Neural Networks:
Artificial Intelligence and Industrial Applications

Springer
Berlin
Heidelberg
New York
Barcelona
Budapest
Hong Kong
London
Milan
Paris
Tokyo

Bert Kappen and Stan Gielen (Eds)

Neural Networks: Artificial Intelligence and Industrial Applications

Proceedings of the Third Annual SNN Symposium on Neural Networks, Nijmegen, The Netherlands, 14-15 September 1995

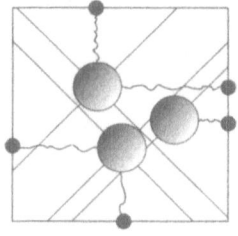

 Springer

Bert Kappen and Stan Gielen
Dutch Foundation for Neural Networks (SNN)
Geert Grooteplein Noord 21
6525 EZ Nijmegen
The Netherlands

ISBN-13:978-3-540-19992-2   e- ISBN-13:978-1-4471-3087-1
DOI: 10.1007/978-1-4471-3087-1

British Library Cataloguing in Publication Data
Neural Networks : Artificial Intelligence
and Industrial Applications - Proceedings
of the Third Annual SNN Symposium on
Neural Networks, Nijmegen, The
Netherlands, 14-15 September 1995
  I. Kappen, Bert II. Gielen, Stan
  006.3
ISBN-13:978-3-540-19992-2

Library of Congress Cataloging-in-Publication Data
A catalog record for this book is available from the Library of Congress

Typesetting: Camera ready by contributors

34/3830-543210 Printed on acid-free paper

Preface

This volume contains a collection of papers that were presented during the Third Annual Symposium on Neural Networks in Nijmegen, The Netherlands. The symposium was held on September 14 and 15 1995 and was organised by the Dutch Foundation for Neural Networks (SNN). The symposium consisted of two parallel tracks.

The scientific track is entitled Neural Networks and Artificial Intelligence. The term "Artificial Intelligence" is often associated with "traditional AI" methodology. Here it is used in its literal sense, indicating the problem to create intelligence by artificial means.

When considering the possibility of artificial intelligence, it is important to realise that we would never consider such a thing without the example of natural intelligence. Therefore, design of artificial intelligence should take advantage of the biological solutions. This is the research field called neural networks.

The aim of the scientific track is two-fold: to give an overview of new developments in neuro-biology and the cognitive sciences. These insights may lead to novel computational paradigms for artificial intelligence. Secondly, to give an overview of recent technical and theoretical achievements in robotics, vision and data modeling.

Until recently the use of neural networks has been restricted to the academic world. However, over the last few years industries and businesses have started to become aware of the benefits of neural networks for commercial use. Their ability to learn by examples, to deal with noisy data and with nonlinear structures makes neural networks suitable to deal with problems where conventional computing fails. These abilities of neural networks result in various commercial advantages: such as for instance improved accuracy and system performance, the ability to automate previously manual processes, and reduced development effort. The ultimate proof of the commercial benefits of neural computing is given by the rapidly increasing number of applications.

The industrial track is entitled Neural Networks in Practice. As is emphasised by this title, the industrial track presents working neural network solutions to real industrial problems. The aim of the track is to convince industry and business that neural networks provide solutions where other methods fail. To illustrate that neural computation is not restricted to any specific industrial area, the presentations have been selected from a broad range of application areas, varying from industrial process control to marketing and finance. Finally we would like to stress that most presentations concern applications which are at this moment in commercial use.

This industrial track is one of the activities of the Stimulation Initiative for European Neural Applications (SIENA). SIENA is an Esprit project (EP 9811) whose objective is to accelerate the take-up of Neural Network technology in Europe through a program of activities with a specifically industrial orientation.

SNN is a partner in SIENA for the Benelux, in a multinational consortium with other partners in the United Kingdom, Spain, France and Germany.

We would like to thank the industrial program committee and our colleagues in the SIENA project for their invaluable assistance in the design of the industrial program. We would like to thank the scientific program committee for their help in the review of the scientific program. Finally, we would like to thank Elma Burg for her great help in the preparation of these proceedings.

Bert Kappen
Stan Gielen

Organisation

Scientific Program Committee:

A. Aertsen (Jerusalem), S.I. Amari (Tokyo), J. Buhmann (Bonn), B. van Dijk (Amsterdam), R. Eckhorn (Marburg), S. Gielen (Nijmegen), T. Heskes (Nijmegen), A. Herz (Oxford), B. Kappen (Nijmegen), B. Kröse (Amsterdam), V. Lopez (Madrid), D. Mackay (Cambridge), T. Martinetz (München), J. Taylor (London), W. von Seelen (Bochum).

Industrial Program Committee:

E. Auée (Arnhem), S. Hafner (Stuttgart), T. Harris (Egham, Surrey), B. Kappen (Nijmegen), W. Wiegerinck (Nijmegen), T. Willems (Oisterwijk).

This symposium is organised by the Foundation for Neural Networks (SNN). InnovatieCentrum Midden-en Zuid-Gelderland (IC) and Vereniging Artificiele Neurale Netwerken (VANN) collaborated in the organisation of the industrial track.

The Foundation for Neural Networks (SNN) is a university-based non-profit-making organisation that stimulates fundamental and applied research on neural networks in the Netherlands. The research program investigates neural information processing strategies for artificial behaviour, vision, pattern recognition and cognitive systems and various application areas. SNN collaborates with industry on neural solutions for their specific industrial applications. Currently, several groups from the universities of Nijmegen, Utrecht, Amsterdam, Delft and Groningen are participating. *(Contact address: Geert Grooteplein 21, 6525 EZ Nijmegen, The Netherlands, tel +31 80 614245, fax +31 80 541435, e-mail snn@mbfys.kun.nl. After 10 October 1995: tel + 31 24 3614245, fax +31 24 3541435.)*

InnovatieCentrum Midden-en Zuid-Gelderland (IC) is an initiative of the Dutch Ministry of Economic Affairs to make knowledge accessible and applicable for small and medium-sized enterprises. IC operates on the basis of independent and individual advice. Also in the field of neural networks IC helps entrepreneurs of SMEs to investigate whether this technology can be useful in their specific situation. *(Contact address: E. Auée, Bergstraat 35-4, 6811 LC Arnhem, The Netherlands, tel +31 85 458948, fax +31 85 459311, e-mail Ed_Auee.ICNN.ICNN-EXTERN@NOTES.compuserve.com. After 10 October 1995: tel +31 26 4458948, fax +31 26 4459311.)*

De Vereniging Artificiele Neurale Netwerken (VANN) is an independent association. It offers its members (academics and industrials) an informal forum to discuss neural networks and their applications. The VANN is the Dutch Special

Interest Group of the International Neural Network Society and yearly organises 5 mini-symposia on practical applications of neural networks. To keep its members and others interested a year-book is published. *(Contact address: D.J.N. Egberts, Julianalaan 35, 6721 ED Bennekom, The Netherlands, e-mail egberts@biologica.nl).*

Contents

NEURAL NETWORKS IN PRACTICE

Applications of Neural Networks - Orals

Applications of Neural Networks - Posters

NEURAL NETWORKS AND ARTIFICIAL INTELLIGENCE

Neurobiology - Orals

Segmentation Coding by the Visual System
Neural Signals that Possibly Support Scene Segmentation

Reinhard Eckhorn

Physics Department, Neurophysics Group, Philipps-University
D-35032 Marburg, Germany

Abstract

Synchronized oscillatory activity in the visual system has been proposed as a temporal label of feature-linking, by assuming that *spatial segmentation coding* of a visual scene is based on synchronized oscillations within a segment's neural representation and on desynchronized signals among different segments. In an extended hypothesis it was stated that segmentation may be based on any type of synchronization and desynchronization, being rhythmical or non-rhythmical, stimulus-dominated or internally generated. This was supported by experimental evidence obtained by multiple microelectrode recordings from the visual cortex of cats and monkeys. It includes stimulus specific synchronization and facilitation at zero-delay correlation during short stimulus-locked responses and during sustained oscillations (35-90 Hz). We now extend the synchronization hypothesis to *spatio-temporal segmentation coding* by suggesting that the synchronized single or repetitive interruptions of excitations define temporally precise internal representations. By applying this to natural vision, we can argue that during phases of slowly changing retinal images (like ocular fixation and smooth pursuit) oscillations may prevent perceptual "smearing" by interrupting the flow of visual information repetetively. However, when a visual object suddenly changes its position, ongoing oscillations should immediately be interrupted by the single excitation-inhibition cycle evoked by the objects displacement. Such stimulus-locked cycles might acts as "reset" signals of prior, and evoke new, segmentations of an object against the remaining scene. Experimental evidence in support of this hypothesis includes our observations on suppression of oscillatory by stimulus-locked responses and on changes of oscillation frequencies with velocity, size and intensity of a stimulus.

6

1 Introduction

A visual scene is a dynamic process projected onto the retinae of the moving eyes. The immense flow of information has to be structured by the visual system in space and time for obtaining a relevant estimate of the relations among the various local and global visual features. This requires a flexible preprocessing system capable of extracting and binding the visual features such that a meaningful scene of relevant visual objects and actions can be perceived. At present, these tasks of feature extraction, feature binding and scene segmentation can neither be executed satisfactory on computers, nor are the corresponding neural mechanisms and codes known that are utilized by the visual system. However, several recent investigations of multiple site recordings from the visual cortex of cat and monkey suggested that oscillatory activities of high frequencies (35-90 Hz) may play a role in segmentation coding /1-6/. This induced one of the most active discussions in visual neuroscience by asking whether feature-binding is defined by neural synchronization and feature separation by desynchronization. The present paper combines previous, and adds new, proposals of how stimulus-specific processes of synchronization, desynchronization and inhibition may interact in the visual cortex in order to cope with the difficult task of segmenting dynamic visual scenes.

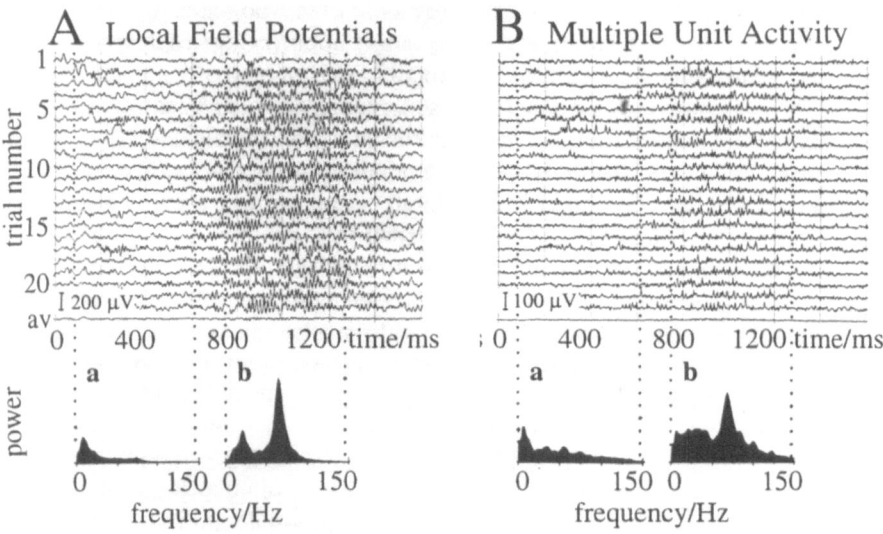

Fig. 1: High frequency (60-90 Hz) oscillations in primary visual cortex of an awake monkey. **Upper panels:** 22 responses to identical stimulus repetitions (of slowly moving light bar). When the stimulus was in the receptive fields (600-1400 ms after stimulus start) oscillations occurred at high amplitudes. **Lower panels:** spectral averages of the responses in upper panels. A: Local field potentials (10-150 Hz). B: Multiple unit activity (Fig. modified from /10/).

2 Segmentation by Synchronized and Desynchronized Neural Activities ?

For the visual system a central question of segmentation coding is how the local feature detectors, characterized by receptive field properties of visual neurons, are combined such that they are associated into coherent perceptual objects and events. The "synchronization hypothesis" seems a promising approach: It states that those neurons participate in the representation of a visual object whose activities engage in a common synchronized state in response to stimulation by that object /7/. This hypothesis attracted attention when stimulus-specific synchronized oscillations of 35-90 Hz were found in the visual cortex of anesthetized cats /1-6,8/ and awake monkeys /9-12/ (Fig.1). Later the synchronization hypothesis was extended /13/. Two types of synchronized cortical signals were proposed as candidates for feature association: (1) stimulus-locked signals, evoked by transient retinal stimulation, and typically non-rhythmic; (2) oscillatory signals, induced by sustained stimuli, and typically not locked in their oscillation phases to stimulus events. Both types of signals can occur synchronously in those neurons that are activated by a common stimulus. Oscillatory activities in particular have extensively been studied. They were found synchronized in paired recordings within vertical cortex columns, in separate columns of the same cortical area, and even between different cortical areas and hemispheres. The average phase difference between oscillatory events was typically close to zero /1,4/. Oscillatory events in any two assemblies were more closely correlated the more similar were their receptive field properties, and the better a common stimulus activated the assemblies simultaneously.

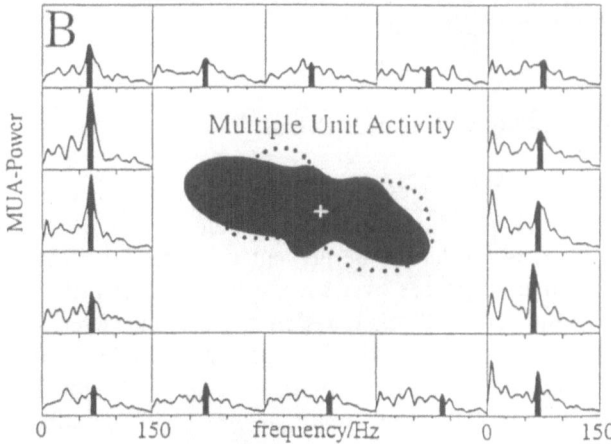

Fig. 2: Orientation - direction tuning of oscillations were generally sharper than of conventional response measures (summated responses). An example of multiple unit activity is shown. Shaded areas indicate oscillatory components while dotted line indicates tuning curve of summated responses. (Fig. modified from /10/).

2.1 Orientation Tuning is Sharper with Oscillatory Activity Compared With Conventional Response Measures

The validity of the synchronization hypothesis requires that synchronized oscillations occur preferentially among coupled neighboring neurons with similar receptive field properties. This should be visible in the tuning characteristcs of the neurons' preferred stimuli if mutual facilitation occurs among synchronized neurons. Tuning curves of oscillatory signals were therefore expected to have sharper characteristics than those obtained for the conventional response measures (e.g., summated response). Indeed, our Marburg group could recently show this for orientation and direction characteristics in the visual cortex of an awake monkey (Fig.2) /10/.

2.2 Zero-Delay Correlations Among Separate Cortical Locations

Average phase differences between oscillations at separate locations of the same cortical area and among different visual cortical areas were narrowly distributed around zero. Such in-phase oscillations were recently demonstrated between visual areas V1 and V2 of an awake monkey /11/. This was unexpected because V1 and V2 are known as serially arranged suggesting a delay of V2 against V1 oscillations. An explanation of zero delay correlations would be, for example, a common oscillatory input to V1 and V2 from a subcortical source. It is more probable, however, that zero-delay is mainly due to temporally symmetric feedback interactions among the cortical areas /11,14/. In conclusion, associations of features the representations of which are dispersed among different cortical areas could be efficiently supported by oscillations at zero phase difference.

2.3 Rhythmic Temporal Segmentation ?

Temporal segmentation seems necessary for clear percepts, because retinal images move permanently. Even under conditions of rest and ocular fixation microsaccades and drifts of the eyes can hardly be prevented. Cortical neurons, on the other hand, can persist in their responses over several hundreds of milliseconds even to very short visual stimuli of 10-20 ms duration /e.g.15/. However, if a figure with sharp contours has evoked cortical activations persisting while the retinal image of that figure shifts across the retina the contour's perception might be smeared, because it activates successively many other neurons with neighboring receptive fields. Smearing might be prevented if the continuously varying stream of afferent visual information would be "chopped" by the cortex into series of consecutive short "image frames". Optimal "frame rates", however, should then be adapted to the rate of retinal image displacement (velocity) and the retinal size of the object's image. Faster framing would be desirable with higher movement velocities and smaller size (according to the sampling theorem of signal theory).

2.3.1 Frequency of Oscillation Depends on Stimulus Velocity and Size

Astonishingly, stimulus specific oscillations (35-90 Hz) observed recently in the visual cortex seem to fulfill these requirements of image sampling. Oscillatory events in the visual cortex of anesthetized cats and especially in awake monkeys had on average, higher oscillation frequencies with faster stimulus movements or with smaller stimulus features compared with slower movements or larger features /16/.

A simple calculation shows that spatial precision of perception based on such a code is well in the range known from psychophysical experiments. In a typical recording experiment a stimulus object of 2° (visual angle) moving at 1°/s induced a dominant oscillation frequency in the awake monkey visual cortex of about 60 Hz. This corresponds to a retinal image displacement of one arc minute visual angle during a single activation-inhibition cycle (17ms). An even higher precision in spatio-temporal representation would exist during periods of ocular fixation. Assuming residual drift movements of the eyes of about 0.1°/s during fixation tasks and typical oscillation frequencies around 40 Hz results in image displacements of 9 arc seconds during a single oscillation cycle (25 ms "frame rate"). This is obviously better than the perceptual resolution at sharp contrast borders.

2.3.2 Frequency of Oscillation Depends on Stimulus Contrast

We found the average oscillation frequency also depends on the contrast of stimuli (intensity, stimulus strength; Fig.3C). High oscillation frequencies occurred at low contrasts whereas frequencies slowed down at higher contrasts /16/. In addition, the lower frequency oscillations were generally of larger amplitudes. My suggestion for a functional role of this effect for scene segmentation is speculative and assumes that low frequencies and high amplitudes have a potential of establishing larger cortical "synchronization fields" /13/ compared with small amplitudes and high frequencies. This seems plausible because larger distances will require longer spike propagation delays than short ones, and mutual synchronization of separate assemblies is only possible if the activation delays are shorter than about a third of the oscillation period /14/. Hence, strong stimulus features (according to their high intensity or contrast) would enable feature linking over more extended ranges of the visual field than weak stimuli. This corresponds to our everyday experience.

2.4 Associations Among Parts of an Object may be Coded by Coupled Oscillations at Different Frequencies

As larger objects and higher contrasts induced lower oscillation frequencies, and as often a broad range of frequencies were simultaneously induced by single stimuli, we asked whether the features to be linked have to do so at the same frequency or whether this could also be possible at different frequencies. We found, indeed stimulus related phase coupling among oscillations at different frequencies in about 60% of recordings from cat and monkey visual cortex, including couplings at harmonic and incommensurate frequencies /17/: 1.) within the gamma-frequency band

(30-90Hz), 2.) between gamma and low frequency (1-30 Hz) rhythms, and 3.) among rhythms at lower frequencies (1-30Hz).

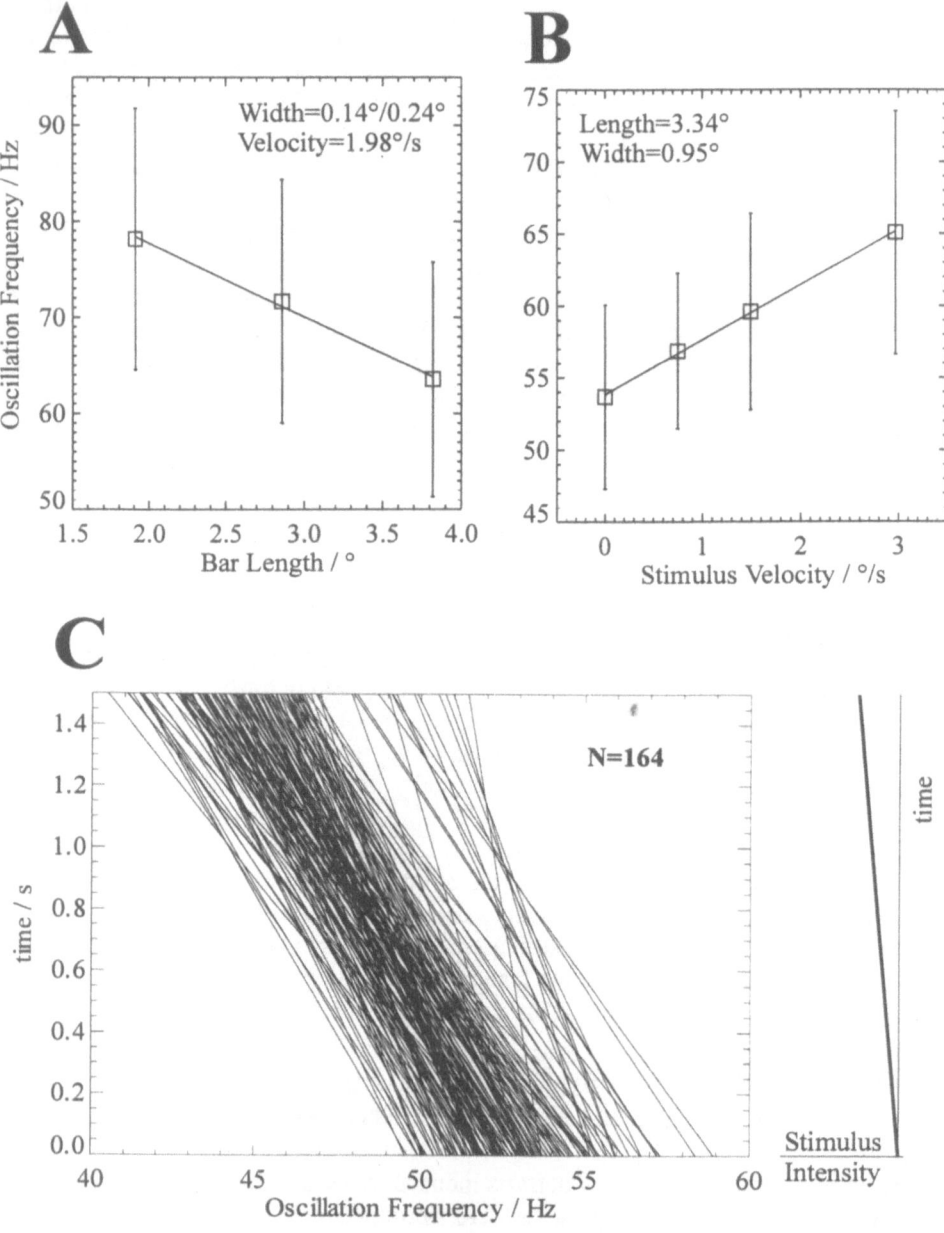

Fig.3: Dependencies of average oscillation frequencies on stimulus size, velocity and contrast. The average oscillation frequencies shifted to lower values accompanied by higher amplitudes if the length of a moving stimulus (light bar) was elongated (**A**), or if its width was broadened, or if its velocity was decreased (**B**), or if its contrast was increased (**C**). (Modified from /16/.)

It seems plausible for us to assume that neurons can synchronize their oscillations over broader visual areas at low frequencies than at higher ones. Accordingly, the activities evoked by the outer borders of an object may therefore synchronize at the lowest frequency, whereas those of the object's subfields may engage at higher frequencies. However, the different frequency components induced by the same object may loosely be coupled in order to code its entity. In summary, the coding of feature associations may not be restricted to phase-coupling of oscillatory events at a common high frequency (35-90 Hz), but may be implemented by coupled rhythmic processes at different time scales.

2.5 Segmentations due to Sudden Retinal Image Shifts

While oscillations of high amplitudes are generally induced during epochs of slow retinal drift rates in which cortical cells are activated by sustained thalamic inputs, fast transient components were primarily evoked by stimulus epochs of changing high velocity and direction. In the visual cortical areas 17 and 18 of the anesthetized cat oscillations were gradually reduced in amplitude and finally fully suppressed with increasing amplitudes of fast transient stimulus movements /6/.

Comparable measurements were recently made in the awake monkey supporting the observations in the anesthetized cat /A. Guettler & R. Eckhorn, unpublished/. In monkeys rapid changes of the retinal image due to (micro) saccades or sudden switchings of an object in its contrast or position could immediately disturb ongoing oscillations. The latter were, as in the cat, generated only during states of stable or slowly moving retinal images.

Applying these observations to natural vision, we can argue that during phases of slowly changing retinal images (like ocular fixation and smooth pursuit) oscillations may prevent perceptual "smearing" by interrupting the flow of visual information repetetively. However, when a visual object suddenly changes its position, ongoing oscillations are immediately interrupted by a single excitation-inhibition cycle evoked by the object's displacement. Hence, the stimulus-locked cycle might acts as "reset" signal of the prior, and evoke a new, segmentation of the object against the remaining scene.

Acknowledgements: This article could not have been written without the numerous discussions with my colleagues and their extensive help in experiments and data processing, as credited in the text and figure captions. Expert help in experimental techniques came from U. Thomas and financial support from Deutsche Forschungsgemeinschaft (Ec 53/6, 53/7).

References

1. Eckhorn R, Bauer R, Jordan W et al. Coherent oscillations: a mechanism of feature linking in visual cortex? Multiple electrode and correlation analysis in the cat. Biol Cybern 1988; 60:121-130
2. Gray CM, König P, Engel AK, Singer W. Oscillatory responses in cat visual cortex exhibit inter-columnar synchronization which reflects global stimulus properties Nature 1989; 338:334--337

12

3. Eckhorn R. Principles of global visual processing of local features can be investigated with parallel single-cell and group-recordings from the visual cortex. In: Aertsen A, Braitenberg V (eds), Information Processing in the Cortex. Springer, Berlin New York London 1992; pp 385--420

4. Engel AK, König P, Gray CM et al. Stimulus-dependent neuronal oscillations in cat visual cortex: Inter-columnar interaction as determined by cross-correlation analysis. Europ J Neurosci 1990; 2: 588--606

5. Engel AK, König P, Singer W. Direct physiological evidence for scene segmentation by temporal coding. Proc Natl Acad Sci USA 1991; 88: 9136--9140

6. Kruse W, Eckhorn R, Schanze T et al. Stimulus-induced oscillatory synchronization is inhibited by stimulus-locked non-oscillatory synchronization in cat visual cortex: two modes that might suppoort feature linking. Soc Neurosci Abstr 1992; 18:131.3

7. Reitboeck HJ. A multi-electrode matrix for studies of temporal signal correlations within neural assemblies. In: Basar E, Flohr H, Haken H, Mandell A (eds), Synergetics of the Brain, Springer-Verlag, Berlin Heidelberg New-York, 1983; pp 174--182

8. Gray CM, Engel AK, König P et al. Stimulus-dependent neuronal oscillations in cat visual cortex: Receptive field properties and feature dependence. Europ J Neurosci 1990; 2: 607-619

9. Kreiter AK, Singer W. Oscillatory neuronal responses in the visual cortex of the awake Macaque monkey. Europ J Neurosci 1992; 4: 369--375

10. Eckhorn R, Frien A, Bauer R et al. High frequency (60-90Hz) oscillations in primary visual cortex of awake monkey. NeuroReport 1993; 4:243--246

11. Frien A, Eckhorn R, Bauer R, Woelbern T, Kehr H (1994) Stimulus-specific fast oscillations at zero phase between visual areas V1 and V2 of awake monkey. NeuroReport 5: 2273-2277

12. Frien A, Eckhorn R, Woelbern T et al. Oscillatory group activity reveals sharper orientation tuning than conventional measure in primary visual cortex of awake monkey. Soc Neurosci Abstr 1995 (in press)

13. Eckhorn R, Reitboeck HJ, Arndt M et al. Feature linking among distributed assemblies: Simulations and results from cat visual cortex. Neural Computation 1990; 2:293-306

14. Gerstner W, Ritz R, van Hemmen JL. A biologically motivated and analytically soluble model of collective oscillations in the cortex. Biol Cybern 1993; 68: 363-347

15. Rolls ET, Tovee MJ. Processing speed in the cerebral cortex and the neurophysiology of visual masking. Proc R Soc Lond B 1994; 257:9-15

16. Eckhorn R, Frien A, Bauer R et al. Oscillation frequencies (40-90Hz) in monkey visual cortex depend on stimulus size and velocity. Europ J Neurosci Suppl 1994; 7: 90.04

17. Schanze T, Eckhorn R. Phase coupling between oscillations at different frequencies in the visual cortex of cat and monkey. In: Elsner N, Heisenberg M H (eds), Gene, Brain, Behaviour. Thieme, Stuttgart New York 1993, p442

Synchrony and Fast Plasticity in the Visual Cortex

Bob W van Dijk

Graduateschool Neurosciences Amsterdam, Laboratory for Medical Physics and
the Netherlands Ophthalmic Research Institute
Amsterdam, the Netherlands

1. Introduction

The traditional view of sensory cortical processing is that of static parallel local cortical circuits detecting or signalling specific local features or feature gradients (e.g. [1]). Global processing in this view occurs by somehow comparing "linking features" in areas that lie more central in the cortical hierarchy. It is more and more realized, however, that this view is oversimplified and possibly not correct. On the one hand the hierarchical scheme is contradicted by growing evidence that even at the presumed-lowest level in the hierarchy, such as the primary auditory or visual cortices, global stimulus features are of importance to the activity patterns [2-4], and even behavioral characteristics influence responses [5-7]. On the other hand there is growing evidence that neuronal connections are far from being static.

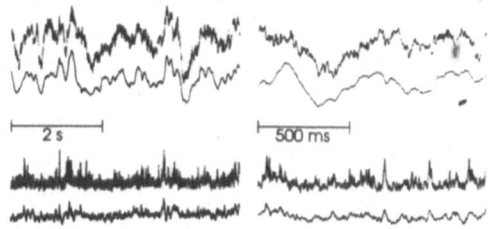

Figure 1. Recorded (thick lines) and predicted (thin lines) activities in cat cortical area 18. Top two traces: synaptic activity: bottom two traces spiking activity.

There is also accumulating experimental evidence that spatio-temporal patterns of active neurons play a central role in cortical processing. Such patterns of precisely synchronized active neurons, showing bursting or oscillatory firing, are distributed over large portions of the cortical mass (a.o. [5,8-10]). On the basis of these observations many of these authors have put forward the hypothesis that the cortex uses "synchrony" as a linking carrier to associate features extracted by neurons in different sensory areas. A cortical neuron in such a scheme would be part of many of these neuron assemblies and would be linked in one of these dependent on stimulus and behaviour. This further stresses the need for dynamic functional connections in the cortex. However, so far there has been no experimental evidence that directly demonstrated that the cortex indeed employs synchrony and coherence of the activities of distributed neurons for sensory, associative, motor or memory tasks. In this paper I will show experimental evidence that functional connections in the cortex between ensembles of neurons can change at short time scales and I will explain the role synchronous activity may play in the cortex by describing the dynamics of an artificial neural network that has both biological firing rules and biological plasticity at comparable time scales.

2. Methods

Experiments were done on adult female cats under Urethane anaesthesia. The procedures for surgery and preparation of the animal, as well as visual stimulation are described in detail elsewhere [11,12]. We recorded the laminar profiles of intracortical neuronal activity by inserting a linear array of electrodes radially into area 18, one of the primary visual cortical areas of the cat. This array consists of 16 electrodes each 50x50μm in size and 150μm apart. Signals were spatially and temporally filtered to provide a set of spiking activities and set of synaptic activities. The spiking activity at each electrode was determined by measuring the envelope of the high frequency content (1k-10kHz) of the signal at the tip (this type of signal is referred to as multiple unit activity by Eckhorn, e.g. [13]). The synaptic activity was determined by taking the second spatial derivative - using a three point formula - of the low temporal frequency (0.5 - 70Hz) content of the signal at the tips. Continuous data segments were stored while three different visual stimuli were presented to the animal: a non-modulated homogeneous TV monitor (which will be referred to as the no-stimulus condition), sinusoidally moving black and white gratings or bars, of which the velocity, the orientation, the contrast, and the number of bars were adjusted to optimally evoke time locked responses (this will be referred to as the moving bar condition) and a 100% modulated homogeneous flashing stimulus (referred to as the flash condition). The different stimuli were presented in random sequences in time epochs of 1 to a few minutes, contiguous recordings typically lasted about 30 minutes.

We define the *functional connectivity* to be identical to the *transferfunction* between the activity at one site in the cortex and the activity at a second site. We therefore developed a method to estimate this transferfunction from the two sets of simultaneously recorded data. This method finds the equivalent linear time invariant system, that transfers the data from the independent set of signals $\{x(t)\}$ to the dependent set of signals $\{y(t)\}$. This linear transfer can be written as:

$$y_i(t)=\sum_{j=1}^{q} \int_{\tau=-\infty}^{\infty} x_j(t-\tau)h_{ij}(\tau)d\tau +r_i(t)$$

Here h_{ij} is the linear system kernel, and $r_i(t)$ the residual signal.

Our method solves the "best" kernel, i.e. such that the full least square residual is minimal. This optimization problem is solved simultaneously for all signals from the set of dependent signals, using repeated partialization in the frequency domain to achieve regularization. The method requires that two conditions are met: i. the signal transfer is nearly linear; ii. the transfer is stationary over the time interval during which the signals were recorded. This was tested in two ways: i. the quality of the prediction of $\{y(t)\}$ based on $\{x(t)\}$ was tested by graphically comparing selected epochs of $\{y(t)\}$ with the model prediction; and ii. by computing the multiple regression coefficient between the recorded and estimated signals.

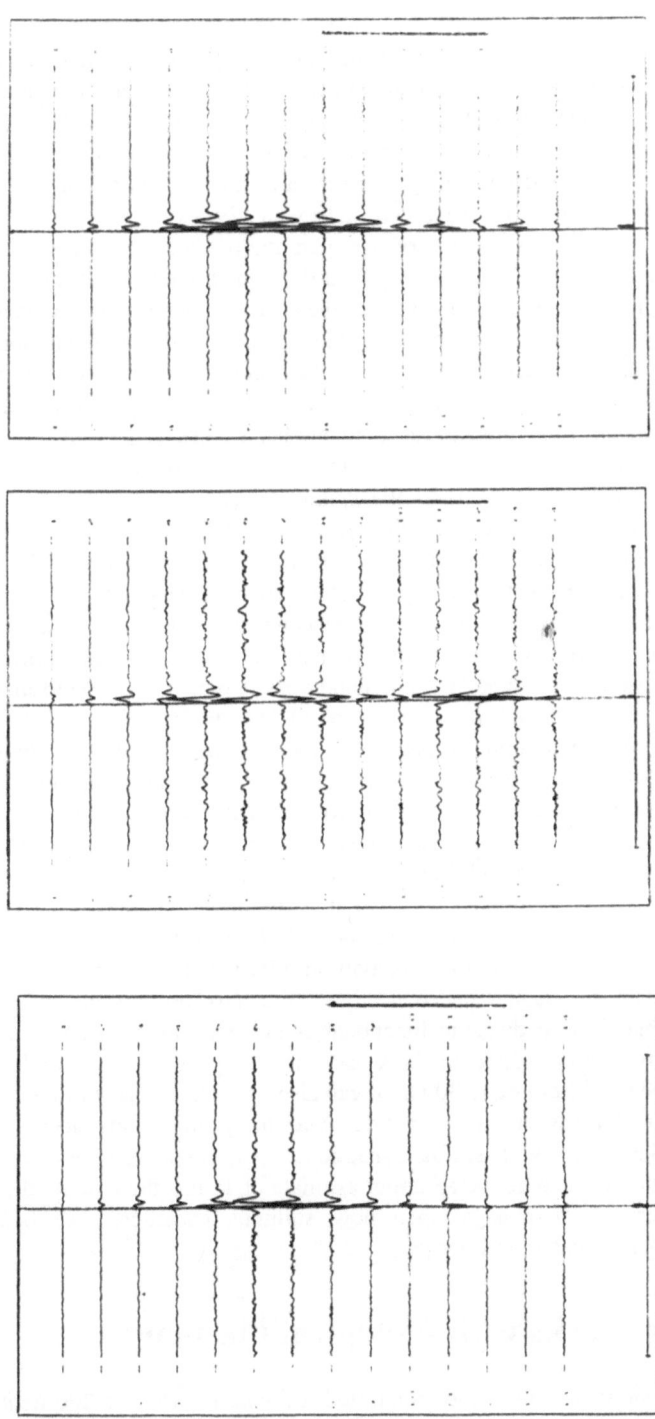

Figure 2. Compound transfer function from spiking activities at electrodes 8-12 to synaptic activities at all electrodes under three stimulus conditions: left no stimulus; middle flashes; right moving bars.

3. Results

Figure 1 shows typical data epochs at two different time scales in the left and right halves of the figure. Thick lines are the recorded data epochs, thin lines are the corresponding predictions based on our linear model. That is in the two sets in the top of the figure recorded and predicted synaptic activities are shown, while in the two sets at the bottom of the figure recorded and predicted spiking activities are shown. Data as shown in this figure learned us that our linear model worked quite well. Even though there evidently are differences in the precise wave shapes of recorded and predicted data, the two data sets were very similar. In fact most of the observed differences could be attributed to a static non-linearity. This is easily understood, since the synaptic activities have a gaussian amplitude distribution, while the spiking activities have a poisson amplitude distribution. These graphical comparisons also learned us that the connectivity is stable when the visual stimulation is unchanged. The transfer kernels were derived from long records, 20 to 30 minutes in duration, and could be used to predict signals in arbitrarily chosen individual epochs from this record of 0.5 to 5 s duration. Multiple regression coefficients were in the range from 0.6 to 0.8 when the spiking activities were chosen as the independent variable; they were smaller for the reverse correlations ranging from 0.5 to 0.75. Since not every synaptic potential change results in a spike discharge, while every spike results in synaptic potentials this finding can intuitively be understood.

Figure 2 shows the changes that occur in the effective connections within a single column of cat area 18. This figure consists of three panels each depicting the functional connections recorded during the three stimulus conditions. Compound transferfunctions are shown over the time interval from -1.18 to +1.18s. From left to right these transferfunctions were recorded during respectively the no-stimulus condition, the flash condition and the moving bar condition. Each panel shows predicted synaptic activity signals at the 14 recording positions; from top to bottom of the cortex corresponding to from top to bottom in each panel. The predicted responses in figure 2 are those that result from a unit pulse at spiking activity channels 8 to 12 simultaneously at $t = 0$. From the figure it can be seen that the compound signal transfer in the column has changed significantly between the different stimulus conditions. Of the 11 cats that we studied and analyzed, 6 showed different functional connectivity in different stimulus conditions; the other 5 showed virtually no changes in connectivity. We have no explanation for the dichotomy. In cats that showed dynamic functional connectivity the transfer functions obtained during the moving bar condition show oscillatory responses. These oscillations were broadband in the band from 40 to 80 Hz. It can also be seen that most of the power in the transferfunction for any stimulus condition was for positive time delays. This suggests that the spiking activity causes the synaptic activity, rather than the other way around. From our data we could derive an upper estimate of 1s for the time it takes for the functional connections to change upon new stimulus conditions; we believe, however, that these changes occur even faster.

4. Artificial Networks with Biological Plasticity.

We studied the behaviour of a randomly connected artificial network of 200 neurons. Each of these neurons consisting of three compartments, representing dendrites, soma and axon. These neurons could be inhibitory or excitatory. Each neuron had spiking dynamics, governed by the transmembrane potential and the channel dynamics of each compartment. Synapses in our simulations are modeled by channels of which the conductivity changes after input according to an alpha function. To put fast synaptic plasticity in our model, we imposed additional dynamics on the maximal conductances of the synaptic channels. We assumed a simple hypothesis [15] proposed as a model of long term potentiation and depression. According to this hypothesis a synapse becomes stronger if it is activated when the transmembrane potential is above the potentiation threshold, it will become weaker if it is below this threshold but still above a second, the depression threshold, while it will not change at all if it is below both thresholds or when the synapse is never activated. In our model we further assume a sigmoid dependence of the synaptic efficiency change on the distance between membrane potential and potentiation or depression thresholds. The slope of this curve will be called the plasticity rate (a).

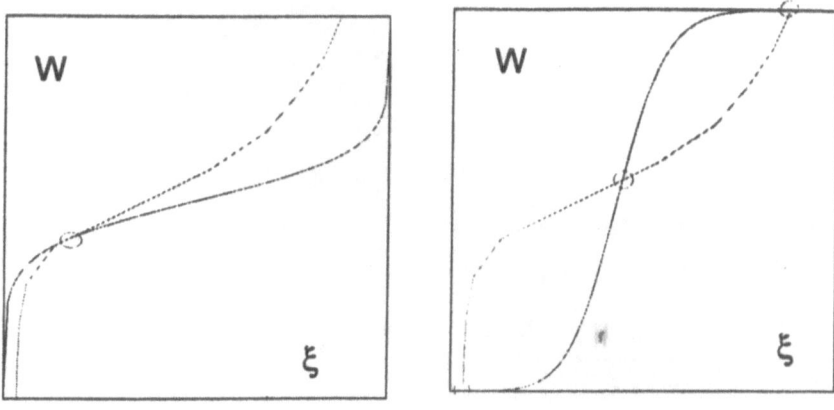

Figure 3. Hardware and software of the mean field. Left for low a; Right for high a.

To study the behaviour of such a network with mixed dynamics, i.e. the dynamics of the spiking activity and the dynamics of the synaptic changes, we used the mean field description. Thus we assumed that all neurons had a similar firing rate that only depended on the network's average excitation and inhibition. The behaviour of the network can now be expressed by a small set of order parameters. One of these ξ expresses the average drive of synaptic change by the activity given a collection of synaptic efficiencies. The other is the average synaptic efficiency W. The function $\xi(W)$ represents the hardware of the network and is not linked to the plasticity rules (except by the plasticity threshold values); the function $W(\xi)$ represents the software of the system, i.e. the asymptotic efficiencies that will occur for different hardwares. Figure 3 shows these two functions for two different plasticity rates. In the left panel a is low in the right panel a is high. The crossing points indicate where the mean field system would be in equilibrium. For low a there is only one equilibrium and it is stable; for high a there are three equilibria two of which, one at high mean connectivity the other at low mean connectivity, are stable.

W

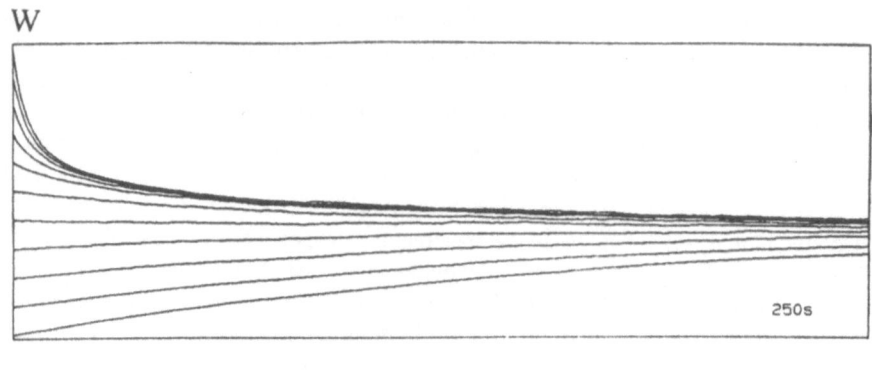

250s

W

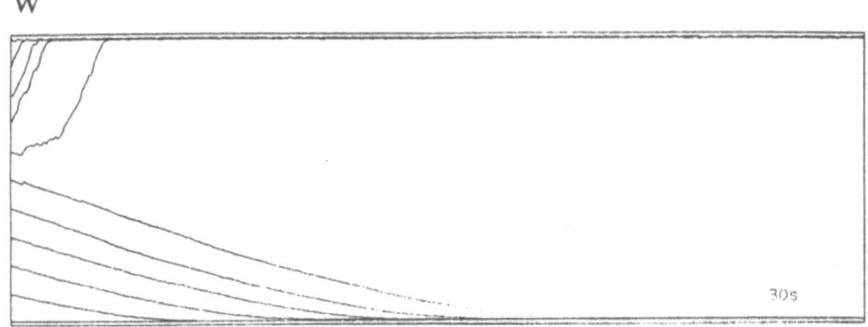

30s

Figure 4. The evolution of the connectivity of the mean field.

Figure 4 shows the behaviour of the average connectivity in the network at these parameter settings for different initial conditions. The panel in the top for low a indeed shows that the network goes to one and the same level of connectivity independent of initial conditions, while with high a the network goes to either a highly connected or a weakly connected state. These data show that synaptic plasticity not only may play a role in learning and memorizing, it also may have an important role in stabilizing networks. Although not evident from the behaviour of the mean field, our simple one-layered network with plasticity showed a rich repertoire of behaviours, including fixed points and limit cycles both in stable states of low and of high mean connectivity. These behaviours depended mostly on the character of the external input, and on the ratio of inhibition and excitation within the network.

Figure 5 explains the relation between synchrony and connectivity in our model network. The top panel shows the normalized cross-correlation functions of the average spiking activity of the network for different levels of connectivity. It shows that when the network is strongly connected, the neurons fire in synchrony. The two bottom panels show changes of the average connectivity of the network induced by external input. In the bottom panel the network was driven by synchronous diffuse input when it was in a stable low-connected state. As a result of this input the network changed into a new stable state of high connectivity. The middle panel shows the reverse. The network was in a stable high connected state, but by repeated asynchronous input was brought back into a low-connected state.

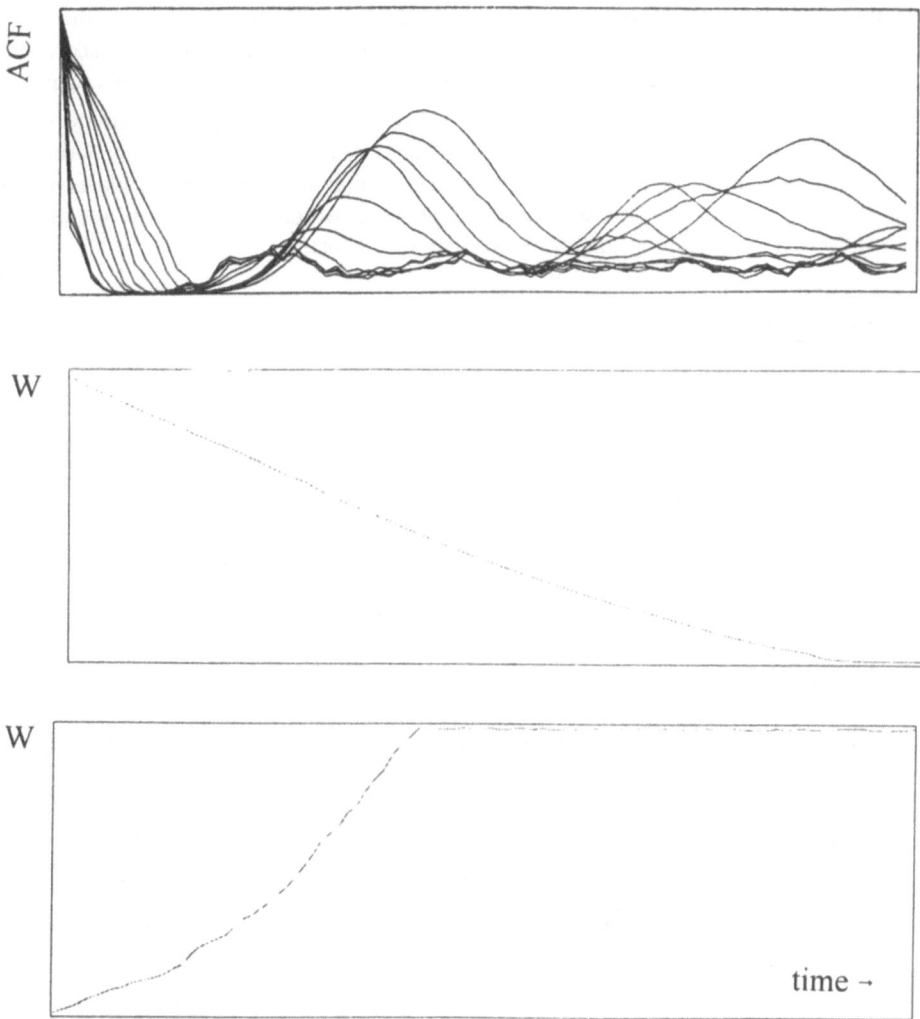

Figure 5. Top: Autocorrelation functions of the number of spikes produced by the network at various levels of mean connectivity; Middle: change of the mean connectivity induced by anti-synchronous diffuse external input; Bottom change of the mean connectivity induced by synchronous diffuse external input. The time axes in these three panels differ in scale.

5. Conclusions

Our data have shown that the visual cortex of an anaesthetized cat changes its connectivity when different visual stimuli are presented. Based on these data we suggest that the state of connections in the cortex signals different sensory percepts. To show that synchrony and changes of synchrony might be the relevant cues to signal such connectivity changes, we studied a simple one-layered and fully connected neural network with "biological" plasticity and biological firing rules. These simulations showed that this network has a rich repertoire of dynamic behaviours, including multiple equilibrium states. Synchronous or anti-synchronous input to the network can drive it to different stable connectivity states. The plasticity plays a double role, it makes state changes possible and it makes the equilibrium states stable. On the other hand the connectivity of the network determines the synchrony of the spiking activity of the neurons of the network. In other words in our model network synchrony writes the connections, which can be memorized because of the stabilizing role of plasticity, the connections are read out from the network from the synchrony of its activity. We propose a similar role for synchrony and plasticity in the visual cortex.

6. References

1. Barlow, H.B. Single units and sensation: A neuron doctrine for perceptual psychology? Perception 1972; 1: 371-394.
2. Knierim, J.J. and van Essen, D.C. Spatial organization of suppressive surround effects in neurons of area V1 in alert monkeys. Soc. Neurosc. Abstr., 16:#523.6.
3. Fox, J.M., Delbruck, T., Gallant, J.L. et al., Modulation of classical receptive field responses by moving texture backgrounds in monkey striate cortex: spatial and temporal interactions. Soc. Neurosc. Abstr. 1990; 16:#523.5.
4. Lamme,V.A.F., van Dijk, B.W. and Spekreijse, H. Contour from motion processing occurs in primary visual cortex. Nature 1993; 363:541-543.
5. Freeman, W.J. and Skarda, C.A. Spatial EEG patterns, non-linear dynamics and perception: the neo-Sherringtonian view. Brain Res. Rev. 1986; 10: 147-175.
6. Freeman, W.J. and van Dijk, B.W. Spatial patterns of visual cortical fast EEG during conditioned reflex in a Rhesus monkey. Brain Res. 87; 422: 267-276.
7. Vaadia, E., Bergman, H. and Abeles, M. Neuronal activities related to higher brain functions - theoretical and experimental implications. IEEE trans. BME 1989; 36: 25-35.
8. Gray, C.M., Konig, P., Engel, A.K. and Singer, W. Oscillatory responses in cat visual cortex exhibit inter-columnar synchronization which reflects global stimulus properties. Nature 1989; 338: 334-337.
9. Abeles, M. Local cortical circuits. Studies of brain function vol. 6. Springer, London, 1982.
10. Edelman, G. The remembered present; a biological theory of consciousness. Basic Books Inc., 1989.
11. van Dijk, B.W., Vijn, P.C.M. and Spekreijse, H. Low temporal frequency desynchronization and high temporal frequency synchronization accompany processing of visual stimuli in anaesthetized cat visual cortex. In: Pantev et al. (eds) Oscillatory event related braon dynamics. Plenum Press, New York, 1994, pp 183-204.
12. Vijn, P.C.M., van Dijk, B.W. and Spekreijse, H. Topography of occipital EEG reduction upon visual stimulation. Brain Topogr. 1992; 5: 177-181.
13. Eckhorn, R., Bauer, R., Jordan, W. et al. Coherent oscillations: a mechanism of feature linking in the visual cortex? Biol. Cybern. 1988; 60: 121-130.
14. Vijn, P.C.M., van Dijk, B.W., Slopsema, J.S. and Spekreijse, H. Inter-laminar signal transfer in cat visual cortex: a multi-variate analysis of multiple-unit synaptic and spiking activities. (in press)
15. Artola, A. and Singer, W. Long term depression of excitatory synaptic transmission and its relationship to long-term potentiation. TINS 1993; 16: 480-487.

Rapid Neural Synchronization: From Spiking Cells to Synfire Webs

Andreas V.M. Herz

Department of Zoology, University of Oxford

Oxford, OX1 3PS, England

1 Introduction

Various important cognitive tasks such as the recognition and behavioral response to complex natural scenes are accomplished within a few hundred milliseconds after stimulus onset [1]. Since at least a dozen successive cortical areas are involved in this task, on average no more than 30 milliseconds are available per processing level.

Firing frequencies in sensory and motor cortices are typically in the 30 to 100 Hertz range and lower in associative areas. This means that at least in some brain regions, a significant part of neural information processing is completed within a few firing cycles, possibly already when the first action potentials are generated in response to a stimulus.

It is often argued that this result implies that major cortical processing is primarily done in a feedforward manner. However, the above argument does not rule out that 'results' of neural computations are held in a (dynamic) short-time memory — to be probed by 'down-stream' processing stages through long-range feedback loops. Thus, although the first wave of excitation after stimulus onset may indeed be processed by feedforward connections, feedback circuits may nevertheless play an important role for rapidly solving cognitive tasks.

The observations trigger an important question: What kind of network processing can be done with the first action potentials after stimulus onset?

A theoretical investigation of this question requires a description of neural activity in terms of interconnected spiking model neurons — approaches based on a firing-rate approximation cannot capture the relevant temporal details. Furthermore, the analysis should focus on the short-time dynamics of such networks, consider network topologies beyond uniform all-to-all coupling schemes so as to allow for biologically significant computations, and should not rely on mean-field approximations for the same reason.

A first step in that direction has been taken in refs. [2] and [3]. The present contribution summarizes the main results of this study. For concreteness, only very simple integrate-and-fire neurons are discussed. It is shown that these models rapidly relax to phase-locked solutions — complex but cyclic attractors are reached as soon as every neuron has fired its first spike. Simulations of more elaborate integrate-and-fire models display almost identical dynamical behavior in the short-time regime, while deviating at longer times.

The results shed new light on the computational role of lateral connections. The models may also help to obtain a better understanding of stimulus-dependent synchronization of action potentials [4, 5] and provide a mathematical environment to investigate the 'binding' of stimulus features through synchronized neural assemblies [6]. The models may finally be used to extend the concept of purely feedforward 'synfire chains' [7] to richer architectures.

2   The Model

'Integrate-and-fire neurons' capture the essence of spiking nerve cells. Below the firing threshold $u_{thresh} = 1$, each neuron i, $1 \leq i \leq N$, operates as a leaky integrator,

$$C(du_i/dt) = -[u_i(t) - u_0]/R + I_i(t) \ . \tag{1}$$

In the absence of the input current $I_i(t)$, the cell potential $u_i(t)$ relaxes to a fixed rest value u_0 with a time constant that depends on both the input capacitance C and the transmembrane resistance R. The term u_0/R can be absorbed in $I_i(t)$ and for convenience one may thus set $u_0 = 0$. When the potential of cell i reaches the threshold u_{thresh}, the cell produces an action potential and instantaneously resets its potential to $u_{reset}=0$.

When an action potential arrives at a synapse from cell j to cell i, a synaptic current briefly flows. For simplicity, the duration of this current is set to zero. By assuming vanishing signal delays and a linear summation of synaptic inputs, the current $I_i(t)$ into cell i is then given by

$$I_i(t) = \sum_{j,n} T_{ij}\delta(t - t_j^n) + I_i^{ext}(t) \ . \tag{2}$$

The t_j^n are the times at which neuron j generates an action potential, T_{ij} represents the strength of the synaptic connection from neuron j to neuron i, and $I_i^{ext}(t)$ denotes an external input current.

If synaptic (or dendritic) time constants of the biological neurons to be modeled are much longer than the duration of action potentials, a cell triggered by an action potential of another cell fires as soon as its membrane potential reaches the threshold; the remaining synaptic current is integrated afterward. If, however, synaptic (and dendritic) time constants are comparable with or shorter than the duration of action potentials, synaptic currents that arrive during spike generation are partially lost. Thus if, for example, cell j is triggered by cell i at time t, the two firing events should be described by

$$u_i(t^+) = T_{ij} \tag{3}$$

and

$$u_j(t^+) = \gamma[u_j(t^-) + T_{ji} - 1] \tag{4}$$

where γ, $0 \leq \gamma \leq 1$, denotes the level of charge conservation.

Because the duration of action potentials and synaptic currents has been set to zero, the firing of one neuron can set off instantaneous firings of others and generate events in which many neurons are active simultaneously. Notice that in what follows, terms such as "synchronized neurons" always refer to the time of spike generation. According to this definition, a periodic network state (also called a "phase locked solution") may or may not be "globally synchronized."

In passing, let me note that with (3) replaced by $u_i(t^+) = 0$, uniform all-to-all couplings of equal strength, $I_i^{\text{ext}}(t) = \text{const}$, and $\gamma = 0$, the network gradually approaches a state where all cells fire in unison [8]. On the other hand, stable periodic solutions that are *not* globally synchronized for short-range couplings exist if eq. (3) is used [3]. This demonstrates that model details have a pronounced effect on the long-time and stationary behavior of coupled integrate-and-fire neurons. They do, however, only play a minor role for the biologically relevant short-time behavior.

Numerical simulations of networks with integrate-and-fire neurons often show a rapid convergence to approximately phase locked solutions, normally followed by a slow reorganization of the (locally) synchronized clusters. A mathematical proof of the first phenomenon has been obtained in the limiting case $R \to \infty$ of perfectly integrating cells and uniform input currents $I_i^{\text{ext}} = I > 0$ [2, 3].

In this situation, information coming from the periphery or other areas of the brain is encoded in the initial values for the u_i, not in time-dependent input currents. This choice resembles the experimental situation of "stimulus-induced oscillations" whereas the alternative scheme would model "stimulus-locked oscillations" [4].

Due to the constant positive input current I, each model cell fires regularly (with frequency I^{-1}) in this case if there is no further synaptic input from other cells. Thus I^{-1} represents the spontaneous firing rate of an isolated neuron. By rescaling time, the capacitance C and input I in (1) can be taken as unity. The overall dynamics may then be summarized by the following update rules:

(i) Initialize the $u_i(t = 0)$ in $[0, 1]$ according to the external stimulus.

(ii) If $u_i \geq 1$ and if neuron i is next in the update scheme then

$$u_i \to u_i' = \gamma(u_i - 1) \tag{5}$$

and

$$u_j \to u_j' = u_j + T_{ji} \ . \tag{6}$$

(iii) Repeat step (ii) until $u_i < 1$ for all i.

(iv) If the condition of step (ii) does not apply then

$$\frac{d}{dt} u_i = 1 \quad \text{for all } i \ . \tag{7}$$

3 Rapid Synchronization

For the simple model (5) – (7), the following theorem holds [2, 3]:

Theorem. Assume that the synapses satisfy $T_{ij} \geq 0$ and

$$\sum_j T_{ij} = A < 1 . \tag{8}$$

Then all solutions of (5) – (7) converge to cyclic oscillations with period $P = 1 - A$. The attractor is reached as soon as every neuron has fired once.

Notice that synaptic symmetry, $T_{ij} = T_{ji}$ is *not* required. In order to proof the proposition, let us first show that no neuron fires more than once in any interval of length P.

Lemma. Let $n_i(t, t')$ denote the number of times neuron i fires in $[t, t')$. If the conditions of the theorem hold then $n_i(t, t + P) \leq 1$.

Proof of the lemma. Starting at time t, let i denote the first neuron that fires twice and let t' denote the time when it does so. The total change in u_i from t to t' due to synaptic currents and the external input must be equal to or greater than 1,

$$(t' - t)(1 - A)/P + \sum_j T_{ij} n_j(t, t') \geq 1 . \tag{9}$$

Since neuron i is the first to fire twice, the number of firings of each of the other neurons up to t' is less than or equal to 1. For T_{ij} nonnegative, the lefthand side of (9) is, therefore, less then or equal to $(t' - t)(1 - A)/P + A$. This implies that (9) can only be satisfied if $t' - t \geq P$.

Returning to the proof of the theorem, let $t_{\max}$ denote the first time where every neuron has fired at least once. Some cells may have fired repeatedly before $t_{\max}$, depending on the parameter values and initial conditions. Let t_i denote the last time when neuron i fires before $t_{\max}$, $t_{\min}$ the minimum of all these times t_i, and k a cell that fires at $t_{\min}$ for the last time without being triggered by other cells at that time.

By definition, every cell discharges at least once in the interval $[t_{\min}, t_{\max}]$. This implies in particular that every neuron j from which cell k receives synaptic input emits one or more action potentials in that interval. Each spike adds T_{kj} to u_k. The total change of u_k in $[t_{\min}, t_{\max}]$ is thus equal or greater than $A + t_{\max} - t_{\min}$. This number has to be smaller than 1 because otherwise, neuron k would fire a second time in the interval $[t_{\min}, t_{\max}]$ in contradiction to the assumption. It follows that $t_{\max} - t_{\min} < P$.

Evaluated at time $t = t_{\max} - P$ and combined with the above result, the lemma implies that every cell fires exactly once in $[t_{\min}, t_{\max}]$ and no cell fires in $(t_{\max} - P, t_{\min})$. This proves that in finite time $t_{\max} \leq 1$, a limit cycle is approached in the sense that $u_i(t) = u_i(t - P)$ for $t \geq t_{\max}$. The argument also shows that the attractor is reached as soon as every neuron has fired once.

4   Collective Computation

The periodic solutions of the last section are stable under small perturbations but not asymptotically stable. Thus strictly speaking, a limit set of P-periodic solutions is approached. From a neurocomputational point of view, this demonstrates that although the network functions as an attractor neural network, individual 'limit cycles' could easily be left once simple desynchronizing mechanisms such as neural fatigue are added to the model.

The spatiotemporal output characteristics of the model systems in response to spatially structured inputs, for example grey-valued (visual) patterns, are readily understood if one first considers random initial conditions drawn from a uniform distribution with mean $\bar{u}$ and width w. Simple natural scenes may then be thought of as arrays of noisy grey-valued plateaus, separated by strong brightness variations at object boundaries.

To simplify the discussion, two-dimensional networks with nearest-neighbor interactions of strength α and periodic boundary conditions are chosen. The average network response is then governed by the two parameters α and w. The former characterizes the dependence upon the internal dynamics, the latter describes the dependence upon the statistics of the initial conditions.

At low noise level, that is for $w \leq \alpha$, the very first action potential in the network triggers all cells and leads to a globally synchronized oscillation. At high noise level, that is for large w and small enough α, such an avalanche is impossible. There is also not enough time for slow long-range ordering because the convergence takes place in finite time. This implies that the probability distribution for events decreases exponentially as a function of their size.

The computational network properties are therefore as follows. The systems always relax to periodic oscillations where every neuron fires with the same rate P^{-1}, independent of the structure of the initial condition. The spatiotemporal characteristics of the network oscillation, however, strongly reflect the statistics of the initial pattern. Regions with high similarity are smoothed out and represented by synchronized clusters of neurons. Regions with high variability are represented by spatially uncorrelated firing patterns. Integrate-and-fire networks thus operate like spatiotemporal extensions of resistive grids with line-breaking elements. More general network architectures that include longer-range connections and inhibitory synapses can be 'programmed' to perform specific computations as demonstrated in Figure 1.

Our results prove that already simple locally coupled integrate-and-fire neurons are able to rapidly perform nontrivial computations. During the first wave of neural activity after stimulus presentation, noisy input patterns are locally smoothed. The stimulus quality is encoded by the relative firing times of the synchronized clusters. The firing patterns themselves are stored dynamically in periodic neural reverberations to be used for later processing. The findings substantiate the hypothesis that networks of spiking neurons are capable to bind and represent objects through synchronized neural assemblies [6].

Note that in the first, rapid phase, connections operate in a 'lateral' mode — because the entire computation is accomplished once the last neuron fires for

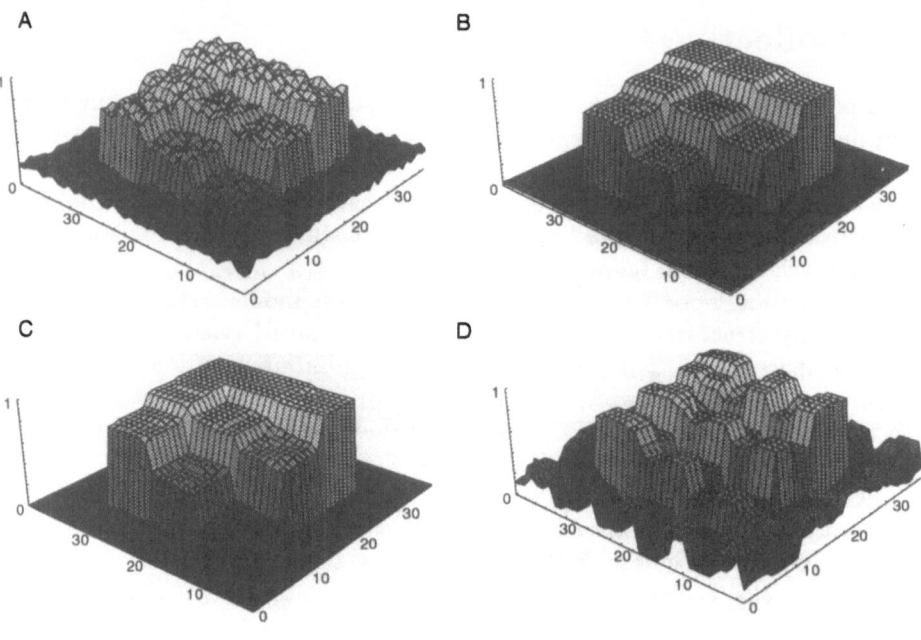

Figure 1: Collective computation with spiking neurons. Input pattern (A) and output of three networks (B - D) with 40 × 40 cells (x and y axis) and periodic boundary conditions. The initial values for the membrane potentials u_i (z axis in A) are noisy plateaus ($w = 0.1$) with means $\bar{u} = 0.9$, 0.85, 0.8; 0.8, 0.7, 0.6; 0.7, 0.5, 0.3, and a background of 0.2. The spatiotemporal network response is represented by the time elapsed since the last firing of a neuron (z axis in B - D). (B and C) Noise reduction through moderate local excitation; the output patterns closely resemble the original (noiseless) image, that is, the stimulus quality (grey-level) has been encoded in the relative firing times of the model cells. In (B) each neuron is connected to its four nearest neighbors with strength $\alpha = 0.06$. Additional couplings in (C) to the four next nearest neighbors with efficacy $\beta = 0.03$ give rise to an overfitting of areas with little stimulus differences. Contrast enhancement is possible through long-range inhibition if interaction distances match the typical length scale of the stimulus pattern. Such inhibitory long-range couplings ($\gamma = -0.03$) are included in (D), joining each neuron i with 16 cells (out of 32) along the boundary of a square of size 9 × 9, centered at i. In addition, local excitatory couplings with $\alpha = 0.05$ and $\beta = 0.02$ are used.

the first time, at least this neuron does not experience true feedback. Feedback loops are, however, of prime importance for the periodic oscillations exhibited by the network afterward. The example demonstrates that lateral cortical connections may serve a double purpose — they allow rapid computations as well as dynamic short-time memory.

5 Toward Synfire Webs

The output of one network with integrate-and-fire neurons can be used as input for a second system of the same type. This procedure may be iterated so as to build 'synfire webs' of arbitrary complexity. These systems are characterized by 'lateral' connections within elementary networks, feedforward connections between networks, and feedback loops from 'higher' areas to 'lower' processing levels.

What can be learned from the previous analysis about the dynamics possible in such network structures?

As a simple toy example, assume for the moment that there are no driving currents, that is, $I_i = 0$. In this case, the membrane potentials stay constant and no neuron will fire. If, however, the neurons receive pulsed external inputs of size P with a frequency P^{-1}, they will again relax to periodic solutions with the same firing frequency. (This is because the assumption of constant input current is not necessary for the proof in section 3. All what is needed is that the accumulated input over certain time intervals is the same for every neuron.) Once the stimuli cease, the network will go back to the silent mode.

Similar results can be obtained for more general pulsed input patterns and nonzero I_i. These considerations may open a new route to understand stimulus-dependent temporal correlations in frontal cortical areas. Previous theoretical approaches [9, 10] have shown that pure feedforward systems such as 'synfire chains' [7] are capable to generate self-focusing traveling waves of synchronized neural activity. The present work might help to extend these results to a network theory including lateral and feedback connections. At the same time, ad-hoc assumptions for a discrete-time description of the collective dynamics [11] might be replaced by a microscopic explanation. It seems also possible to extend the present framework to more realistic models of firing nerve cells [12] including axonal, synaptic, and dendritic delays [13].

Acknowledgment Except from the (more speculative) arguments of the last section, all results presented in this article were obtained in a collaboration with John J. Hopfield. The author's work has been supported by a Beckman Institute Fellowship and the Commission of the European Communities under the Human Capital and Mobility Programme.

References

[1] Thorpe, S.J. (1993) Temporal synchrony and the speed of visual processing. *Behavioral and Brain Sciences* **16**, 473-474.

[2] Herz, A.V.M. & Hopfield, J.J. (1995) Earthquake cycles and neural reverberations: Collective oscillations in systems with coupled threshold elements. To appear in *Physical Review Letters*.

[3] Hopfield, J.J. & Herz, A.V.M. (1995) Rapid local synchronization of action potentials: Toward computation with coupled integrate-and-fire neurons. *Proc. Natl. Acad. Sci. USA* **92**, 6655-6662.

[4] Eckhorn, R., Bauer, R., Jordan, W., Brosch, M., Kruse, W., Munk, M., & Reitboeck, H.J., (1988) Coherent oscillations: A mechanism of feature linking in the visual cortex? *Biological Cybernetics* **60**, 121-130.

[5] Gray, C.M. & Singer, W., (1989) Stimulus-specific neuronal oscillations in orientation columns of cat visual cortex. *Proc. Natl. Acad. Sci. USA* **86**, 1698-1702.

[6] von der Malsburg, C. (1981) *The Correlation Theory of Brain Function* (MPI for Biophysical Chemistry, Göttingen).

[7] Abeles, M. (1991) *Corticonics* (Cambridge Univ. Press, Cambridge, U.K.).

[8] Mirollo, R.E. & Strogatz, S.H. (1990) Synchronization of pulse-coupled biological oscillators. *SIAM J. Appl. Math.* **50**, 1645-1662.

[9] Gewaltig, M.-O., Diesmann, M. & Aertsen, A. (1995) Propagation of synfire activity in cortical networks: A statistical approach. This Volume.

[10] Herrmann, M., Hertz, J.A. & Prügel-Bennett, A. (1995) Analysis of synfire chains. Nordita preprint 95/5 S.

[11] Bienenstock, E. (1995) A model of neocortex. To appear in *Network*.

[12] Gerstner, W. & van Hemmen, J.L. (1992) Associative memory in a network of 'spiking' neurons. *Network* **3**, 139-164.

[13] Gerstner, W. & Herz, A.V.M. (1995) Work in progress.

Dynamic Representations Provide the Gradual Specification of Movement Parameters

Klaus Kopecz[1], Wolfram Erlhagen[2] and Gregor Schöner[3]

[1] Dept. of Medical & Biophysics, Univ. of Nijmegen, The Netherlands
[2] Institut für Neuroinformatik, University of Bochum, Germany
[3] CNRS-LNC, Marseille, France

Abstract: In an attempt to give a theoretical foundation of the abstract notion of "motor programming", neural representations of movement parameters are introduced on which activity dynamics is subject to sensory and preinformation as well as to intentional factors. A motor program is defined as a sufficiently developed cluster of activity, the location of which codes for parameters of the upcoming movement. Simulation results closely reproduce experimental findings. They show that motor programs can be prestructured by the stimulus set used during sessions and that the motor program can adapt gradually over time to a visual specification. Testable predictions are derived from the theory. As a concrete example, the programming of visually guided saccades is considered.

1 Introduction

The behavioral account to the notion of "motor programming" is based on observations of reaction times and parameters like movement amplitude, direction, force, etc. Sufficient experimental evidence has been accumulated to view a motor program (MP) as evolving from a dynamic process guided by sensory information and knowledge about future events (preinformation). This view is based on results from discrimination tasks, precue paradigms and double-step studies. Important for this study, MPs are prestructured by the stimulus set subjects are exposed to. This has been demonstrated in experiments where subjects have to synchronize movement initiation to an acoustic tone within a regular sequence. A visual target is presented at a variable time preceding movement initiation. This paradigm is depicted in Fig. 1. The target-tone interval (TT) is typically varied from 0 to 300 ms. Importantly, with vanishing or small TT, sensory information cannot be used any more to guide the movement. The observation is that subjects then show a default response depending on the target set used. With a set of two targets specifying two movement parameters with equal probability, this default response is largely scattered around the average of the two specifications. For increasing TT, the mean observed movement parameter gradually approaches the actual target specification with concurrently decreasing variability. Experimental results are described in [1, 2] for arm movements and in [3] for saccadic eye movements.

The aim of this study is to present a theoretical language which provides a natural framework for the integration of sensory and preinformation for purposes of motor programming. Movement parameters are specified by locations of activity on a neural representation. Thus, the principle of space coding found throughout the nervous system is used. The theory is applied in a concrete model architecture for the programming of visually guided saccades. Simulation results show in how far the above reviewed properties are reproduced and that testable predictions can be made.

2 Model architecture

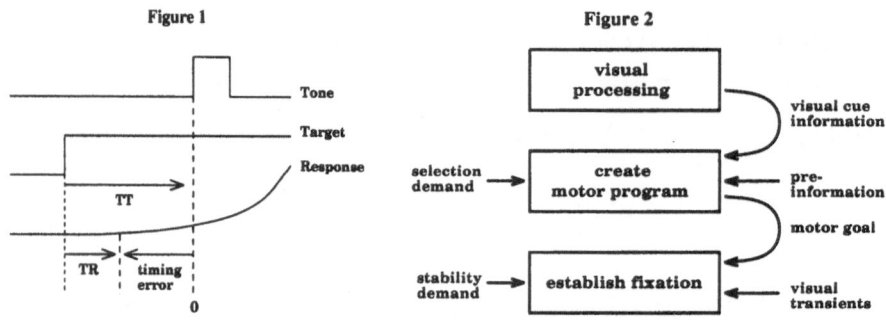

To refer to an exemplary behavior, we consider the saccadic eye movement system. Fig. 2 shows the model architecture together with its relation to several types of information. The visual processing level is introduced to provide a transformation on a spatial scale which decreases over time. On the programming level, information on possible movement targets is integrated (visual and preinformation). The location of created activity codes for the programmed parameter value. To allow for non-stimulus driven movements, the *selection demand* is introduced as an intentional factor with which activity can be created without visual information. To implement the potential for self-generating activity, dynamics has to be introduced, which led to the notion of *dynamic representations*. The creation of activity on the programming level is governed by the following equation:

$$\dot{u}(x,t) = -u(x,t) + h_{sel} + \int_R w(x - x')S[u(x',t)]dx' + f_{vis}(x,t) + f_{pre}(x)$$

The movement parameter is depicted by x (here, direction or amplitude). $u(x,t)$ is called the "field variable" from which "activity", $S(u)$, is derived through the sigmoid transfer function S. $w(x)$ represents the interaction kernel, which is of an excitatory-center inhibitory-surround type. These dynamics are able to generate one unique connected cluster of activity, providing decision making with ambiguous input and the self-generation of activity. The excitability of the field is controlled by the selection demand, h_{sel}. Voluntary movements are initiated by increasing h_{sel}. Activity is then generated around the location of maximal input determined by the superposition of visual information, f_{vis}, and preinformation, f_{pre}. In extreme, when a movement is required without visual information, the motor program is based on f_{pre} alone. To account for fluctuations in reaction times (RTs) and movement endpoints , Gaussian white noise has been added to the field dynamics.

The task of the fixation level is to stabilize gaze during fixations. In brief, fixation related activity stabilizes gaze against perturbations on a retinocentric representation governed by the same type of dynamics as used on the programming level. The balance between fixation related and peripheral activity determines the time of movement onset and thus the RT. The location where peripheral activity is created determines the saccadic endpoint. The *stability demand* has been introduced as an intentional factor which controls fixation stability and thus RTs.

In [4, 5] it has been shown that this model architecture together with its implementation in terms of dynamic representations successfully reproduces and predicts findings on gap/overlap effects and speed-discrimination trade-offs, the behavior in double-step conditions and the occurrence of anticipations. A similar theory for the programming of arm movements is proposed and analyzed in [6]. Here, the model behavior in the synchronized saccade paradigm illustrated in Fig. 1 is studied in more detail.

3 Simulation results

To start a movement at the prescribed time t_0, the process of activity generation must be initiated at $t = t_0 - t_p$, where t_p is the internal processing time, composed of the time needed to create activity on the programming level, which in turn must create peripheral activity on the fixation level, as well as the efferent delay. Importantly, in the model, the mean t_p depends on the set of TTs used, because visual information speeds-up activity creation. In the model we found a mean processing time of 175 ms for the used TT set.

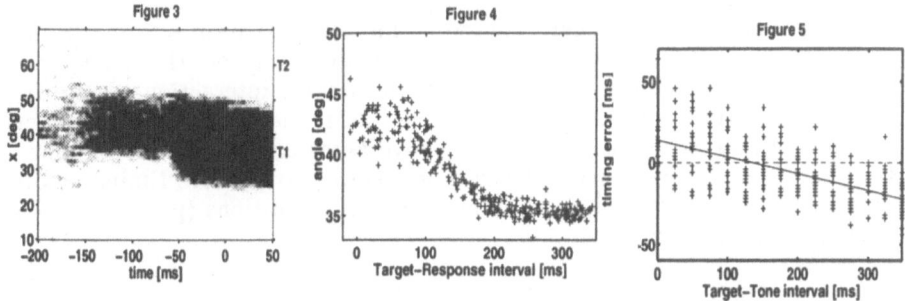

Fig. 3 illustrates the time course of activity on the programming level for TT= 100 ms. The possible target positions are indicated by T1 and T2, where T1 is actually presented. The time of required movement initiation is set to zero. The continuously acting preinformation is centered around the average of T1 and T2, based on the assumptions that the two individual prespecifications are sufficiently broad and overlapping to yield a maximum at the average location. At $t = -175$ ms, the selection demand is increased to create a cluster of activity, the location of which is here completely based on preinformation. Preceding this initiation, preinformation is reflected in an increased level of spontaneous activity. When visual information becomes effective, it results in a continuous modification of the motor program in direction to the target. Depending on the time when the motor program becomes effective in creating peripheral activity on the fixation level, the movement will be more or less directed to the target. This is illustrated in Fig. 4, where the stochastic dependence of movement endpoints on the TT-interval is shown. Further, movements based solely on preinformation are more variable. This increased variability is caused by the shape of preinformation, which is taken to be much broader than visual information. As a result, the motor program purely based on preinformation shows a large location variability, which can also be observed in the activity in Fig. 3 during the period from -170 to -60 ms, when compared to the later period where visual information acts additionally. The simulated data in Fig. 4 reproduces the main findings described in [1, 2, 3].

Two predictions arise from the here exposed view of the motor programming process. First, although we adjusted the internal processing time to meet the

32

required begin of the movement on the average, the model predicts a systematic dependence of the timing error (cf. Fig. 1) on TT. Because visual information contributes to programming activity, it speeds-up movement initiation, resulting in a negative correlation between timing error and TT as depicted in Fig. 5. The relation is in a good approximation linear for the used range of TT intervals, with a correlation coefficient of −0.7. Such a correlation has been indeed described in [1]. The second prediction arises from the assumption that preinformation shows averaging due to broad, overlapping individual specifications. Thus, one expects a transition from the monomodal average specification to a bimodal shape of preinformation, whenever the two possible targets become sufficiently separated. The model behavior with such a bimodal prestructure is shown in Figs. 6,7. Now, for small TT, programming dynamics selects randomly one of the modes, so that the probability of movement errors decreases from 50% to 0% as the TR interval increases. This transition from averaging to bistable behavior of movement errors as target separation increases, has been described in [2].

Fig. 6 depicts a case where the wrong location has been selected from the prestructure so that visual information must first "cancel" the incorrect motor program. This suppression through the lateral interaction results in an increased internal processing time, so that the model predicts a difference in the distribution of timing errors between averaging and bistable conditions. As shown in Fig. 8, this distribution is more skewed to positive timing errors in bistable conditions (panel B) than in averaging conditions (panel A), or may even show a separate mode.

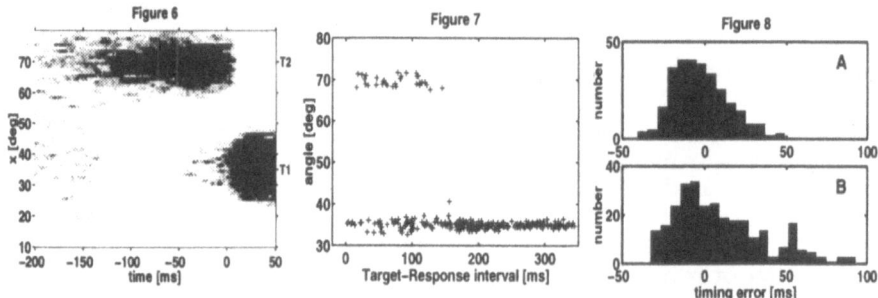

References

[1] W Hening, M Favilla, C Ghez. *Exp. Brain Res.* **71**:116-128, 1988
[2] M Favilla, J Gordon, MF Ghilardi, C Ghez. *Soc. Neurosci. Abst.* **16**:1089, 1990
[3] TR Stanford, LH Carney, DL Sparks. *Soc. Neurosci. Abst.* **16**:901, 1990
[4] K Kopecz. Saccadic reaction times in gap overlap paradigms. *Vision Research*, in press, 1995
[5] K Kopecz. Dynamic neural representations provide a theoretical framework to understand saccadic motor programming. *Submitted*, 1995
[6] G Schöner and W Erlhagen. Dynamic neural field theory of motor programming. *Submitted*, 1995

Acknowledgements: We acknowledge grants from the European Communities (ERB4001GT931013) and the German Science Foundation (DFG, KOGNET).

Recording from Foveal Striate Cortex while the Monkey is Looking at the Stimulus

J. Krüger, I. Bondar, H. Haan

AG Hirnforschung, Hansastr. 9a, D-79104 Freiburg, Germany

Abstract

We recorded from the foveal parts of area 17 where the stimuli are processed to which the monkey is directing its gaze. Most procedures used in the past did not allow for this [1,2, but see 3]. We observed high frequency oscillations in local field potentials whenever colours (30-60 Hz, monkey Z.) or other targets (60-120 Hz, monkey W.) were shown. However, central processes were influential as well.

A bundle of 32 microelectrodes was implanted in the foveal projection of area 17 of each of two java monkeys with the tips distributed within about 1 mm^3. Filtering isolated local field potentials (10-500 Hz) and spikes (.5-10 kHz). Cardboard stimuli and a variety of objects were presented by hand to the untrained animals. The animals were allowed to touch and grasp them. Video recordings were taken showing the animal, the stimuli and the computer clock. The heads of the animals were only loosely restrained.

In monkey Z. oscillations of local field potentials in the 30-60 Hz range (Fig. 1) appeared whenever it looked at a coloured object. Red objects, some of them subtending 1° of visual angle or less, were most efective, followed by blue and green. A great variety of uncoloured (or yellow) objects did not yield convincing evidence of oscillations. Thus, colour saturation appeared to be the relevant parameter.

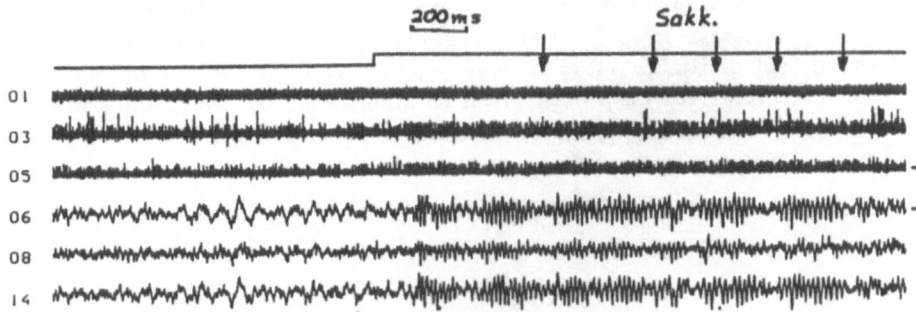

Fig. 1. Recordings from foveal area 17 of monkey Z. Large red stimulus. The step trace indicates the instant of a saccade towards the stimulus. The upper three recording traces show spikes, the remaining ones local field potentials. Channel numbers 5 and 6 (left) are from the same electrode.

With the evaluation of the data of monkey Z. we demonstrate that monitoring the gaze direction can be replaced by recognizing neural events[4]: After an experiment in which a variety of stimuli were presented to the monkey, we first identified oscillatory events above 30 Hz in printouts of neuronal recordings by eye. The events were ranked by magnitude for which printed templates were used. In Fig. 1 electrodes #6, 8 and 14 show rank-1, -2 and -1 oscillations, respectively.

The important point was that the oscillatory sections were identified in the raw data without relying on knowledge about any experimental circumstances. Thereafter we independently analysed the video recordings, and identified "presentation periods" in which a stimulus was presented, excluding only those intervals during which the monkey's gaze clearly was far removed from the direction of the stimulus. The 198 presentation periods formed 12% of the total recording time (4% colour stimuli, 8% other).

There was a total of 115 rank-1 and -2 oscillatory events, 82 of which occurred within the presentation period of a colour. Taking the 66 rank-1 events alone, the result was still clearer: All except 2 of these events occurred during a presentation period of a colour which in this case was mostly red or blue. It is unlikely that this can have occurred by chance since the colour presentation periods occupy only 4% of the total recording time. In the remaining cases (oscillations without colour presentation), in general no stimulus at all was presented, and the monkey's gaze was dispersed in a variety of directions. However, in one rank-1 and eleven rank-2 cases the gaze direction was consistent with the location of a small red object in the background which was visible from the monkey's position.

We conclude that the size of the "receptive field" of the oscillatory potential was of the order of 1° or less, since when small stimuli were used most changes of oscillatory activity were accompanied by changes of spike activity recorded simultaneously. At these instances minute eye movements could often be detected in the video recording.

Thus, based on consistent observations over a time span of nine days we can conclude from a large rank-1 oscillation that monkey Z. was looking at a a red or blue target. Probably, it also was aroused by it, without relying on a previous training paradigm: it sometimes became agitated and disconcerted upon presentation of the objects. We believe that the oscillations are a colour-dependent message from a central evaluation process. An adapted version of the model of Niebur et al.[5] can possibly account for our observations. Other interpretations have been given to high-frequency oscillations observed (or explicitly not observed) under different circumstances [6-12].

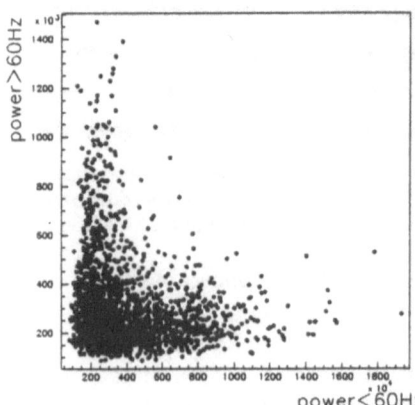

Fig. 2. Oscillatory power calculated in 512-ms intervals. Monkey W. Each interval is a point in the diagram. Abscissa: power in the frequency range < 60 Hz (relative units). Ordinate: power in the > 60 Hz-range. Note that high values never occur in both ranges simultaneously.

Monkey W. showed a different picture: Often large waves in the 20-Hz range were seen in the local field potentials of most electrodes (Fig. 3). When objects were shown to the animal, and it turned its gaze in that direction, often the large waves disappeared and oscillations of a smaller amplitude in the frequency range of 60-120 Hz appeared. It was surprising that the appearance of high power was mutually exclusive in the "<60-Hz" and the ">60-Hz" ranges (Fig. 2).

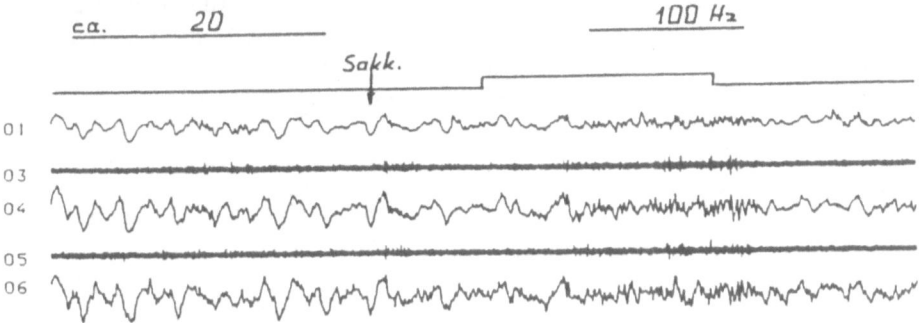

Fig. 3. Recordings from foveal area 17 of monkey W. Striped yellow-black stimulus. At left in the recording W. looks straight upwards to the ceiling. At "Sakk." it makes a large saccade, and a head movement, directed towards the vicinity of the stimulus. The step trace indicates further saccades towards and away from the stimulus. Recording traces #3 and 5 show spikes, the remaining ones local field potentials. Channel numbers (left) 3/4 and 5/6 are from the same electrodes.

So far, we could not find out a particular stimulus requirement to evoke these oscillations (one was consistently over several days that the monkey looked straight into the video camera). However, it was remarkable that on all electrodes showing spike activity most spikes were elicited during the high-frequency periods. Since the monkey could direct its gaze into a richly structured laboratory environment, it was unlikely that the sudden appearance of spikes on many electrodes, the disappearance of large 20-Hz waves, and the onset of high frequency oscillations, were merely due to particular optical stimulus parameters. We rather believe that a central process is involved since we observed firstly that the pupil widened during the high frequency periods, and secondly that these periods occurred more often, and increased in duration, as the end of the recording session was approached.

It was noteworthy that oscillations and spike activity apparently did not correlate in neither animal. Possibly the oscillations arise from central disinhibitory influences, and longer-lasting IPSPs are intercalated between them and their expression in spike activity of area 17 cells.

The differences between the observations in our two animals may on the one hand arise from differences in electrode location with respect to layers and columns, and on the other hand from differences in social rank-order, and from their different reactions to the stimuli

We are interested in developing our special procedures because
— they enable us to study the process of seeing in the most relevant foveal projection areas, without using training procedures for mere methodological reasons.

— we can avoid unnecessary coercive measures to the animal (gaze direction training, rigid head fixation), and

— an access to higher brain functions may be gained.

The principle is to repeatedly recognize neural events, and to take records of accompanying outer circumstances, such as stimuli or behaviour. The two essential requirements are that the neural events are sufficiently distinct so that they can be recognized as single events (multisite recording will be required for this in general), and long-term recording stability so that confidence can gradually accumulate about the significances of the patterns observed. The more stringent we are in the realm of recognizing neural events, the more relaxed we can be with the exact specification of external events.

We thank E. Schottmann for help.

Literature

1. Hubel, D. H. & Wiesel, T. N. Receptive fields and functional architecture of monkey striate cortex. J. Physiol. London 195: 215-243 (1968)
2. Wurtz, R. H. Comparison of effects of eye movements and stimulus movements on striate cortex neurons of the monkey. J. Neurophysiol. 32: 987-994 (1969)
3. Gallant, J. L. Connor, C. E. & Van Essen, D. C. Responses of visual cortical neurons in a monkey freely viewing natural scenes. Soc. Neurosci. Abstracts 20(1): 838, #349.11
4. Krüger, J. & Becker, J. D. Recognizing the visual stimulus from neuronal discharges. Trends in Neurosci 14: 282- 286 (1991)
5. Niebur, E., Koch, C. & Rosin, C. An oscillation-based model for the neuronal basis of attention. Vision Res. 33: 2789-2802 (1993)
6. Eckhorn, R., Bauer, R., Jordan, W., Brosch, M., Kruse, W., Munk, M. & Reitboeck, H. J. Coherent oscillations: a mechanism of feature linking in the visual cortex. Biol. Cybern. 60: 121-130 (1988)
7. Gray, C. M. & Singer, W. Stimulus-specific neuronal oscillations in orientation columns of cat visual cortex. Proc. Natl. Acad. Sci. USA 86: 1698-1702 (1989)
8. R. Eckhorn, A. Frien, R. Bauer, T. Woelbern, H. Kehr (1993) High frequency (60-90 Hz) oscillations in primary visual cortex of awake monkey. Neuroreport 4: 243-246
9. Von der Malsburg, C. Nervous structures with dynamical links. Ber. Bunsen- Ges. Phys. Chem. 89: 703-710 (1985)
10. Livingstone, M. S. Visually-evoked oscillations in monkey striate cortex. Soc. Neurosci. Abstracts 17: 176, #73.3 (1991)
11. Kreiter, A. & Singer W. Oscillatory neuronal responses in the visual cortex of the awake macaque monkey. Europ. J. Neurosci 4: 369-375 (1992)
12. Young, M. P., Tanaka, K. & Yamane, S. On oscillating neuronal responses in the visual cortex of the monkey. J. Neurophysiol. 67, 1464-1474 (1992)

Propagation of Synfire Activity in Cortical Networks: a Statistical Approach

Marc-Oliver Gewaltig[1], Markus Diesmann[2], and Ad Aertsen[2]

[1]Institut für Neuroinformatik, Ruhr-Universität Bochum, Germany,

[2]Center for Brain Research, The Weizmann Institute of Science, Rehovot, Israel

1 Introduction

Recently it was demonstrated that the activity of frontal cortical neurons in the awake behaving monkey comprises excessive occurrences of highly accurate (~1-3 ms) spatio-temporal firing patterns. Moreover, these patterns can be related to the behavioral state of the animal [1, 10]. On the basis of the characteristic anatomy and physiology of the cortex, it was proposed that *synfire* activity, propagating through the sparsely firing cortical neural network, presents a natural explanation for this phenomenon [2, 1]. In order to test this hypothesis, we investigated the dependence of reliable synfire propagation on the structural and the dynamical properties of a model cortical network, using the newly developed simulation tool SYNOD [6].

2 Synfire Chains

Synfire chains consist of *diverging/converging links* connecting a number of *groups* of neurons. A *diverging/converging link* can be described by two structural parameters, the *width w* and the *multiplicity m*, defining the number of neurons in a group and the minimum number of connections from a member neuron to the next group.

In simulation studies we found, that successful transmission from one group to the next requires two conditions to be fulfilled. First, the number of firing neurons within the sending group has to be larger than some minimum number a_0. Second, their spike time distribution must be narrower than some critical width σ_0. These critical values a_0 and σ_0 depend both on the structural parameters w and m, as well as on the details of the single neuron dynamics [7]. In order to characterize the dynamics, we need to assess the influence of the degree of synchrony in the spike arrival time distribution. Unfortunately, existing measures of neural transmission focus on two limiting cases, full synchrony and random arrival [2, 3]. Intermediate cases with a finite degree of temporal dispersion are not addressed.

3   Pulse-Packets

We introduce here the concept of *pulse packets* [7, 8], in order to overcome these restrictions and to quantify the degree of temporal synchrony in propagating volleys of neural activity. A pulse packet is a probabilistic description of the activity of a group of neurons, represented by a pulse density function $\rho(t)$. This pulse density function is determined by two parameters: the *activity a*, defining the number of active neurons in a group and the *width* σ, defining the temporal dispersion of the group activity (Fig. 1). This parametric description of synfire activity provides a conceptual framework that allows us to derive an appropriate neural transmission function and, thereby, to enhance our analytical insight into the role of the single neuron dynamics. Using this approach, we investigated

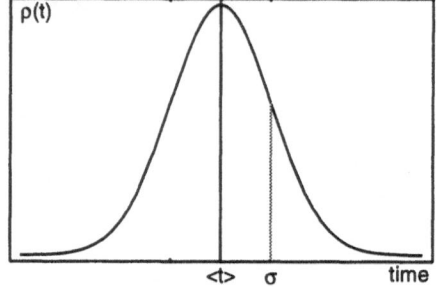

Figure 1: a pulse packet, the area under the graph represents the activity a.

the response of a model neuron [9] to input activity with varying degrees of synchrony. From the model neuron we recorded the response (time of first spike), collected in a PST-histogram over many trials. After normalization for the number of trials, the resulting output distribution was similarly described as a pulse packet, and the associated pulse density $\rho(t)$ along with the values of a and σ were determined. Thus, we could investigate how the output distribution changed, depending on the input distribution. To this end, the input was varied systematically from a sharp synchronous volley of spikes (small σ, large a) to an asynchronous rate variation (large σ, small a). For each pair of input parameters (a_{in}, σ_{in}), we measured the corresponding output pair (a_{out}, σ_{out}). As an alternative approach, we simplified the model neuron such that an analytical treatment was possible. This yielded a relationship for the input-output relation of pulse packets that could be solved numerically. Again, for each pair of input parameters (a_{in}, σ_{in}), we measured the corresponding output pair (a_{out}, σ_{out}).

4   Results

The resulting input-output relation between incoming and outgoing pulse packets can be visualized in so-called iterative maps. These yield a compact characterization of the neuron's firing dynamics. In contrast to earlier approaches where the neuron's firing probability is measured quasi-statically as a function of DC-current, this new transmission function takes full account of the dynamic properties of the input distribution $\rho(t)$. One appropriate way to look at it is to plot σ_{out} versus σ_{in} for constant a_{in}. The result from the numerical study is shown in Fig. 2; observe that for small values of σ_{in}, the outgoing pulse packet is wider than the incoming one. Synchronous input is thereby dispersed in time. With increasing σ_{in}, however, the curve crosses the diagonal and runs below it. Thus, beyond this intersection the neuron exhibits a synchronizing behavior, the intersection itself represents a stable attractor of an invariant pulse packet. These features of the single neuron dynamics can be carried over to describe the behavior of groups of neurons. In the simplest case of completely connected groups, the above distribution, appropriately scaled for the number of neurons in a group, directly describes the group's temporal response. This framework can be extended to the case of incompletely connected groups, where every neuron 'sees' only a fraction of the pulse packet.

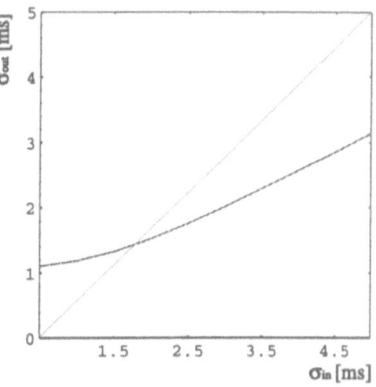

Figure 2: Iterative map for σ

Hence, we can determine the stable point of synfire activity traveling along a chain of such groups, and investigate its dependence on the neuron parameters and those of the chain.

5   Conclusions

The formalism of pulse packets provides the appropriate framework to clarify the notion of *coincident* firing. This yields a natural solution to the question whether the cortical neuron acts as an 'integrator' or as a 'coincidence detector' - a question which was raised many years ago [3] and was revived recently [5, 11]. The notion of pulse packets conveniently embeds these two different modes into one unified concept. Our investigation shows that the neuron may behave as either of the two, depending on the degree of synchrony of the input activity. The temporal structure of the input determines which of the two aspects is emphasized.

The approach outlined here opens the way for a quantitative description of network dynamics beyond the single neuron level. It provides a parametric language to describe the propagation of synchronous activity in networks, that can be characterized as 'locally feed-forward', i.e. locally composed of chains of groups of neurons. At the same time, it provides a conceptual bridge to link the single neuron dynamics to the mechanisms involved in stable transmission of information in such networks. Both aspects accommodate an analytic treatment of the model. An example for such analytic treatment is the stability analysis, using a dynamical systems approach [4] (see also Arndt et al. in this Volume). Finally, and most interestingly from the experimental point of view, the spike time distributions obtained in our simulations can be compared with the spike time statistics in recurring patterns in physiological data, and thus be used to test the synfire hypothesis for activity in the working brain.

40

Acknowledgments We gratefully acknowledge stimulating discussions with Moshe Abeles and Martin Arndt. Partial funding was received from the German Ministry of Science and Technology (BMFT), the Israeli Academy of Science, and the Human Frontier Science Program (HFSP).

References

[1] Abeles M, Bergman H, Margalit E, Vaadia E (1993) Spatio-temporal firing patterns in the frontal cortex of behaving monkeys. Journal of Neurophysiol 70: 1629-1643

[2] Abeles M (1991) Corticonics. Neural circuits in the cerebral cortex, Cambridge University Press, Cambridge, UK

[3] Abeles M (1982) Role of cortical neuron: integrator or coincidence detector? Israel Journal of Medical Sciences 18:83-92

[4] Arndt M, Erlhagen W, Aertsen A (1995) Stability of activity propagation in "synfire chains" - A dynamical systems approach. In: Goettingen Neurobiology Report 1995, p. 542. Menzel R, Elsner N (eds). Stuttgrt, New York: Thieme

[5] Bernander Ö, Koch C, Usher M (1994) The effect of synchronized inputs at the single neuron level. Neural Computation 6:622-641

[6] Diesmann M, Gewaltig M-O, Aertsen A (1995) SYNOD: An Environment for Neural Systems Simulations. Technical Report GC-AA/95-3, The Weizmann Institute of Science, Rehovot Israel

[7] Diesmann M, Gewaltig M-O, Aertsen A, Abeles M (1994) Structural and dynamical aspects of syn-fire activity in cortical networks. In: Proc 17th Ann Meeting European Neurosci Assoc, p 118. Oxford Univ Press, ENA

[8] Gewaltig M-O, Diesmann M, Aertsen A (1995) Role of the cortical neuron: both integrator and coincidence detector. In: Learning and Memory. Menzel R, Elsner N (eds). Stuttgrt, New York: Thieme

[9] Gewaltig M-O, Diesmann M, Aertsen A, Abeles M (1994) A realistic and computationally efficient model of cortical pyramidal nerons. In: Proc 17th Ann Meeting European Neurosci Assoc, p 118. Oxford Univ Press, ENA

[10] Grün S, Aertsen A, Abeles M, Gerstein G, Palm G (1994) Behavior-related neuron group activity in the cortex. In: Proc 17th Ann Meeting Europ Neurosci Assoc, p 11. Oxford Univ Press, ENA

[11] Shadlen M, Newsome W (1994) Noise, neural codes and cortical organization. Current Opinion in Neurobiology 4:569-579

Propagation of Synfire Activity in Cortical Networks- a Dynamical Systems Approach

Martin Arndt[1], Wolfram Erlhagen[1], and Ad Aertsen[2]

[1]Institut für Neuroinformatik, Ruhr-Universität Bochum, Germany
[2]Dept. of Neurobiology, The Weizmann Institute of Science, Rehovot, Israel

1   Introduction

During the last years, several models and related theories discussed the possible functional role of synchronized neuronal activity in cortical function. Here, we focus on recent findings by Abeles and colleagues on the abundance of accurate spatio-temporal spike patterns in the activity of neurons in the prefrontal cortex of awake behaving monkey, and their dependence on stimulus and behavioral context [1,2]. These findings support the hypothesis, that synchronous spike volleys propagate through the cortex in 'reverberating synfire chains' (RSC): feedforward networks with additional feedback connections. Using simulations of simplified, purely feedforward 'synfire chains', Diesmann and Gewaltig could demonstrate [3] that the stability of propagation of 'synfire volleys' in such chains strongly depends on the density of inter-node connectivity. Thus, the stability properties of these systems are described by iterative maps, which exhibit stable and instable fixpoints for the mean activity and the temporal width of the propagating 'pulse packet'. Motivated by these results we set out to develop a theoretical analysis of the stability properties of synfire propagations based on dynamical systems theory.

2   The 'Synfire chain' Model

The theory we present here is designed to describe the propagation of synchronous spike activity in a simple feedforward 'synfire chain' [4] without reverberating connections (Fig. 1). The network consists of a number of layers with w neurons per layer. Each neuron receives m inputs from the neurons of the preceding node. For a fully connected ('complete') chain the multiplicity of interlayer connectivity is equal to the number of neurons per node, i. e. $m=w$.

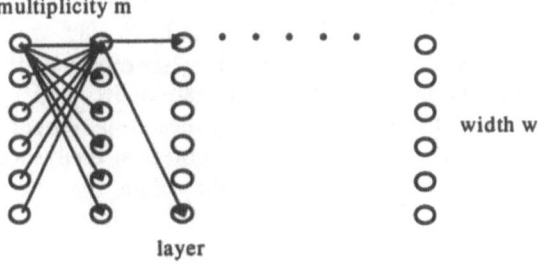

Fig. 1: Graph of a complete feedforward 'synfire chain' with the main structural parameters

For the individual model neuron we chose an 'integrate-and-fire' neuron for which the postsynaptic potential u is modeled by a leaky integrator. The firing probability is described by a sigmoid function (threshold Θ), the refractory dynamics v by a second leaky integrator. This leads to the following set of equations for the single neuron dynamics:

$$\tau_e \cdot \dot{u}_k = -u_k + v_e \cdot \sum_l \left[f(u_l - \Theta) \right] - v_{ie} \cdot v_k$$
$$\tau_i \cdot \dot{v}_k = -v_k + v_{ei} \cdot f(u_k - \Theta)$$

3   Stability of Synfire Propagation

3.1   Results from Simulation

We studied the activity dynamics in such simplified synfire chains by stimulating the first layer of the network with a brief volley of 1 ms . As a result, synchronous spike activity propagates along the network. The parameter varied during the simulation was the threshold level of the single neuron. The dynamics of synfire activity are described by the velocity (Fig. 2) and the temporal width of the propagating volley (Fig. 3) are shown below.

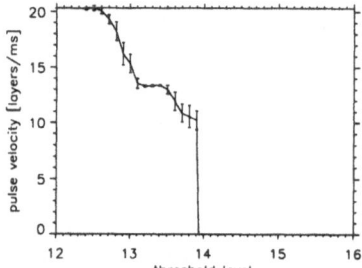
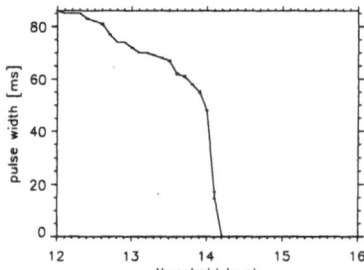

Fig. 2: Volley velocity vs. threshold   **Fig. 3:** Volley width vs. threshold

We found that for increasing threshold the velocity of the volley steadily decreases to zero up to the critical threshold value. Similarly, the width of the volley decreases to zero. Below the critical threshold stable propagations exist. If the threshold value is further increased the propagation becomes instable, i. e. the volley amplitude decreases to zero. Motivated by these separate ranges of stable and instable synfire propagations we developed a theoretical analysis using dynamical systems theory to predict velocity and width of the volley.

3.2 Results from Theory

For the case of a 'complete synfire chain', i. e. with fully connected nodes, the network can be simplified to a chain of single neurons, since every neuron in the complete chain receives the same input. This simplification allows a transition from the description of neuronal activity by discrete firings of single neurons to a continuous mean field description in terms of the interacting dynamics of membrane potential u and threshold v.

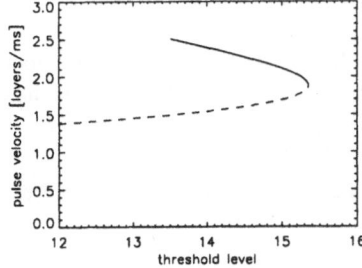

Fig. 4: Graph of a complete feedforward 'synfire chain' reduced to a chain of node neurons

The interaction of the neurons is described by a convolution (∗) between the neuronal output and a gaussian kernel that describes the local neighborhood connectivity This leads to a neural field description as given by Amari [5]:

$$\tau_e \cdot \dot{u} = -u + w(x) * f(u(x) - \Theta) - v_{ie} \cdot v; \qquad w(x) = v_e \cdot \left[e^{-\frac{1}{\sigma} \cdot (x-1)^2} + e^{-\frac{1}{\sigma} \cdot (x+1)^2} \right]$$

$$\tau_i \cdot \dot{v} = -v + v_{ei} \cdot u$$

Transformation to a reaction-diffusion-system yields a system that is similar to Fitzhugh-Nagumo equations [6,7], developed to describe the propagation of action potentials along the axon:

$$\tau_e \cdot \dot{u} = w_2 u'' - u + f(w_0 u) - v_{ie} \cdot v; \qquad w_0 = 2\sqrt{\pi} \cdot v_e \sigma, \qquad w_2 = \frac{\sqrt{\pi}}{2} \cdot v_e \sigma^3,$$

$$\tau_i \cdot \dot{v} = -v + v_{ei} \cdot u$$

Further analysis of the corresponding characteristic equation following the work of Rinzel and Terman [8] yields velocity (Fig. 5) and width (Fig. 6) of the volley solution as a function of the threshold level Θ.

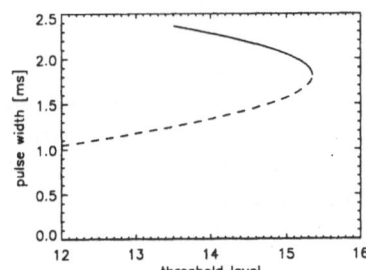

Fig. 5: Volley velocity vs. threshold   **Fig. 6:** Volley width vs. threshold

Together with the stable solutions (solid curves in Figs. 5 and 6) there exist also instable solutions (dashed curves in Figs. 5 and 6) with lower propagation velocities and smaller widths of the volley. The knee of the curves marks the critical threshold above which a stable propagation is no longer possible.

4   Discussion

We developed a theoretical description of the dynamics of activity propagation in simplified 'synfire chains'. The results of the analysis allow the prediction of propagation stability depending on single neuron parameters, i. e. threshold level.

In a related simulation study [3] it was shown that stable propagation requires a minimum amount of coincident activation at the input. In a related paper at this conference [9] they give an alternative analytical approach by using a statistical description for the shape of their puls packets.

The discrete spike dynamics in a discrete network of 'complete synfire chains' was analytically transformed to continuous field dynamics. After further transformation to a reaction-diffusion-system, the propagation of 'synfire' activity could be described by Fitzhugh-Nagumo like dynamics. The range of stable propagation can therefore be predicted. The analysis of the propagation velocities for stable and instable solutions provides a basis for the understanding of synchronization and binding phenomena. We are currently investigating whether this theory can also describe the shape of the stable solution and synchronization effects in incompletely connected chains.

Acknowledgements Special thanks to Marc-Oliver Gewaltig and Markus Diesmann for their expert support using their simulation environment SYNOD. Additional funding was received from the German Ministry of Science and Technology (BMFT), the German Research Foundation (DFG, Ae 10/2-1) and the Human Frontier Science Program (HFSP).

References

[1] Abeles M, Bergman H, Margalit E, Vaadia E (1993) J Neurophysiol 70:1629-1638

[2] Abeles M, Prut Y, Bergman H, Vaadia E, Aertsen A (1993) In: Brain Theory, Aertsen A (ed) Elsevier, Amsterdam

[3] Gewaltig M-O, Diesmann M, Aertsen A, Abeles M (1994) In: Goettingen Neurobiol Rep, p559; Diesmann M, Gewaltig M-O, Aertsen A, Abeles M (1994) In: Goettingen Neurobiol Rep, p560. Elsner N, Breer H (eds). Thieme, Stuttgart

[4] Abeles M (1991) Corticonics, Cambridge University Press

[5] Amari S, Arbib M A (1977) In: Systems Neuroscience, Metzler J (ed), Academic Press

[6] Fitzhugh R (1961) Biophys J 1:445-466

[7] Nagumo JS, Arimoto S, Yoshizawa S (1962) Proceed IRE, 50:2061

[8] Rinzel J, Terman D (1982) SIAM J Appl Math 42, 1111-1137

[9] Gewaltig M-O, Diesmann M, Aertsen A (1995)

Neurobiology - Posters

Aspects of Spatiotemporal Learning in Artificial Neural Networks: Modeling Synaptic Membrane Currents Using SPICE Simulations

Jaap Hoekstra and Petra Mantel
Delft University of Technology, Dept. Electrical Engineering,
P.O. Box 5031, 2600GA Delft, The Netherlands
e-mail: jaap@neuron.et.tudelft.nl

Most of todays artificial neural network models disregard the spatial structure of the neuron's extensive dendritic system. There are a couple of reasons, however, to take such a structure into account. First, as a consequence of the membrane properties of dendrites, this spatial structure influences the temporal processing of synaptic inputs. The time course of the membrane potential response at any place in the dendrite depends on the location of the synapse and the arrival time of an impulse. Second, the mutual distance of synapses are important, for instance, in case excitatory and nearby inhibitory synapses are considered, or in case of non-hebbian local learning.

The spatial structure of dendrites can be taken into account by the use of compartmental modeling techniques. In compartmental modeling the dendrite is subdivided into sufficiently small segments (or compartments), in which the physical properties (e.g. dendrite diameter, specific electrical properties) are spatial uniform and the potential is taken constant. The differences in potential occur *between* compartments rather than within them [2, 4]. The advantage of this modeling is that it places no restrictions on the membrane properties of each compartment.

Recently compartmental modeling received much attention again. A new generation of artificial neural networks extends its models with more (neuro-) biological knowledge.

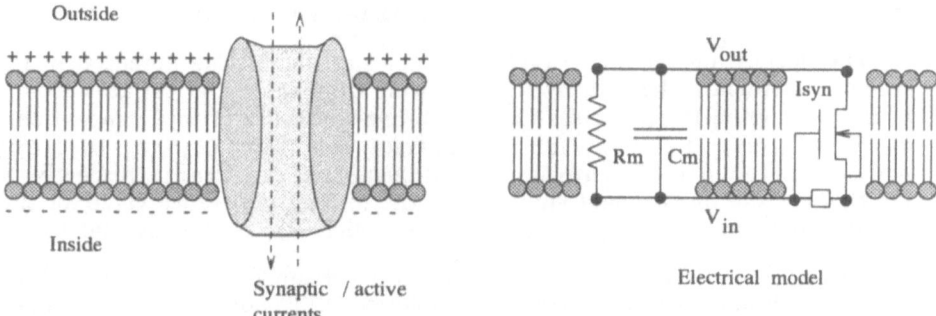

Figure 1: Schematic biological view and electrical model of a voltage controlled ion channel in a single compartment.

The modeling with compartments is done in three different ways:

- by deriving analytically expressions for the potential between the compartments [3];

- by developing programs in which the potential is derived numerically;

- by using electrical circuit simulation programs, like SPICE, to simulate dendritic structures [2] and references therein.

The basis for the modeling is the equivalent (electrical) circuit of a compartment. It models the three basic classes of ionic channels that are found in the membrane: *passive* or *leak*, and the *synaptic* and *active* channels. This circuit, see Fig. 1, a leaky capacitantance representing the leakage current and a transistor representing the voltage controlled behavior. First, the constant conductance, g_{leak}, in series with a constant battery (E_{leak}) through which the passive ionic current flows. Second, the synaptic (chemically gated) channels that are represented by a time-varying conductance ($g_{syn}(t)$), which is in series with a battery (E_{syn}) representing the (reverse) potential of the synaptic processes. Third, the active channels, which represent active membrane activity, a time- and voltage-dependent conductance ($g_{act}(t, V)$) in series with a battery (E_{act}) whose value is the equilibrium potential of the ionic species involved. The electrical potential V across the membrane is measured with respect to the resting potential, that is, the potential when there is no current flowing across the membrane. The compartment is joint to its neighbors by junctional resistors.

Many compartmental models only quantitatively consider passive dendritic membranes [1, 2, 4]. Although, some of the (commercial) compartmental neural modeling programs can cope with active membranes, it is, however, assumed, that there is no one modeling approach or program that is better than all the rest, each having its own advantages and disadvantages.

Most SPICE simulations of compartmental methods deal with passive dendritic membranes [2], or use very precise detailed descriptions, and thus complex and relatively large circuits, to model active membranes. From a point of view of implementability, the most simple possible circuitry having enough functional resemblance should be used in order to realize large on-chip networks. In this paper we construct a SPICE model for active dendritic membranes. After an sketch of how active channels can be incorporated, we construct simple electrical models for signal repetition, and the synapse, along with simulation results.

Figure 2 shows an equivalent circuit for a compartmental model of a ladder of segments of passive dendritic membrane. The structure resembles the structure of an electrical transmission cable.

The dotted line encloses exactly one compartment. The potential in the ith compartment is denoted by $V[i]$. Viewing from the center towards ground of the ith compartment we see the membrane resistance of this compartment $r_m[i] = R_m/\Delta x$ and the membrane capacitance $c_m[i] = C_m \cdot \Delta x$. Here, R_m and C_m are the membrane resistance and membrane capacitance, respectively. On the left and on the right side, half of the internal resistance value for this compartment appears; $r_i[i] = R_{in} \cdot \Delta x$, in which R_{in} is the internal lateral resistance.

In case of an active membrane, we model the channels to open if the voltage in the membrane exceeds a certain threshold value. The open channels conduct ionic currents. The channels can thus be modeled by voltage-gated current sources, or transconductances. As a suitable electrical equivalent we choose

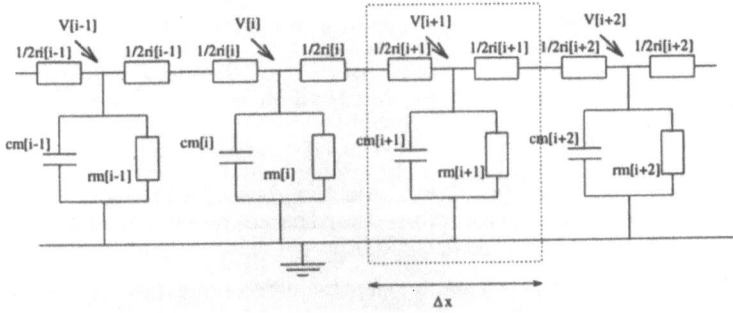

Figure 2: Circuit model transmission cable used in the SPICE simulations.

the MOS-transistor. The current through the transistor (from the source, s, to the drain, d) only depends on the voltage on the gate, g.

By connecting a voltage source to the transistor's drain and by coupling its source to the transmission cable, a voltage on the gate can direct the current towards the transmission cable. This voltage on the gate can be thought to originate from another neuron or from the transmission cable itself. In the first case the transistor could, in principle, model a synapse; in the second case the transistor could, in principle, model an active membrane.

In the models we also incorporate simple notions of the *refractory* periods, that is: the *absolutely* refractory period, a period after activation during which it is impossible to elicit a response to a second stimulus regardless of its strength, and the *relatively* refractory period during which the membrane potential is resuming its normal (resting) value. Figure 4(a) shows the response of the (transmission) cable on a voltage pulse on the transistor's gate. The pulses on the gate cause a short current pulse, which size and duration are independent of the duration of the pulse on the gate. The second pulse is less in amplitude showing the decrease in excitability of the membrane.

To consider a coupling of the gate of the transistor to the cable itself, we notice that, in case of an active membrane, the transistor has to conduct only if the voltage on the cable exceeds a threshold value. A circuit capable of performing this action, meets the following requirements:

- the circuit's input is the cable voltage;

- the circuit's output is a voltage exceeding the cable voltage to direct the

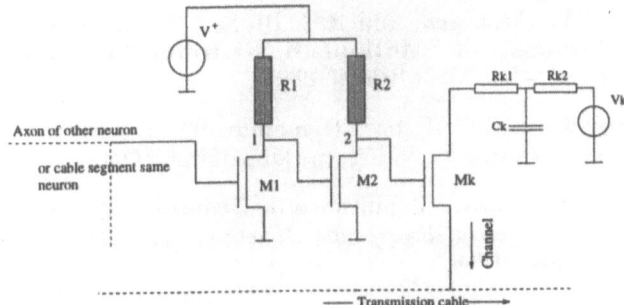

Figure 3: Circuit model used for the simulations of active membrane.

transistor's gate, in case the circuit's input exceeds the threshold value;

- the circuit's output is a voltage lower than the cable voltage, in case the input voltage is less than the threshold value;

- the circuit's input should not withdraw current from the cable.

The circuit depicted in Fig. 3 exhibits the desired behavior. The input of the circuit can, of course, also be used for the connection from other artificial neurons.

The SPICE simulation of a cable, representing an active membrane, is shown in Fig. 4(b). In this case a synaptic pulse (the most left pulse) causes the cable potential at the next compartment to exceed the threshold value, producing an action potential. This potential is passed to the next compartment. The figure shows the potentials in between the first five compartments.

The conclusion of our research is that it is possible to model some aspects of active membranes, within the framework of compartmental modeling, with relatively simple circuits. On going research should indicate if it is possible to model small artificial neural networks and to develop learning rules for these networks.

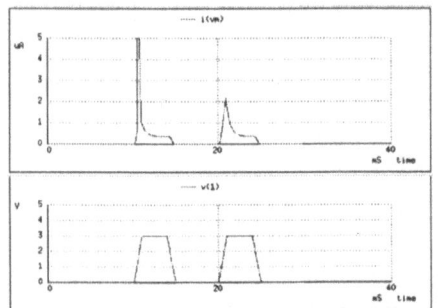 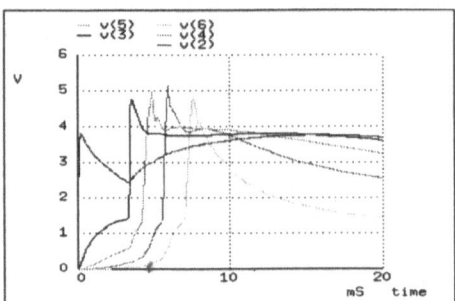

Figure 4: SPICE simulation of (a) the refraction periods; (b) an active membrane

References

[1] C. Mead. **Analog VLSI and Neural Systems**, Addison-Wesley Publ. Corp. Inc., 1989.

[2] I. Segev, J.W. Fleshman, and R.E. Burke. "Compartmental Models of Complex Neurons", in: **Methods in Neuronal Modeling** C. Koch and I. Segev (Eds.), The MIT Press, 1989.

[3] P.C. Bressloff and J.G. Taylor. "Dynamics of Compartimental Model Neurons", *Neural Networks*, Vol. 7, pp. 1153-1165, 1994.

[4] W. Rall. "Theoretical Significance of Dendritic Tree for Input-Output Relation", In: *Neural theory and Modeling*, R.F. Reiss (Ed.), Stanford University Press, 1964.

Possible Functional Roles of the Bipartite Dendrites of Pyramidal Cells

Ralf Möller   Horst–Michael Groß

Fachgebiet Neuroinformatik, Technische Universität Ilmenau
Ilmenau, FR Germany
e-mail: Ralf.Moeller@Informatik.TU-Ilmenau.DE

Abstract

A theory about the possible functional roles of the bipartite dendrites of cortical pyramidal cells is presented that tries to fuse aspects of both Cardinal Cell Theory and Assembly Theory. The article treats resulting functional differences of both dendritic pathways in conjunction with an hypothesis concerning the existence of two activity ranges. The model includes a three–rule system of self–organization.

1 Cortical architecture

The main cortical cell type are pyramidal cells. These cells show a typical dendritic morphology. An *apical* dendrite originates from the top of the cell body and runs perpendicular to the cortical surface into direction of the pia mater. These dendrites are contacted by synapses from cortico–cortical fibres that enter an areal from the white matter. *Basal* dendrites form a local plexus around the cell body. Synapses at basal dendrites are mainly set up by pyramidal cells in the neighbourhood. *Figure 1* shows this simplified picture of the cerebral cortex called 'skeleton cortex' according to BRAITENBERG, who also introduced the terms A- and B–system for the non–local apical path and the local basal path [2].

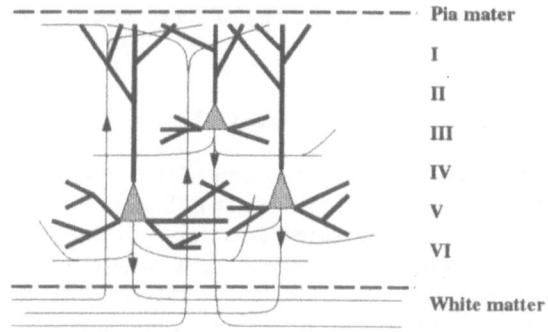

Figure 1: *A(pical)- and B(asal) system of pyramidal cells.*

2   Cardinal Cell and Assembly Theory

Figure 2 compares Cardinal Cell Theory and Assembly Theory. *Cardinal Cell Theory* as proposed by BARLOW [1] assumes a purely *unidirectional hierarchical structure* of neural systems. Low level neurons coding sensory signals feed higher level neurons that respond to complex features up to complete objects in the visual case. The neural system probably forms an inverse pyramide, because there has to be a vast number of complex cells at the top level of the hierarchy. Only few active cells represent a sensory situation. The function used to combine responses of low level neurons into the activity of high level neurons should be *conjunctive* in order to preserve the specialization of cells. Combinatorial explosion is one of the major drawbacks of this theory. *Assembly Theory* that goes back to HEBB [3] on the contrary concentrates on *lateral connections* between neurons. It avoids combinatorial explosion by 'cutting' the hierarchy at a much lower level — hierarchical structure is neglected in most cases. Sensory situations are represented by simultaneous activity of a large number of cells each coding a rather simple feature. Lateral connections between neurons of one level serve for the formation of cell assemblies, i.e. groups of cells that are active due to excitatory interactions. If one neuron can participate in more than one cell assembly, coding space enlarges drastically [5]. Therefore, (weak) *disjunctive* effect of lateral presynaptic cells is necessary, so different cell groups are able to excite one neuron.

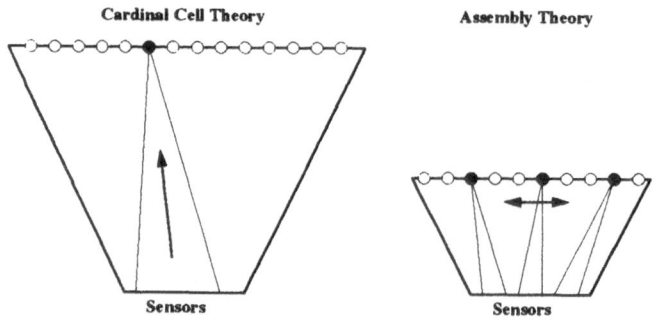

Figure 2: *Comparision of Cardinal Cell Theory and Assembly Theory.*

3   Functional interpretation

In difference to BRAITENBERGs interpretation of the two dendritic pathways of pyramidal cells [2] we try to fuse Cardinal Cell Theory and Assembly Theory by the assumption, that the *hierarchical structure* of the Cardinal Cell Theory corresponds with the *A–system* of pyramidal cells, whereas the *lateral structure* of the Assembly Theory finds its biological counterpart in the *B–system*. A neuron can either be activated from a specific input combination through the hierarchical A–system or can take part in different assemblies through the B–system. Following this, synaptic effects of *apical synapses* onto the postsynaptic neuron have to be *conjunctive* and effects of *basal synapses disjunctive*.

We further assume *two separate ranges of activity*: *high firing frequencies* for signals originating from 'real' *external inputs*, and *low frequencies* that code *hypotheses* about signals. Real signals are only mediated by the A–system, hypotheses by both systems. A pyramidal cell signals real activity, if enough parts of the apical represented conjunction are activated, otherwise the neuron fires at the hypotheses level. Independent of the activity level of presynaptic cells the basal dendrite is only able to cause hypotheses activity in the postsynaptic neuron. This can be explained from the point of learning. Lateral connections reflect statistical relations between conjunctions of signals. High basal weights between two neurons that resulted from frequent coincident activation via the A–system, could never be reduced if the basal pathway could evoke 'real' activity in the case of a 'real' conjunction of the basal partner neuron.

4   Self–organization

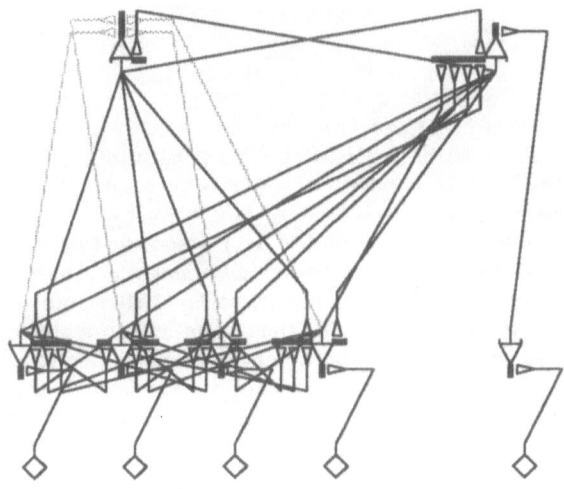

Figure 3: *Diagram of all connections of a simple test network.*

Self–organization of synaptic weights in both the A- and B–system requires a learning rule system of three rules. The *apical learning rule* is HEBBian with normalization of the weight sum reflecting conjunctive properties of the apical path. The rule connects 'real' pre- and postsynaptic activity and a term expressing real activity at a number of *basal* partner neurons. In the consequence, a conjunction is only represented in apical weights, if the conjunction often takes part in one ore more assemblies in the basal path. A *sensitivation rule* temporarily dissolves the conjunctive character of the apical path to allow for the alteration of the represented conjunction — without this rule, no other conjunction could ever excite the neuron up to the level necessary for learning. A 'pre–not–post–LTD' rule [4] has to be used for the *basal learning rule*, because asymmetric relations between two channels should create weights

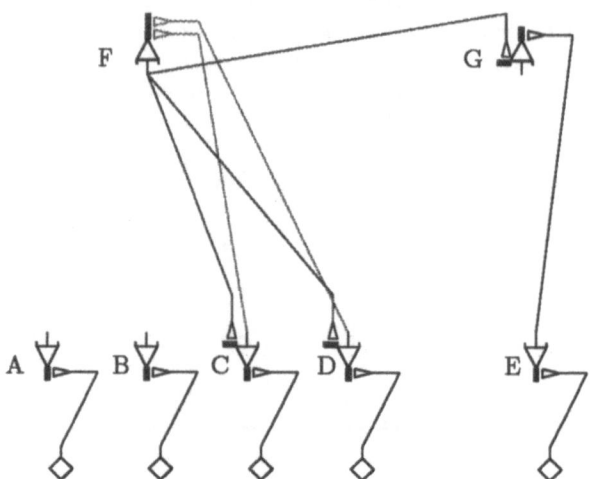

Figure 4: *Final weight state of the test network. Gray levels depict weights.*

in the direction from premise to conclusion. Related problems stated in [4] are avoided by the assumption of separate activity ranges. All learning rules are influenced by 'real' activity only.

Figures 3 and *4* show the effect of the learning rule system with a simple network. An asymmetric complex coincidence between the conjunction of channels C and D on one side and channel E on the other side was presented together with random activity on all channels. The learning rule system leads to the specialization of neuron F to the apical conjunction of C and D, and F and G are connected by asymmetric basal weights.

References

[1] H.B. Barlow. Single unit and sensation: A neuron doctrine for perceptual psychology. *Perception*, 1:371–394, 1972.

[2] V. Braitenberg. Cortical architectonics: general and areal. In M.A.B. Brazier and H. Petsche, editors, *Architectonics of the cerebral cortex*, pages 443–465. Raven Press, New York, 1978.

[3] D.O. Hebb. *The Organization of Behaviour*. Wiley, New York, 1949.

[4] P.A. Hetherington and M.L. Shapiro. Simulating Hebb Cell Assemblies: the necessity for partitioned dendritic trees and a post-not-pre LTD rule. *Network*, 4:135–153, 1993.

[5] G. Palm. On the internal structure of cell assemblies. In A. Aertsen, editor, *Brain Theory*, pages 261–271. Elsevier Science Publishers, 1993.

The Role of Cerebellum in Motor Control

Franz Mechsner

Department of Neural Information Processing, University of Ulm

D–89069 Ulm/ Germany

May 30, 1995

1   Introduction

The cerebellum is involved in training and performing of complex movements. The neurons of the cerebellar network and their interactions are well known. Clinical symptoms are well identified. Yet, in spite of the large body of knowledge accumulated up to now there is still need for a comprehensive, coherent theory that elucidates the role of cerebellum in acquisition and use of motor skills.

The theory presented here is an attempt to formulate such a synthetic view. We begin by defining a fundamental operation that seems powerful enough as to explain cerebellar function in motor control. Then we sketch a network model that is able to perform the fundamental operation in the context of motor activity. The model appears to be consistent with the known physiological, anatomical and clinical data. A more detailed description of the theory is given in [1].

2   The fundamental operation

For the sake of simplicity assume the motor system, without cerebellum, to be composed of many "motor control units" - groups of neurons of uniform behaviour. The nervous system starts and controls motions by exciting selected "motor control units". The activity of a particular motor control unit is called "elementary motor activity" (EMA). It is assumed that - even without cerebellum - the nervous system can combine EMAs rather freely, depending on the task.

However, due to the bounded control capacity of the brain, there is an upper limit to the number of EMAs that can be carried out simultaneously. The faster the coordinated motion the lower this maximum. Suppose there are 100 motor control units and therefore 100 possible EMAs. Imagine a particular motion being executed by simultaneously activating 5 EMAs labelled A, B, C, D, E for, let's say, 200 ms with a certain frequency. Assume furthermore that 5 EMAs is the upper limit in the case of the movement in question. This means: First, if performed faster the coordinated motion becomes unmanageable. Second, it is impossible to add any further EMAs and keep control.

According to the present theory the nervous system is able to train higher coordinated motions if the cerebellum comes into play. Its fundamental operation is to "couple" EMAs by means of a learning process according to the following rule:

- *If a set of EMAs has often been performed simultaneously the participating EMAs exert a strong excitatory influence on each other when performed together.*

It is claimed that this fundamental operation is sufficient to explain the role of cerebellum in the aquisition and use of motor skills. To appreciate this consider the nervous system driving the EMAs A, B, C, D, E with roughly the same control pattern as before training. The effect is: Since those 5 EMAs activate each other the coordinated movement is performed stronger and faster. However, by reducing the intensity of the control pattern the nervous system can perform the movement at a slower pace as well. While saving some handling capacity in this way the nervous system can simultaneously add new EMAs and keep control. The first effect allows training of higher motion speed - also in the case of sequences -, whereas the second effect allows exercising of higher spatial coordination. Apart from the mentioned effects cerebellar "coupling" allows better reproducibility and in some cases better synchronization of coordinated movements.

An important point is still lacking in this oversimplified picture: It is by no means necessary to assume here that a trained motion is a fixed pattern of muscle activity, performed always in the same, unalterable way. This would render the theory unplausible because real motions can be executed in a rather flexible way, and usually a particular motion cannot be performed at higher speed by simply multiplying the vector of control activity as suggested in the previous paragraph. It must be realized that the cerebellar "couplings" allow for many variations in the activity pattern of simultaneous EMAs. The nervous system can, for instance, adaptively tune necessary variations in the control pattern to account for mechanical disturbances or for changes in dynamical properties of the interacting body parts with varying motion speed. The existence of cerebellar "couplings" does not guarantee by itself improved motor skills - it is the nervous system as a whole that can improve its motor control capacity by making use of the fundamental operation.

Complicated and highly coordinated motions can be learned as described and performed in a flexible fashion. This holds true for totally synchronous muscle activities as well as for overlapping activities and sequences, fast ones in particular: Due to temporal distributivity in motor planning and control, to axon conductance time spans and delayed feedback, the cerebellum can "couple" EMAs that influence muscle activities which are not synchronous. Note that the described picture is meant just for illustrating the essence of cerebellar function - the notion of "EMAs" has to be much refined and adapted when it comes to explain realistic motor learning and control, especially if learning processes in the cerebral cortex come into play.

3   The network model

The detailed network model is inspired by the theories of cerebellum put forward by Marr [2], Albus [3], and Houk [4]. However, it is different in many respects allowing for the establishment and flexible use of "couplings" as outlined.

Its main features are described in short: Imagine a simplified model cerebellum consisting of a cerebellar cortex folded around a single nucleus. Suppose for the time being that the neurons of the nucleus are quasi-spontaneously active with a certain frequency. This frequency can only be varied by inhibitory action of Purkinje cells of the cerebellar cortex. It is assumed that the parallel fiber activity in the cerebellar cortex is - by means of Golgi cells - restricted to 1% of the possible activity, the level of activity remaining constant all the time over the cerebellum. There is an "initial state" of the cerebellum where all synapses from parallel fibers to Purkinje cells have maximum weight. The resulting constant activity of Purkinje cells is claimed to inhibit the constant spontaneous activity of all nuclear neurons exactly down to zero.

Only due to learning processes this situation can change and the cerebellum can have effects. As first suggested by Albus [3] it is assumed that active synapses from parallel fibers to a particular Purkinje cell are slightly weakened if a climbing fiber signal occurs at roughly the same time. However, one single Purkinje cell is not assumed to have significant effects. It is claimed that the cerebellum is composed of modules, and that each of them consists of a huge number of Purkinje cells exclusively projecting to a particular "compartment" in the nucleus - a group of neurons of uniform behaviour. Each nuclear compartment can exert exitatory influence on a particular EMA. Every EMA can be influenced by a corresponding cerebellar module whose output is provided by a particular nuclear compartment.

Only if a great portion of the Purkinje cells in a particular module receive input by means of weakened parallel fiber synapses, as described, a significant reduction of inhibition in the corresponding compartment takes place. Thus the nuclear neurons become active. It is assumed: all Purkinje cells in a module receive the same sort of climbing fiber input. They are excited synchronously by these fibers if the particular EMA influenced by this module - let's call it "D" - is carried out. It is assumed furthermore that parallel fibers signal the execution of EMAs to Purkinje cells. This implies: If a particular EMA, let's say "F", is often performed together with EMA "D", those parallel fiber synapses signalling "F" to "D"-Purkinje cells are weakened. The same holds true for those parallel fiber synapses signalling "D" to "F"-Purkinje cells. Therefore a reciprocal "coupling" of "D" and "F" takes place.

This is a rough idea of how the "coupling" operation is performed by the cerebellar network. However, it is obvious that a "coupling" of two EMAs as described can be a mostly undesired effect - imagine for instance a thumb bending that invariably leads to a middle finger bending. But the reciprocal "coupling" of only two or three EMAs is avoided in the model - the cerebellum is only able to effectively couple sets of many EMAs. This takes place because not all Purkinje cells of a particular module receive the same parallel fiber input: Purkinje cells of one module are placed together in, let's say 10, groups called microzones. These microzones are distributed all over the cortex, thus each of them receives different parallel fiber input and therefore information about another subset of EMAs. Only if several EMAs are executed together they can influence sufficiently many microzones of a module to exert a significant effect to the corresponding nuclear compartment. Altogether the consequence is: The cerebellum can only influence complex movements not simple ones. Experiments performed by Thach [5] support the view that this is indeed the case.

References

[1] F. Mechsner, " A new hypothesis on cerebellar function", in preparation

[2] D. Marr, "A theory of cerebellar cortex", J. Physiol. 202: 437-470 (1969)

[3] J.S. Albus, "A theory of cerebellar function", Math. Biosci. 10: 25-61 (1971)

[4] J.C. Houk, " Model of the cerebellum as an array of adjustable pattern generators", in: M. Glickstein et al (eds.), Cerebellum and neuronal plasticity, Plenum Press New York, 249-260 (1987)

[5] W.T. Thach, "The cerebellum and the adaptive coordination of movement", Annu. Rev. Neurosci. 15: 403-442 (1992)

A Point Process Approach to Cortical Networks

Stefan Rotter

Max-Planck-Institut für Entwicklungsbiologie
Spemannstraße 35/IV
72076 Tübingen, Germany

Ad Aertsen

Center for Research of Higher Brain Functions
Weizmann Institute of Science – Dept. of Neurobiology
Rehovot 76100, Israel

Abstract

The "cognitive" properties of some artificial neuronal networks have introduced attractive models for cortical function. We discuss an extended framework for the description of biological nerve nets such that a direct comparison with the signals from electrophysiological recordings on the level of individual nerve cells becomes feasible. The mathematical analysis of these models leads to explicit conditions on their biophysical parameters giving rise to unexpected conclusions. We demonstrate that the "dynamic repertory" of a system of interacting spiking neurons is dramatically enhanced, if signals are admitted to have a time structure. Some possibilities of a spatio-temporal code in presence of plastic synapses and an appropriate learning rule are discussed.

1 Spatio-temporal patterns of action potentials

There is wide consensus that neurons in the neocortex exchange information by means of action potentials. In contrast to the highly reproducible waveform of spikes from an individual neuron, however, the timing of their occurrence is far from deterministic. As is known from single cell recordings both in slice preparations and in intact brains of behaving animals, there is a considerable variabilty in neuronal responses to repeated presentations of identical stimuli. Any theory of cortical function must therefore account for the specific properties of a stochastic code relying on identical point-like events distributed in space and time.

We explore the spatio-temporal dynamics of cortical activity by using an abstract network model, which is compatible with the notion of interacting stochastic point processes. Any such system may be viewed as a Markov process, whose "state" at a given time instance is the spatio-temporal pattern of all previously generated spikes. The "transition probabilities" specify how a pattern gradually evolves in time. The infinitesimal parameters of the process, usually called point process "intensities", are given by the spike probability normalized for an infinitesimal time interval, conditional on the pattern of previously generated action potentials. If these parameters exist, the resulting master equation is readily solved. This general framework sets the stage for a quantitative assessment of otherwise untractable network phenomena.

By taking account of redundancies in the state space, one arrives at a point process model for neuronal function which is of "integrate-and-fire" type. Each cell has its own state space with a "linear" structure and a "nonlinearity" transforming states into rates. The rates, in contrast to the situation in many ad hoc models in use, do not specify averages over a longer period of time, but are considered as instantaneous measures of neuronal excitation. In simple cases, the physical nature of the states can be identified with the help of experiments.

2   Physical states and system identification

By all available evidence, the class of point processes having infinitesimal parameters is sufficiently rich to comprise appropriate models for cortical function on different levels of abstraction. Each specific model amounts to an interpretation of the infinitesimal parameters in terms of biophysical properties of nerve cells. We use the solutions of the general master equation to obtain numerical fits of the parameters of such models to experimental observations. Further analysis of the fitted model's behavior can be expected to yield information on the role of subthreshold processes for the emergence of collective behavior.

Adopting this aproach, we can account for the details of spike generation upon direct current injection in regular spiking neurons, which constitute the major cell type in neocortex [3]. Using systematic variation of injected current in a slice preparation as an experimental paradigm, we identified membrane current, rather than membrane potential, as the significant physical state variable. The model parameters correspond to post-spike hyperpolarization currents, the constant current injected through the electrode, and a characteristic function translating the total membrane current into the point process intensity [5]. There is strong experimental evidence that this current-based description extends to synaptic interaction in terms of excitatory postsynaptic currents [4].

3   Analysis of the network model

In a model network of regular spiking neurons, a stable low level of activity is maintained whenever the post-spike hyperpolarization exceeds the total amount of excitation and inhibition received by each neuron. This hypothesis and its consequence are both in accordance with our knowledge of physiology in the cortex, where synaptic couplings are sparse and weak. Surprisingly though, inhibitory neurons are not required for stability, all rate control might be taken care of by "auto-inhibition". Yet another interesting aspect of this parameter constellation is that the coefficient of variation for the inter-spike interval distribution of single neuron spike trains increases with the total amount of recurrent excitation in the network. Again, no inhibitory neurons are necessary to achieve coefficients as obtained from *in vivo* data. This result is in contrast to some recent observations in the literature [6, 7].

It may be useful to characterize a single cortical pyramidal cell as a stochastic oscillator. The reason for this is not only its regular firing behavior upon DC stimulation, but also its response to transient inputs. In fact, the susceptibility of the model neuron to synaptic inputs shows a distinct phase preference, leading to pronounced resonance phenomena. In addition, the characteristic function translating current into neuronal excitation turns out to be convex, causing the neuron to prefer synchroneous inputs over asynchroneous ones. From this one correctly predicts that transient synchronization of groups of neurons should be part of the dynamic repertory for recurrent networks of regular spiking neurons.

Computer simulations of sparsely connected networks with random topology show that one can indeed distinguish two time scales of synchronization phenomena. At the macroscopic time scale (tens to hundreds of milliseconds), subgroups of neurons may exhibit transient states of "loose" spike synchronization, without appreciable effects on the average rates. Such "assemblies" desynchronize and reorganize periodically upon persistent stimulation. Within the periods of enhanced group activity, one can observe complex spatio-temporal patterns on a millisecond time scale reoccurring with great precision. Similar findings, both at the macroscopic time scale [8] and at the level of millisecond spike patterns [1, 2], have been reported from multi-neuron recordings in the prefrontal cortex of the behaving monkey.

4   Plasticity of time structure

Adaptive properties of a nervous system are most valuable for contolling the agent's behavior in a highly variable environment. We began to investigate the possibilities of plasticity in time structure of neuronal signals by introducing physiologically inspired Hebb-like synapses into our stochastic networks. In the context of point process models with physiological parameters, this leads to a learning rule of covariance type. From numerical simulations we conclude that the weakening of synapses connecting out-of-phase neurons can serve to maintain a stable total amount of excitation within such networks. No artificial normalization of synaptic strengths is necessary. By employing a measure for the "distance" between spatio-temporal patterns which naturally emerges from point process theory we can demonstrate that the mere presence of plastic synapses can have useful consequences. Among other things, one observes the imprinting of temporally extended stimulus patterns, as well as the automatic generation of discriminating representations for distinct stimuli which are presented sequentially.

5   Conclusions

We show that neuronal network models using generalized "integrate-and-fire" dynamics are a mathematical consequence of the assumption that neurons communicate by the use of action potentials. The corresponding dynamical equations are completely solved thus enabling a fit of the model parameters to electrophysiological recordings. A simple parametric characterization of single neuron function is achieved by fitting the model to the regular spiking behavior of cortical pyramidal cells. A number of fundamental properties of recurrent cortex-like networks assembled from such neurons can be predicted, most notably their ability to maintain stable low rates of activity without the help of inhibitory neurons.

Computer simulations indicate that high precision spatio-temporal patterns, embedded in periods of enhanced cooperative group activity, may play a role for coding and computation in such networks. Plasticity of the temporal structure of such patterns is achieved by introducing Hebb-like synapses into the network. The resulting properties brings the point process model close to what general abstract neuronal networks are known to be capable of. By further exploiting our model system's known mathematical structure we expect to derive quantitative predictions to be applied to more complex experimental paradigms involving neural structures with nonrandom topology and plastic properties.

Acknowledgements

Partial funding was received from the German Ministry of Science and Technology (BMFT), the Israeli Academy of Science, and the Human Frontier Science Program (HFSP).

References

[1] M. Abeles, H. Bergman, E. Margalit, and E. Vaadia. Spatiotemporal firing patterns in the frontal cortex of behaving monkeys. *Journal of Neurophysiology*, 70:1629–1638, 1993.

[2] M. Abeles, Y. Prut, H. Bergman, E. Vaadia, and A. Aertsen. Integration, synchronicity and periodicity. In A. Aertsen, editor, *Brain Theory: Spatio-Temporal Aspects of Brain Function*, pages 149–181. Elsevier, Amsterdam, New York, London, Tokyo, 1993.

[3] D. A. McCormick, B. W. Connors, J. W. Lighthall, and D. A. Prince. Comparative electrophysiology of pyramidal and sparsely spiny stellate neurons of the neocortex. *Journal of Neurophysiology*, 54:782–806, Oct. 1985.

[4] A. D. Reyes and E. E. Fetz. Two modes of interspike interval shortening by brief transient depolarizations in cat neocortical neurons. *Journal of Neurophysiology*, 69:1661–1672, 1993.

[5] S. Rotter. *Wechselwirkende stochastische Punktprozesse als Modell für neuronale Aktivität im Neocortex der Säugetiere*, volume 21 of *Reihe Physik*. Harri Deutsch, Frankfurt, 1994.

[6] M. N. Shadlen and W. T. Newsome. Noise, neural codes and cortical organization. *Current Opinion in Neurobiology*, 4:569–579, 1994.

[7] W. R. Softky and C. Koch. The highly irregular firing of cortical cells is inconsistent with temporal integration of random epsps. *The Journal of Neuroscience*, 13:334–350, 1993.

[8] E. Vaadia, I. Haalman, M. Abeles, H. Bergman, Y. Prut, H. Slovin, and A. Aertsen. Dynamics of neuronal interactions in the monkey cortex in relation to behavioral events. *Nature*, 373:515–518, 1995.

Stochastic Resonance and Multi-modal Firing Patterns in Single-neuron Models

Danny Linders

Bert Kappen

Department of Medical Physics and Biophysics, University of Nijmegen

RWCP* Novel Functions SNN† Laboratory

Nijmegen, The Netherlands

Abstract

In analyzing data from experiments on periodically stimulated neurons, the interspike interval histogram shows a multi-modal distribution and under certain conditions, stochastic resonance. We discuss a simple integrate and fire model of a single neuron which is capable of describing these two features. As input for the neuron we use a periodic signal combined with poisson noise, this gives results that resemble data from experiments.

1   Introduction

There has been a great number of experiments in the past involving measurements on single neurons in the presence of some external stimulus. In the special case of a periodic and excitatory stimulus a typical phenomenon is observed: the neuron of interest tends to fire with the same period as the period of the stimulus it receives. We call this the coherent signal.

It depends on the strength of the stimulus whether a spike occurs every cycle or not. In the latter case we observe multi-modality in interspike interval histograms (ISIHs): peaks at integer multiples of the driving period.

Changing the noise level at the input of the neuron gives a different picture of the ISIH. At a specific noise level a maximum occurs in the coherent power at the output of the neuron (this is the power of the coherent signal, see above). This is called stochastic resonance (SR).

It was back in 1960 that Gerstein (Gerstein and Kiang, 1960) observed a multi-modal ISIH when they discussed several ways to analyze data from experiments on single neurons. They used anesthetized cats whom they presented an auditory stimulus, being clicks at various rates and intensities. Siegel (Siegel, 1990) did similar experiments with light instead of sound. Sensory experiments on monkeys were done by Mountcastle (Mountcastle et al., 1969). In all these experiments the phenomenon of multi-modal firing patterns was observed. This effect was discussed by Gerstein and Mandelbrot (Gerstein and Mandelbrot, 1964). They represented the neuron by a random walk model, with a reflecting

---

*Real World Computing Partnership
†Foundation for Neural Networks

barrier at a resting potential and an absorbing barrier at the threshold. Adding a drift term, the strength of which was periodically varied in time, they obtained ISIHs like the ones found in the experiments of Gerstein and Kiang.

Douglas (Douglass et al., 1993), too, observed multi-modal ISIHs in experiments on the mechanoreceptors of the crayfish. In addition, he observed SR in varying the noise level of the stimulus. (Although he used a slightly different definition of SR than we will discuss below.) This phenomenon was discussed by Longtin (Longtin, 1993). He treated SR in two models: a double well potential and an excitable neuron model following the Fitzhugh-Nagumo equations.

Our goal is to demonstrate that SR can be observed in a neuron model with much more simple dynamics, but still biologically plausible.

2   The Model

A suitable model seems to be an integrate and fire model in which the soma potential $v(t)$ plays a central role. Initially the potential starts at zero. It is influenced by two inputs and a decay. The neuron input is divided in two parts: the noise and the signal. The noise is modeled as poisson distributed spikes with an average interval τ_{noise} of constant strength ϵ_{noise}, while the signal has a steady period τ_{sig} and strength ϵ_{sig}. The potential is increased by the noise and the signal and decreased at a constant rate by the decay with strength γ. The change in soma potential can thus be written as

$$\dot{v}(t) = \epsilon_{sig} \sum_{k=1}^{\infty} \delta(t - k\tau_{sig}) + \epsilon_{noise}\, \xi(\frac{t}{\tau_{noise}}) - \gamma, \tag{1}$$

where $\xi(.)$ denotes a poisson process. At the time the potential reaches a threshold θ, the neuron fires and the potential resets to zero.

The choice of a linear decay as an approximation of an exponential decay is motivated by our aim to find an analytical expression for the SR in this model. In the case of exponential decay (and noise from a standard Wiener process) the model turns into an Ornstein-Uhlenbeck process, for which no expression can be found for first passage times. This means that one cannot calculate the probability of firing at a specific time after the last spike. Adding a periodic input will only make things more complicated, thus this model is better left behind. In the model we describe here, if no signal is applied, it is possible to write down the probability that no spike has occurred during a time interval $0 \le t \le \tau$. The following result is due to Pyke (Pyke, 1959).

$$P\{ \sup_{0 \le t \le \tau} (v(t)) \le \theta \} = e^{-\tau/\tau_{noise}} \sum_{n=0}^{\lfloor \gamma\tau + \theta \rfloor} \frac{\gamma\tau + \theta - n}{n!} (\gamma\tau_{noise})^{-n}$$

$$\times \sum_{j=0}^{\lfloor \theta \rfloor} \binom{n}{j} (j - \theta)^j (\gamma\tau + \theta - j)^{n-j-1} \tag{2}$$

For notational convenience this formula has been written for the case the process has been scaled so that ϵ_{noise} equals one, thus $\gamma \to \gamma/\epsilon_{noise}$ and $\theta \to \theta/\epsilon_{noise}$.

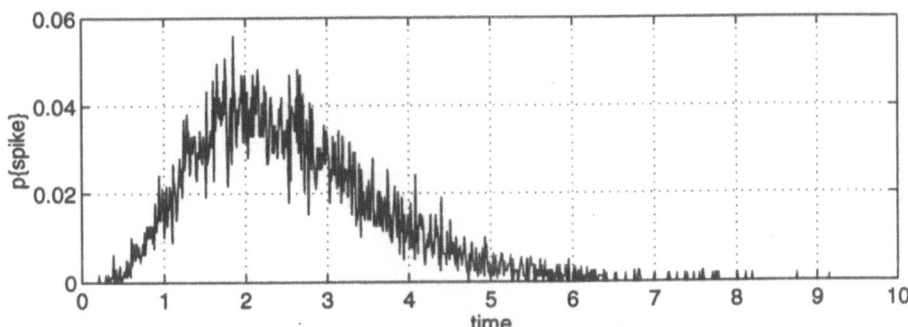

Figure 1: The ISIH of spontaneous activity. Parameters $\theta = 4$, $\gamma = 0.05$, $\tau_{noise} = 0.5$ and $\epsilon_{noise} = 1$.

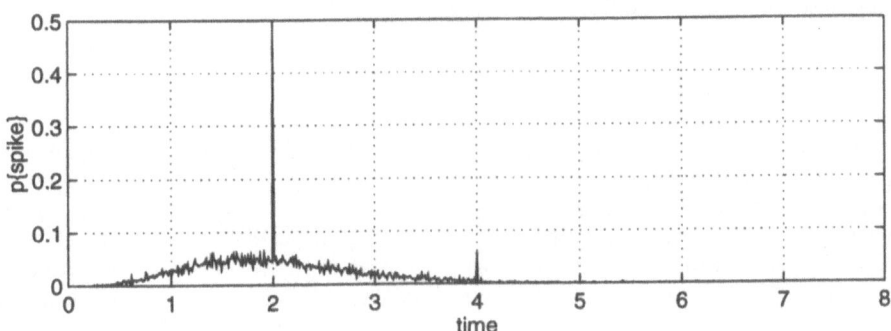

Figure 2: The multi-modal ISIH with a driving force. Parameters $\theta = 4$, $\gamma = 0.05$, $\tau_{sig} = 2$, $\epsilon_{sig} = 0.7$, $\tau_{noise} = 0.5$ and $\epsilon_{noise} = 1$.

3   Results

First we treat the case that no signal is given to the neuron, only noise. The output the neuron then produces, can be viewed as spontaneous activity. Whether the input noise must be seen as a collection of random inputs from other neurons or as the result of unknown processes in the neuron itself, or perhaps both, is not argued. The ISIH then looks like Fig. 1.

Adding a periodic and excitatory stimulus makes the occurrence of a spike at the moment of the stimulus more probable. This results in peaks at integer multiples of the driving period. The multi-modal ISIH is plotted in Fig. 2.

The first peak in this figure represents spikes that have occurred with the same frequency as the stimulus. We called this the coherent part of the output signal. By varying the noise level we can observe a change in the height of this peak: from a small peak at low noise level, going through a maximum at a medium noise level and decreasing again at a high noise level. Thus SR is observed in this model, it is plotted in Fig. 3.

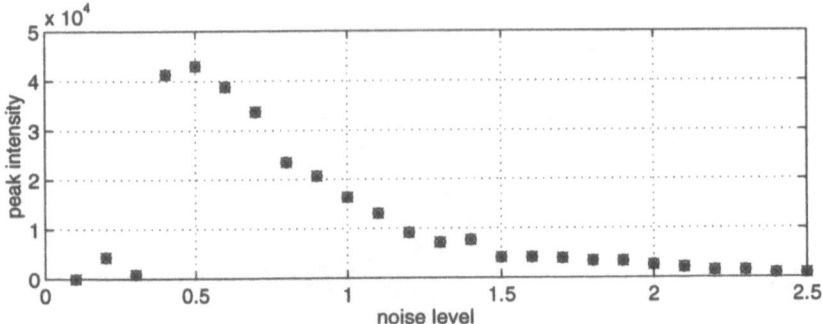

Figure 3: The peak intensity (vertical axis) in the ISIH with the same period as the stimulus when the noise level τ_{noise} (horizontal axis) is varied. At a specific noise level a maximum in the coherent output occurs: this is the point of SR. Parameters $\theta = 4$, $\gamma = 0.05$, $\tau_{sig} = 2$, $\epsilon_{sig} = 0.7$ and $\epsilon_{noise} = 1$.

4 Discussion

We have seen that the simple model we used is capable of describing multi-modal firing patterns and SR. The reason to choose this model lies in its simplicity. Several other models, that may seem more realistic in a biological sense, are that complex that finding an analytical description for the ISIH turns out to be very difficult, if not impossible. Although we haven't been able to write down such an expression for our model yet, it looks as if we are making progress.

References

Douglass, J. K., Wilkens, L., Pantazelou, E., and Moss, F. (1993). Noise enhancement of information transfer in crayfish mechanoreceptors by stochastic resonance. *Nature*, 365:337–340.

Gerstein, G. L. and Kiang, N. Y.-S. (1960). An approach to the quantitative analysis of electrophysiological data from single neurons. *Biophysical Journal*, 1:15–28.

Gerstein, G. L. and Mandelbrot, B. (1964). Randow walk models for the spike activity of a single neuron. *Biophysical Journal*, 4:41–68.

Longtin, A. (1993). Stochastic resonance in neuron models. *Journal of Statistical Physics*, 70:309–327.

Mountcastle, V. B., Talbot, W. H., Sakata, H., and Hyvärinen, J. (1969). Cortical neuronal mechanisms in flutter-vibration studied in unanesthetized monkeys neuronal periodicity and frequency discrimination. *Neurophysiological journal*, 32:452–484.

Pyke, R. (1959). The supremum and infimum of the poisson process. *Ann. Math. Statist.*, 30:568–576.

Siegel, R. M. (1990). Non-linear dynamical system theory and primary visual cortical processing. *Physica D*, 42:385–395.

Cognitive Modelling and Rule Extraction - Orals

The BB Neural Network Rule Extraction Method

F.R. Wiersma
University of Twente
the Netherlands

M. Poel
University of Twente
the Netherlands

A.M. Oudshoff
KPN Research
the Netherlands

3 May 1995

Abstract

In this paper a rule extraction method for a multilayered neural network is proposed. The method is not only able to extract rules from a neural network trained with boolean valued inputs but also from a neural network trained with continuous valued inputs.

Simulations using test data of which the rules are known a priori demonstrate the ability of this method to extract comprehensible and valid rules.

1 The BB Method

This paper adresses the problem of extracting rules from neural networks. Specifically a method referred to as Black Box method (BB method) is decribed. The BB method exploits the confidence principle of a neural network. With the confidence principle is meant the way a neural network uses his continuous valued output(s) to express the level of confidence in a certain output. In the BB method responses for which the neural network expresses a low level of confidence, are used to simplify the rules that have been extracted making these rules more comprehensible.

The BB method consists of three steps. In the first step (section 1.1) the input fields which are not relevant for the neural network are determined. In the second step (section 1.2) rules are generated for all the possible non conflicting [1] input combinations. In the third step (section 1.3) these generated rules are pruned using the confidence principle. The three steps are described using the following simple example:

A three layered (11 input, 12 hidden and 1 output unit) fully connected neural network was trained with the boolean function

$g(a, b, c, d, e, f, g, h, i, j, k) = (a \wedge b) \vee (g \wedge h \wedge i) \vee (\neg a \wedge g \wedge h) \vee (b \wedge c \wedge \neg f)$

After training all 2000 (1800 recall, 200 test) input-output combinations were classified correctly.

1.1 Step 1: Identifying Useful Input Values

To reduce the size of the search space of the second step, input fields that are not relevant for the neural network are determined in this step.

---

[1] Input values are conflicting if e.g. they are used to encode a set of values for one atrribute

This is done by rating all input fields for how much a change from 0 to 1 of an input field can maximally change an output field. The rating for input field I_k with respect to unit u from layer l is given by equation:

$$R(k, u, l) = \begin{cases} 2 \times f(\frac{1}{2} \times \sum_t |w_{t,u,l}| \times R(k,t,l-1)) - 1 & \text{if } l > i\_l \\ 2 \times f(\frac{1}{2} \times |w_{k,u,l}|) - 1 & \text{if } l = i\_l \end{cases} \quad (1)$$

In equation 1, f denotes the transfer function that is used in the neural network, $w_{t,u,l}$ denotes the weight connecting unit u in layer l to unit t in layer $l-1$ and l_i denotes layer $input_layer$.

For the trained network from the simple example the rating values are: $Rating(\{a, \ldots, k\}, g, 2) = \{1.0, 1.0, 1.0, 0.0, 0.0, 1.0, 1.0, 1.0, 0.9, 0.0, 0.0\}$. The input fields $\{d, e, j, k\}$ have a low value for $Rating$ and these input fields are therefore not relevant for the neural network.

1.2   Step 2: Generating the Rules

In this step, a production rule is generated for each possible not conflicting input combination that produces an output larger than a certain predefined threshold θ.

The input fields that have a low value for $Rating$ are ignored to reduce the search space. These low rated input fields are kept[2] $\frac{1}{2}$ during the search.

For each not conflicting input combination $I_{1 \ldots n}$ with output field k larger than θ the following production rule is generated:

$$\text{IF } A_1 \text{ AND } \ldots \text{ AND } A_n \text{ THEN Out\_k}, A_k = \begin{cases} \text{Inp\_k} & \text{if } I_k == 1 \\ \text{NOT Inp\_k} & \text{if } I_k == 0 \end{cases}$$

In general this creates a large set of rules, which will be pruned in the next step. In the simple example the value for θ was $\frac{1}{2}$ and a set of 58 rules was generated in this step.

1.3   Step 3: Pruning the Rules

In this step the set of rules generated in step 2 is pruned, thus making the rules more comprehensible. This pruning is done as follows:

1. Find a rule R that has not been pruned yet.

2. Prune arguments from the rule R as follows: the pruning is done from the first argument A_1 to the last argument A_n. Argument A_k is pruned from the rule R if the following condition holds:

   The sum of the icdvs (input combination density values) of -input combinations for which rule R without argument A_k can be applied[3] and for which the neural networks output value is not the domain $[\theta, 1 - \theta]$- is

---

[2] Instead of $\frac{1}{2}$ any other value from the domain $[0, 1]$ would also be adequate.

[3] A rule can be applied for a input combination if all the arguments from the rule are true for this input combination

less than or equal to $\beta\%$ of the sum of the icdvs of -input combinations for which rule R without argument A_k can be applied-.

The variable θ is the same as the one used in section 1.2. The variable β is the fault tolerance in this pruning step and should not be too large, e.g. 5%. The icdf of an input combination should be an estimate of the density of this input combination in the data set.

3. As long as arguments can be pruned from rule R continue with step 2.

4. Remove all the rules that are implied by the rule that just has been pruned.

   For example: Rule 'IF A THEN C' implies rule 'IF A AND B THEN C'.

5. As long as there are rules that have not been pruned continue with step 1.

6. Remove all the rules that are covered by other rules in the set of rules.

   For example: Rule 'IF B AND NOT C THEN T' is covered by the rules 'IF NOT A AND NOT C THEN T', 'IF A AND B THEN T'.

In the simple example all input combinations have the same density in the data set, and therefore all icdfs could be set to 1. The value for *beta* in the example was 10%. The set of 58 rules generated in step 2 was reduced in this step to 4 rules. These 4 rules were identical to the function g.

The BB-method can also be used if not all inputs fields are boolean valued. This is done by adding hidden rule units after the input layer. The hidden rule units are used to map continuous valued inputs to boolean valued outputs. Hidden rule units are translated into arguments for the rules generated by the BB-method.

For example: Hidden rule unit u with weights $w_1 = 2$, $w_2 = -1$ and bias = 3 and the sigmoid as transfer function can be translated into $Input\_u = (2 \times Hid\_Input\_1 - Hid\_Input\_2 > 3)$.

2 Empirical Results

2.1 Application with Continous Valued Input Fields

Table 1 contains the rules extracted by the BB method from a neural network trained with function: $h(w, x, y, z, a, b) = (w \geq 3 \wedge w \leq 5) \vee (y \geq z + 4) \vee (a \wedge \neg b)$
Both $w < 3$ and $w > 5$ is impossible. Therefore is 'a AND NOT b' simulated by rules 6,7. This makes the set of rules identical to the original function h.

2.2 Application on a More Natural Problem

A neural network was trained to classify 101 animals identified by 18 attributes into the 7 classes: Mammals, Fish, Birds, Crawfish, Insects, Reptiles, Wormlike. The neural network classified 98 animals correctly after training. Table 2 contains the rules for the classes Mammals, Fish, Reptiles that were extracted

Rule nr	Rule	Explanation
1	R1 = (-17.7×y + 17.2×z) > -6.5	R1 is not a rule but an argument
2	R2 = 28.1×w > 15.4	R2 is not a rule but an argument
3	R3 = -26.0×w > -6.2	R3 is not a rule but an argument
4	IF NOT R1	This is the rule: $y > z + 4$
5	IF NOT R2 AND NOT R3	This is the rule: $3 \leq w \leq 5$
6	IF NOT R2 AND a AND NOT b	This is the rule: a AND NOT b
7	IF NOT R3 AND a AND NOT b	This is the rule: a AND NOT b

Table 1: Rules extracted from a neural network trained with function h.

Class	Rule
Mammal	IF NOT eggs
Mammal	IF hair AND 4 legs AND tail
Mammal	IF hair and NOT venomous AND fins AND 4 legs
Fish	IF eggs AND NOT legs AND tail
Fish	IF eggs AND fins AND NOT legs
Reptile	IF eggs AND aquatic AND venomous AND 4 legs AND NOT tail
Reptile	IF eggs AND aquatic AND toothed AND 4 legs

Table 2: Rules extracted from a neural network trained on animal classes.

by the BB method from this neural network. These rules do not only contain information about the animal classes but also about the generalizations the neural network makes. The rule 'IF hair and NOT venomous AND fins AND 4 legs' for example shows clearly how a neural network can generalize. If such an animal existed it would not be strange if it was classified it as a mammal. A total of 23 rules was extracted. This set of rules classified 95 animals correctly.

3 Conclusions

The representations formed by BB method are comprehensible and allow the network to explain itself. The BB method is able to generate a complete set of valid rules that in a lot of cases performs equally to the neural network itself. This set of rules does not only give an explanation of the classification made on a particular case but it also makes new discoveries accessible to human review.

References

[1] Ichiro Enbutsu Kenji Baba and Naoki Hara, *Fuzzy Rule Extraction from a Multilayered Neural Network*, NEW YORK, 1991

[2] LiMin Fu, *Rule Generation from Neural Networks*, IEEE TRANSACTIONS ON SYSTEMS, MAN, AND CYBERNETICS, VOL. 24, NO. 8, August 1994

[3] J. Denker, D. Schwarts, B. Wittner, S. Solla, R. Howard, L. Jackel and J. Hopfield, *Large Automatic Learning, Rule Extraction and Generalization*, Complex Systems 1, pag. 877-922, 1987

IMPLICATIONS OF HADLEY'S DEFINITION OF SYSTEMATICITY

Nico Bakker

e-mail: N.Bakker@uci.kun.nl

June 7, 1995

Abstract

In the ongoing debate between connectionists and proponents of the LOT Hadley [2] has contributed a new definition of systematicity, a term brought into the debate by Fodor [3]. Two important new aspects of this definition are that it provides a precise test of a systems level of systematicity, and that it puts strong emphasis on learning. Both of these aspects have their implications for the debate. This paper will try to make these implications clear.

An implication of the definition being more precise is that the discussion about whether or not a connectionist network behaves systematically can be greatly reduced (...). As a consequence, the definition raises a challenge for connectionists to strengthen their case on empirical grounds. We will discuss a project that aims to meet this challenge [4], and examine the arguments Hadley [1] uses to reason out that it still does not live up to his definition.

An implication of the emphasis Hadleys definition puts on learning is that the claim that the LOT *does* account for systematicity becomes less clear, as the trouble symbolic systems have in learning is notorious. Hadley cites [6] as an example of a symbolic system that exhibits systematicity. We will take a critical view on this example, to see whether the systematicity exhibited falls within the definition of the concept by Hadley.

1 Hadley's definition of systematicity

Hadley [2] gives a more precise definition of systematicity, and provides us with three levels of systematic behaviour, called weak, quasi and strong. Below only the first and last are given, as the middle is not needed in our discussion. The definition of strong systematicity is the one that comes closest to Fodor's conception of the term.

Weak systematicity: The successful processing (by recognising or interpreting) of no other test sentences than those containing syntactic role/word combinations that also appeared in the training corpus.

Strong systematicity: The successful processing of test sentences, both simple and embedded, containing syntactic role/word combinations that did not appear in the training corpus.

The emphasis put on learning shows from the constraints on the training regime. Presently we will slightly extend the argument Hadley gives for the emphasis put on learning. First, we will adapt Fodor's famous illustration of the concept of systematicity to our context as follows. A person who understands "Mary loves John" is able to generalise this and understands "John loves Mary", using both Mary and John in a new syntactic role. Now generalisations can be made in many ways, for example one could also generalise the first sentence to "Mary John loves". Therefor, the correct use of words in novel positions necessitates learning to distinguish nouns from verbs and to come to know the argument structure of specific verbs.

2   The work of Christiansen and Chater

The work of Christiansen and Chater [4], hereafter denoted as CC, consists of a simple recurrent network, which is trained to predict the syntactic category of the next word in the sentence before it is given as input. This is an extension of the work of Elman [5]. The grammar to be learned is quite complex. It contained, among others, left recursive genitives (e.g. Mary's boy's cats) and conjunctions of noun phrases (e.g. John and Mary). A few sample sentences are:

> girl who men chase loves cats.
> man who girls in town love thinks that Mary jumps.
> Mary who loves John thinks that men say that girls chase boys.

The training and test corpus were designed to fall within the constraints of strong systematicity.

Two possible occurrences of strong systematicity were aimed at. One by excluding 'girl' and 'girls' from genitive context (e.g. neither 'girl's cats' nor 'Mary's' girls' were allowed in the training set). Another by excluding 'boy' and 'boys' from noun phrase context (e.g. both 'boys and men' and 'John and boy' were disallowed in the training corpus). As a result of these constraints the network can be tested for strong systematicity, for in both cases it has to process sentences containing words in new syntactic positions.

The results of the test for strong systematicity in genitive context were very limited. For example, after processing 'Mary' the network predicted a genitive marker, among others. After this genitive marker, i.e. 'Mary's', had appeared as input to the network, a noun was predicted. So far, so good. But the activation levels of 'girl' and 'girls' were near zero, while the activation levels of all other nouns lay between 0.1 and 0.2 . This indicated that 'girl' and 'girls' were not expected by the network as the next word. Another example showed only slightly better results.

In contrast to the first result, the test for strong-systematicity in the context of a noun phrase conjunctions were quite successful. For example, after the network had processed 'Mary says that John and ... ' it predicted a noun. And although the activation levels of 'boy' and 'boys' are "minimal compared with the other nouns" [4], the net is able to rightly predict a plural verb after it had processed 'Mary says that John and boy ... '. This prediction is performed despite the fact that the network has only seen singular verbs following 'boy' in the training corpus. It follows that the network has learned to process successfully the noun phrase conjunction for sentences containing words in new positions.

CC end their paper with a discussion of the value of the connectionist approach to syntax acquisition by comparing it with the symbolic approach. They state that most symbolic models have a problem in achieving strong systematicity. As an example they consider the work of Berwick and Weinberg [7], and argue that it does not even come close to systematicity since it does not *learn* the categories of new words from experience —instead, it is provided with information about the syntactic category of each word, including the argument structure of each word.

3   The critique of Hadley

In his response to CC, Hadley states that the ascription to the network the capability to predict categories is a theory-laden construal of the network's response. This construal is stated to be questionable when the network fails to activate crucial words belonging to the given category. What Hadley is aiming at is that part of the claim that the new sentences were processed successfully is based on the fact that the network is judged to expect a noun, where of all the nouns 'boy' and 'boys' have the least activity. So, Hadley argues, one cannot on the basis of this judgement claim that 'boy' and 'boys' have been successfully processed in their new syntactic position, as they seem not to be identified as member of the category noun. The claim made by CC is however that the network successfully processed *new noun phrase conjunctions*. Hadley's definition does not give criteria of what is to be considered 'successful' in the processing of sentences, but one could say that although the network had some trouble with it, it did manage to process the new conjunctions successfully.

Still, Hadley has a point. It seems that the network has learned to perform the conjunction systematicly on basis of the context that the test sentence provides, but has not learned to associate a new role to 'boy' and 'boys'. As a consequence, it has not learned to generalise the use of these words in a context free sense.

4   LOT-based models and Hadley's definition

The second implication of Hadley's definition of systematicity comes forth from the emphasis it puts on learning. Because of this, the behaviour exhibited by symbolic models forms poor evidence for the claim that the LOT is able to explain systematicity. A similar argument has been made in [4]. Hadley has responded by mentioning the work of Anderson [6] as a work that shows systematic behaviour. We will now give a discussion of this model.

The model used by Anderson is called LAS. It learns a language from input consisting of a sentence and a semantic network. This network is parsed to determine the syntactic structure of the sentence. But because in general LAS is able to build more than one parse tree from the semantic network, additional input is given denoting the root of the parse tree to be build. As a result, in an indirect way syntactic tags for all the words in the sentence are given. We could give a technical description of this process to argue that although the syntactic information is provided indirectly, still LAS cannot be said to learn because of the fact that knowledge how to extract the syntactic information is build in the process in advance. But this would take a lot of space. Instead we will make two general remarks which will do the job as well. First, LAS is able to generalise after having processed only one sentence. This strongly indicates that input sentences are more likely to be a trigger to execute build in knowledge than data the model tries to learn from. Second, LAS is not able to adapt to so called discontinuous elements. Examples of these are rare in English (e.g. John and Bill borrowed and returned, respectively, the lawn mower), but are common in Latin. This is explained by Anderson as a consequence of the parsing mechanism used, and this indicates once more that LAN's ability to generalise words to new syntactic roles is build in and not learned. In fact it shows that the syntactic roles that LAS can learn are determined in advance.

In conclusion we state that LAS might only be said to have learned the categories of words, and on the basis of this performs some systematicity. We cannot say that it learned to apply a rule or construct systematically, as it was build in the knowledge which rules to apply to which categories in advance.

5 Conclusion

No one would contradict that syntactic knowledge build into a model in advance or given in addition to the input sentences significantly reduces the learning task of the model. Hadley's definition puts great emphasis on learning. The symbolic model discussed had lots of knowledge build in. So, the symbolic approach has still to meet the definition.

The connectionist model appeared to be able to successfully process a new sentence, but only on basis of the context of the sentence. To meet Hadley's definition, a network is needed that can learn to extract syntactic information from the context of a word, and encode this in the representation of that word. Van Gelder and Niklasson [8] propose an Elman [5] like architecture for this purpose (while, remarkably, at the same time they use syntactic information in the encoding of words). This seems a good way to go.

References

[1] R. F. Hadley, Systematicity Revisited: Reply to Christiansen and Chater and Niklasson and van Gelder, **mind & language** 1994 Vol. 9 No. 4:431-444

[2] R. F. Hadley, Connectionist Language Learning, **mind & language** 1994 Vol. 9 No. 3:247-272

[3] J. A. Fodor and B. P. McLaughlin, Connectionism and Cognitive Architecture: A Critical Analysis, **cognition** 1990 Vol. 28:3-71

[4] M. H. Christiansen and N. Chater, Generalization and Connectionist Language Learning, **mind & language** 1994 Vol. 9 No. 3:273-287

[5] J. L. Elman, Incremental Learning, or The Importance of Starting Small **proceedings from the thirteenth annual conference of the cognitive science society** Chicago, IL, 1991

[6] J. R. Anderson, Induction of Augmented Transition Networks, **cognitive science** 1977 Vol. 1:125-57

[7] R. C. Berwick and A. S. Winberg, **The Grammatical Basis of Linguistic Performance: Language Use and Acquisition** Cambridge, MA.: MIT Press, 1984

[8] L. F. Niklasson and T. van Gelder, On Being Systematically Connectionist, **mind & language** 1994 Vol. 9 No. 3:288-302

THE BINDING PROBLEM IN DISTRIBUTED SYSTEMS

P.H. de Vries, G.J. Dalenoort
Department of Psychology, University of Groningen, The Netherlands

Within the brain there exist multiple maps, each of which represents an elementary property of a perceived object such as its location, form, or colour. At the psychological level we perceive the object as a context-sensitive figure and we can sustain this perception even if the presented object is no longer in sight. This psychological observation implies that at the neural level the excitation patterns in different maps are not separated, but remain coupled, even if the excitation in the sensory systems has decayed.

In the proposed model a memory trace is represented by an excitation pattern carried by a variable group of neurons, an extrapolation of Hebb's long neglected notion of a cell assembly [Hebb, 1949]. A cell-assembly can activate or inhibit other cell-assemblies if some of its neurons also participate in these other assemblies. The neurons of a cell-assembly can therefore not be determined in an absolute sense; one has to take into account which other assemblies are active. i.e. what the context is at the psychological level. In a network representation the nodes are cell-assemblies whereas the interactions between them are reflected by the strength and direction of the connections. Since the cell-assemblies correspond to memory traces or concepts at the psychological level, the network will also be referred to as a conceptual network [Dalenoort,1985].

Under appropriate conditions temporary connections between cell assemblies occur. A necessary condition for the occurrence of a temporary connection is that two memory traces are simultaneously active and belong to the same context. It is assumed that at the neural level the corresponding excitation patterns then have the same *spike resonance* (compatible spatio-temporal characteristics). These are brought about by the participation of both patterns in a larger pattern of activity in the network of cell assemblies. At the psychological level such a global pattern is referred to as context.

Temporary connections can be formed between cell-assemblies since each cell-assembly is assumed to be connected to a switching network. If this system is activated simultaneously by two cell-assemblies that have the same spike resonance then a temporary connection is established between them. In the computer simulation of the network the spike resonance between a pair of assemblies is expressed as a function of their temporal and structural correlation.

Before binding can be understood at the neural level it is necessary to know how the tasks that a system can perform, constrain the parameters and interactions of cell assemblies. A characteristic parameter of a cell-assembly is its critical threshold. When the number of neurons participating in an assembly reaches a certain value, then the overall level of excitation will rise autonomously to a certain maximal level because these neurons start to activate each other more strongly. A related set of parameters refer to the strength of this autonomous growth, its maximal value, and to the strength of the inhibition responsible for its decay.

If cell-assemblies are to perform a task collectively, then they must form a network in which stable excitation patterns can occur. Such a pattern corresponds at the psychological level to the accomplishment of a task or to that of an operation within a task. The simplest form of a stable excitation pattern is an excitation loop. We will therefore represent tasks as a succession of excitation loops in a conceptual network. In order to create loops that propagate through the network in a flexible manner, we introduced a process of backward-inhibition. When a cell-assembly has been activated to a level above that of the critical threshold, all its sources of activation receive back an inhibition that puts them under the critical threshold.

The aim of this paper is to substantiate this network architecture by the analysis and computer simulation of a letter-location identification task. In this task a letter is presented to the network of which it has to report the location as well as the name. Since it cannot be assumed that there exist permanent memory traces for an object that are specific for all its potential locations, these memory traces must be created dynamically. Figure 1 illustrates the conceptual network in which the task is represented. The rectangles in this figure represent structures of cell-assemblies, whereas the lines indicate the flow of excitation. The dotted lines represent temporary connections.

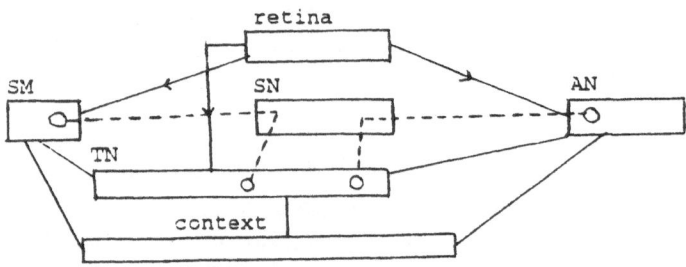

Figure 1, architecture of the conceptual network for letter-location identification (SM:spatial map, SN:switching network, AN:attribute network, TN:task network).

The figure emphasizes the following two points. First there is a separation of the retinal excitation caused by the presented letter. An excitation pattern occurs in the spatial map and another in the attribute map. Secondly, these excitation patterns become re-integrated because both are temporarily connected with excitation patterns in the task network. This re-integration can occur only if the appropriate context is active. The temporary connections produced are a necessary component of the excitation loops that generate the appropriate reaction for the presented letter.

In order to illustrate the evolution of excitation loops a part of conceptual network in Figure 1 has been detailed. Figure 2 shows the subnetwork involved in the binding of the location of the presented letter. The subnetwork involved in the binding of the letter's name is similar. The processing of the various features playing a role in the recognition of a letter have been stylized.

In Figure 2 each circle represents a cell-assembly. The only purpose of the labels of these circles is to explain the network to the reader. They have no function in the processes in the network. The directions of the lines in Figure 2 indicate the flow of excitation or inhibition from one cell-assembly to an another. When cell-assemblies excite each other, i.e. when they form a binary excitation loop, an undirected line is used. The dotted line represents a temporary connection, i.e. a temporary excitation loop.

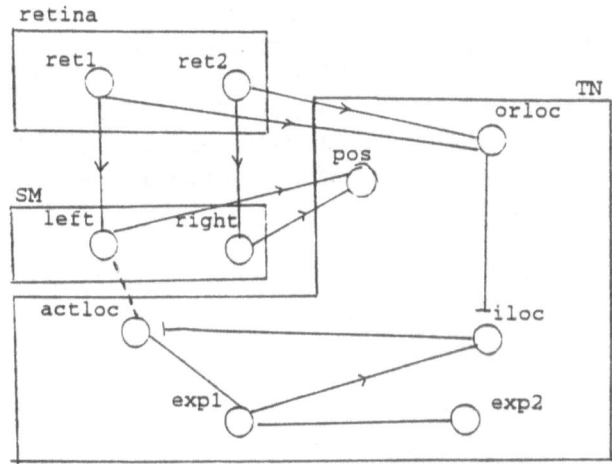

Figure 2, subnetwork for the binding of the location of an object.

The "retina" of the network is connected to two cell-assemblies, each of which can create a specific excitation pattern in the spatial map. In the figure a simple and non-continuous map has been chosen: it represents whether something occurred left or right on the retina. Besides the specific activation of the spatial map, the retina also has a non-specific connection to the task network. Excitation through these connections signals *that* something has happened on the retina, it does not represent *what* has happened nor *where* something has happened.

The task network can be seen as a procedure with two formal arguments, one for the location and one for the name of the letter presented. In Figure 2 the node labeled ACTLOC represents the memory trace for the formal argument for location. The memory trace for this formal argument is assigned an actual value when it becomes temporarily connected to an excitation pattern in the spatial map. For such binding to occur it is necessary that both memory traces are simultaneously active within the same context.

When the network performs the task considered, the activation processes can be represented in a graph with time on the abscissa and the excitation level on the ordinate. The graphs in Figure 3 show the evolution of the excitation patterns for the part of the conceptual network shown in Figure 2, the binding of the location of an object.

Since the network has been instructed about the task, it is in a state of expectation and accordingly the excitation loop connecting EXP1 and EXP2 is permanently active. This produces the characteristic oscillations in Figure 3a. After some time a letter impinges on the network's left part of the retina (Figure 3c). RET1 activates the node for the orientation-reaction ORLOC (Figure 3d) as well as the left part of the spatial map (node LEFT, Figure 3e). The activation of

ORLOC strongly inhibits the intermediate node ILOC, and makes that the excitation level of the latter decays. Consequently, ILOC releases the node ACTLOC, which represents the formal parameter for location. The excitation level of ACTLOC reaches the critical threshold once ILOC has completely decayed (Figure 3f). Since this excitation is simultaneous with that in the left part of the spatial map (the node LEFT, cf. Figure 3e) and both nodes have a comparable role in the context of this task, a temporal connection occurs between the two. Due to this mutual connection the excitation levels in nodes ACTLOC and LEFT will reach the critical threshold. The output node POS (Figure 3h) now receives excitation from the node LEFT strong enough to reach the critical threshold. The supra-threshold activity in this output node combined with that in the spatial map represents the response of the network.

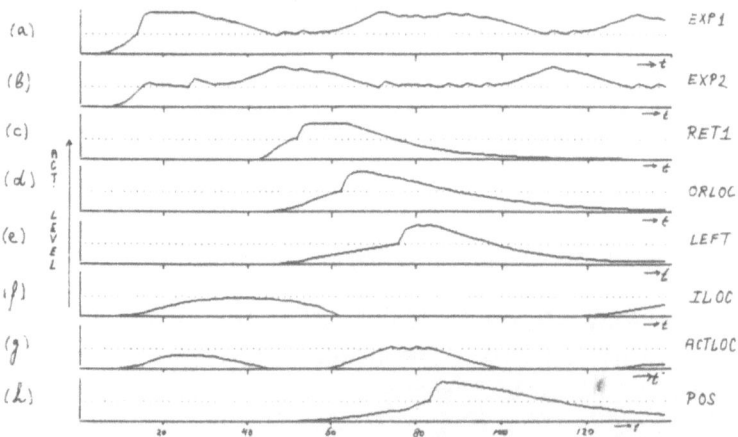

Figure 3, evolution of excitation patterns in the binding of location

The excitation curves presented in Figure 3 are dependent on several parameters. A goal the research is to explore the parameter space in order to find subspaces in which the network can accomplish a certain task. These subspaces must be rather large since we know that the functioning of biological neural networks is robust. The task discussed was relatively simple and did not involve attention. For more complex tasks larger networks will be necessary in which distributed control mechanisms collectively restrict the variety of excitation patterns. These mechanisms, that are independent of content, correspond to attention at the psychological level.

References

Dalenoort, G.J. (1985). The representation of tasks in active cognitive networks, *Cognitive Systems* 1, 253-27.

Hebb, D.O. (1949). *The organization of behaviour*, Wiley, New York.

Integrating Symbolic And Subsymbolic Architectures For Parsing Arithmetic Expressions And Natural Language Sentences

Jonathan A. Tepper, Heather Powell, Dominic Palmer-Brown

Parallel Research Group, Department Of Computing,The Nottingham Trent
University,
Burton Street,Nottingham NG1 4BU
jte@doc.ntu.ac.uk

Extended Abstract

Connectionism is a relatively new approach to language processing and has comparatively few standard methods for syntax analysis and parsing relative to classical symbolic methods. The interest in connectionism has arisen due to its learning capability, tolerance to noisy input, and ability to generalize from previous examples. Classical rule-based techniques are well understood but tend to be intolerant of minor variations that do not strictly adhere to predefined rules.

There have been some successful connectionist approaches to parsing natural language sentences; Selman [1] utilized ideas based upon the Boltzmann machine for syntactic parsing, and Fanty [2] employed localist techniques for context-free parsing. Features such as generalising to the past tense of verbs have also been demonstrated by Rumelhart & McClelland [3]. The majority of research has maintained a degree of symbolic processing but since the criticisms of Fodor & Pylyshyn [4] many connectionists have moved towards a purist approach(e.g Elman[5], Pollack[6], and Smolensky[7]).

This paper presents a novel hybrid AI architecture which is able to parse arithmetic expressions both symbolically and subsymbolically. A method of deterministically parsing simple natural language sentences based on this architecture is also presented. A symbolic element is required for communication and interpretation. A subsymbolic component allows the representation of opaque conceptual information, temporal dynamic structures and complex non-linear structures.

A novel method for adapting the learning rate is also presented. This is suited to training sets with an imbalance between the number of examples of different pattern classes.

Parsing Arithmetic Expressions

The connectionist element of the hybrid system uses a three-layered backpropagation network[8] which consists of 12 input nodes fully interconnected to 6 hidden units. The hidden nodes are fully interconnected to one output unit, which signifies the acceptance or rejection of an expression.

There are eight unique symbols (e.g +, -,*, /,(,),operand,terminator) and the network can be presented with a maximum of 8^4 expressions of length four. Each symbol has its own 3-bit binary representation. A total of 17 expressions are defined as valid and the remaining 4079 expressions are deemed invalid. A number of training set configurations were used to investigate the best training strategy. All training sets consisted of all valid expressions but contained only a percentage of invalid expressions (e.g 1, 10, 20, 30, 40, 50). Due to the high magnitude of invalid expressions relative to valid expressions a method for pattern dependant learning rate adaption was developed.

The network parses expressions from left-to-right by iteratively reducing mathematical terms. Reductions are carried out symbolically. Symbolic structures are used to store intermediate states of the parse and reductions. This approach allows expressions of arbitrary length to be parsed. The input presented to the network at any time step is fixed at four expression constituents and/or symbols from the symbolic structure. A reduction will only be made if the first three symbols are reducible (signified by the connectionist network) and the fourth symbol is a closed parenthesis, operator with lower precedence than that in the expression, or a terminating symbol. A successful parse is encountered if a single symbol taken from the symbolic structure is the only symbol remaining in the input buffer, the terminating symbol has been encountered and the stack is empty.

An unsuccessful parse is reported if there are one or more symbols remaining within the input buffer or stack which are irreducible.

The performance of the network with the various training sets was measured by its generalisation capability against the percentage of valid and invalid expressions contained in the training set. In one case 95% of remaining cases are correctly characterised from 1% of the invalid cases.

The standard backpropagation method does not learn all of the valid cases as they constitute such a small percentage of the overall number in the training set. Two successful experiments to overcome this have been performed. One involved replicating each valid expression in the training set so that valid expressions constitute a substantial proportion of the total training examples. This increased the training set size and the level of success was dependant on the degree of replication. All the network configurations performed well, generalising to greater than 90%. When 50% of the training set consisted of invalid expressions a generalisation rate of 100% was achieved.

The second method involved adapting the learning rate for each pattern and its error. The learning rate for each pattern is calculated as follows

$$\eta = \frac{\Sigma |e_i|}{p} * a + b$$

Where e_i is the error for unit $_i$, p is the total number of hidden or output units and a and b are constants which allow adjustment of the range of η. $a = 0.6$ and $b = 0.2$ were chosen to give $0.2 \leq \eta \leq 0.8$. η is therefore linearly dependent on the average absolute output error within this range.

The training set consisted of all 17 positive examples (no replication) and 50% of the remaining 4079 negative examples as before. All training patterns were learnt after 3000 epochs. The network generalises very well with an accuracy of 97.3%. Further variations of the training set have revealed similarly good results.

Parsing Natural Language Sentences

Classical deterministic parsers such as PARSIFAL[9] are rule-based. They have difficulty in processing ungrammatical and ambiguous sentences and acquiring further language. Ad-hoc extensions (i.e. PARAGRAM[10], ROBIE[11],LPARSIFAL[12]) can provide a limited solution to these problems.

The hybrid architecture proposed provides a framework for deterministic parsing, lexical disambiguation and language acquisition. The connectionist model is based on that used for the arithmetic expression parser. The input tokens consist of sentence constituents and symbols denoting a type of reduction. The output units represent actions to be carried out and positional information to inform the symbolic processor of whereabouts in the input to perform the reduction. There are also association nodes on the output that provide lexical alternatives for ambiguous input. Syntactic rules, derived from a context-free grammar[2], provide a basis for training data so that the network acquires an initial linguistic knowledge structure. Correct input fragments are autoassociated on the output nodes as learning takes place. If an incorrect fragment is present the network may produce a possible alternative. This is used as a means of lexical disambiguation. For example, in the sentence *She wanted to shop,* the word *shop* may initially be incorrectly classified as a noun. The network will produce a verb pattern on the output nodes relative to the position of *shop.*

84

The system parses sentences from right-to-left by iteratively processing sentence constituents and reduced input from the previous time steps. Reductions are carried out symbolically. A symbolic stack is used to store intermediate states of the parse. Unlike many connectionist parsers this approach allows sentences of arbitrary length to be parsed.

The hybrid architecture is able to parse sentences from a simple grammar and disambiguate sentences based on syntactic context. Further work is under way with a large grammar[13] to assess the technique for a realistic subset of natural language.

References

[1]   Selman, B. (1985). A rule-based connectionist parsing system. In *Proc. of 7th Annual Conf. of Cognitive Science*. PP 212-221.

[2]   Fanty, M. A. (1986). Context-free parsing with connectionist networks. In *Proc. of AIP Conf. Neural Networks for Computers*. PP 140-145.

[3]   Rumelhart, D., McClelland J. (1986). On learning the past tenses of English verbs.In *Parallel Distributed Processing, Explorations in the Microstructure of Cognition*, Vol 2. PP 216-271. MIT Press.

[4]   Fodor J.A., Pylyshyn Z.W. (1988). Connectionism and Cognitive Architecture: A critical analysis. In *Cognition*, Vol 28. PP 3-71.

[5]   Elman J.L. (1990). Finding structure in time. In *Cognitive Science*, 14, PP 179-211.

[6]   Pollack J.B. (1990) Recursive Distributed Representations. In *Artificial Intelligence*. Vol 46. PP 77-105.

[7]   Smolensky P. (1990). Tensor Product Variable Binding and the Representation of Symbolic Structures in Connectionist Systems. In *Artificial Intelligence*. Vol 46. PP 159-216.

[8]   Rumelhart, D.E., Hinton, G.E., Williams R. J. (1986). Learning internal representations by error propagation. In *Parallel Distributed Processing, Explorations in the Microstructure of Cognition, Vol 1: Foundations*. PP 318-362. MIT Press.

[9]   Marcus, M. P. (1980). *A Theory of Syntactic Recognition for Natural Language*. Cambridge: MIT Press.

[10]   Charniak, E. (1983). A parser with something for everyone. In *Parsing Natural Language*. New York: Academic Press.

[11]   Milne, R. (1986). Resolving lexical ambiguity in a deterministic parser. In *Computational Linguistics*, 12, PP 1-12.

[12]   Berwick, R. C. (1985). *The Acquisition of Syntactic Knowledge*. Cambridge: MIT Press.

[13]   Noble, H. M. (1988). *Natural Language Processing*. Oxford : Blackwell Scientific.

Bayesian Strategies for Machine Learning: Rule Extraction and Concept Detection

Laura Martignon

Department of Neural Information Processing, University of Ulm

D–89069 Ulm/ Germany

Kathryn Blackmond Laskey

Department of Systems Engineering, George Mason University

Fairfax, Virginia 22030-4444/U.S.A.

June 2, 1995

Abstract

A Markov field represents dependencies in the form of an undirected graph and is a natural representation for specific types of knowledge, such as networks of concepts. A Bayesian learning algorithm is presented for inferring structure and parameters from a database of independent and identically distributed cases from the distribution. We use a conjugate family of prior distributions for model parameters conditional on connectivity structure, and the Bayesian Information Criterion to approximate the posterior probabilities for structures. The procedure is conceived for the general setting of experimental situations, where the correlation structure of a system of attributes has to be learned from examples. Correlations of order two and higher correspond to concepts. The graph induced by drawing cliques corresponding to the detected correlations and/or interactions provides information on the rules to be extracted.

1 Introduction

Any strictly positive probability distribution on the space of binary configurations of a set of nodes induces a Markov graph, where the graph for the Markov field can be recovered from the nonzero coefficients in the energy expansion of the joint probability distribution. Markov fields have been used to represent networks of neurons, both real and artificial. Models of this type have also been suggested as a way to represent concepts. Hinton et al. discuss a distributed representation of concepts that can be encoded in the Boltzmann machine, which uses a Markov field representation [1],[2],[3],[4],[5].

The aim of this paper is to present a Bayesian algorithm that learns correlation structure and correlation strengths based on a database of independent cases arising from the distribution to be learned.

In the realm of knowledge representation **concepts** have been defined in a variety of ways. The common aspect to these definitions is that concepts are formed by gathering sets of attributes, whose simultaneous presence in a given context strikes us as meaningful. When modeling a knowledge base by listing the configurations of binary (0, 1 for "false", "true") states of a fixed set of attributes, the problem is to establish which clusters exhibit the analogue of an "excitatory interaction". In other words, one wants to detect those clusters for which the probability of being simultaneously true is greater than predicted by the simultaneity of strict subsets of the cluster.

2   A family of models for interactions

Let Λ be a set of nodes labeled 1 through k. At a given time each node may be in state 1 or 0. The state of node i at a given time is represented by a random variable X_i. The lowercase letter x_i is used to denote the state of X_i. The state of the set Λ is denoted by the random vector $\mathbf{X} = (X_1, ...X_k)$, which can be in one of 2^k configurations. The actual configuration at a given time is denoted by $\mathbf{x} = (x_1, ..., x_k)$.

Subsets of nodes will be called *clusters* and sets of clusters will be called *structures*. A cluster of nodes in Λ exhibits a positive (or excitatory) interaction when all its nodes *tend* to be active, i.e.,simultaneously in state 1, and a negative (inhibitory) interaction when simultaneous activity tends not to occur.

Let the probability of occurrence of a configuration in a given context be denoted by $p(x)$. The distribution $p(x)$ is assumed to be strictly positive, i.e., all configurations are possible. Let Ξ be a structure and denote by ξ any of its clusters. Define the random variable T_ξ to have value 1 if all nodes in cluster ξ are in state 1 and 0 if any node in cluster ξ is in state 0. That is, T_ξ is the product of the components x_i with i in ξ. By definition T_Φ is 1.

Definition 1.*Let $h(\mathbf{x})$ be the logarithm of the probability $p(\mathbf{x})$ of configuration* $\mathbf{x}$. *The nodes in Λ exhibit* interaction structure Ξ *with* interaction parameters $\{\Theta_\xi\}$ *if $h(\mathbf{x})$ can be written as:*

$$h(\mathbf{x}) = \log p(\mathbf{x}) = \sum_{\xi \in \Xi} \Theta_\xi T_\xi \tag{1}$$

The nodes in a cluster ξ are said to interact *if Θ_ξ is nonzero.*

Given a cluster ξ of j random variables we say that they exhibit a *j-th order correlation* if the marginals on proper subclusters are **not** a sufficient statistic for the marginal distribution restricted to ξ.

It is an important result of multivariate analysis that a cluster ξ exhibits a correlation of order $|\xi|$ if and only if its nodes interact (for details and proofs consult [6],[7],[8],[9].

3   A Bayesian Algorithm for Inferring Structure

Information about $p(x)$ prior to observing any data is represented by a joint probability distribution called the *prior distribution* over the structures Ξ and the interaction strength parameters theta. Empirical observations may be used to update this distribution to obtain a *posterior distribution* that incorporates both prior knowledge and information from the empirical data. The prior distribution is obtained by specifying a prior probability for each structure Ξ and distributions on the parameters $\Theta_x i$ for $\xi \in \Xi$. For a given structure Ξ the model described by (1) is a member of the *exponential family* of probability distributions. Thus a convenient family of prior distributions called the *conjugate family* may be used. By specifying prior probabilities over structures Ξ, we obtain a *mixture distribution*, where the components of the mixture are conjugate prior distributions. Because the conjugate family is closed under sampling, the posterior distribution is also a mixture of conjugate distributions. We estimate posterior probabilities of structures by using the Bayes Information Criterion [10],[11],[12],[13]. The posterior probability of cluster ξ, obtained as the sum of the posterior probabilities of all structures containing ξ, may be interpreted as the probability that the r nodes in cluster ξ exhibit a degree-r interaction. The posterior distribution for $\Theta_x i$ represents information about the magnitude of the interaction. The mean or the mode of the posterior distribution can be used as a point estimate of the interaction strength; the standard deviation of the posterior distribution reflects remaining uncertainty about the interaction strength. Calculating the posterior mixture distribution requires computing, for each possible structure Ξ, the posterior probability of Ξ, the posterior probability of ξ and the conditional distribution of Θ given Ξ. This is clearly infeasible: for k nodes there are 2^k activation clusters, and therefore 2^{2^k} possible structures. In practice, a heuristic search algorithm is used to enumerate a much smaller number of clusters. The posterior distribution is estimated by computing posterior distributions and Bayes factors for enumerated clusters, and normalizing the probabilities of the enumerated clusters to sum to 1.

4   Remarks

Early work in probabilistic expert systems used expert judgment as the primary source of probability information for the models. When available, empirical data was used to estimate probabilities, but experts supplied the structural assumptions. Recently there has been increasing attention devoted to the use of empirical data to learn both structure and parameters. This work has focused primarily on directed graphs [15],[16]. The present work considers Bayesian learning methods for undirected graphical models. When reliable expert judgment about the presence and magnitude of interactions is available, this information may be incorporated in the prior distribution. In previous work [12] we have discussed the specification of prior distributions for use when reliable expert judgment is not available. The class of models proposed in this paper has been used to detect the presence of higher-order correlations and/or interactions in the spiking activity of neurons. We have applied the Bayesian approach discussed in this paper and compared it to conventional statistical significance testings [2],[12],[14]).

The analogy between neurons interacting and attributes forming concepts has a strong intuitive appeal: in a simplified model for knowledge representation attributes can be represented by ideal neurons (or even groups of ideal neurons) and simultaneity (or quasi–simultaneity) of activation turns out to be one of the keys to concept formation and variable binding. It is the core of a popular line of thought in cognitive neuroscience that this analogy is deeper than mere coincidence.

References

[1] J.J.Hopfield, "Neural networks and Physics systems with emergent collective computational abilities, Proceedings of the National Academy of Science, USA (1984):2554–2558

[2] L.Martignon, H.v.Hasseln, S.Grün, A.Aertsen and G.Palm, "Detecting higher order interactions among the spiking events in a group of neurons" to appear in BiologicalCybernetics (1995)

[3] G. Hinton, J.McClelland, D.Rumelhart."Distributed Representations" Parallel distributed processing, D.Rumelhart and J. McClelland (eds) (The MIT Press, Cambridge, 1986)

[4] G. Hinton, T.Sejnowski,"Learning and relearning in Boltzmann machines"in Parallel distributed processing,D. Rumelhart and J. Mc Clelland (eds) (The MIT Press, Cambridge, 1986)

[5] K.B.Laskey, "Adapting connectionist learning to Bayes networks", International Journal of Approximate Reasoning, May (1990)

[6] S. Amari, "Dualistic geometry of the Manifold of Higher–Order Neurons", Neural Networks, 4, (1991), p.443–451

[7] S. Amari, K.Kurata, and H.Nagaoka,"Information Geometry of Boltzmann machines", IEEE Trans. Neural networks, 3, (1992) no.2, p. 260–271

[8] S.Amari and T. SunHan, "Statistical Inference under Multiterminal Rate Restrictions: A Differential Geometric Approach", IEEE Trans.Inf.Theory, 35, (1989), no.2, p. 217–227

[9] Y.Bishop, S.Fienberg, and P.Holland, Discrete multivariate analysis: Theory and Practice, (The MIT Press, Cambridge, 1989)

[10] R.E.Kass, A.Raftery, "Bayes factors and model uncertainty",Carnegie Mellon University Dept. of Statistics, Technical Report Nr.571(1993)

[11] M.H.de Groot, Optimal Statistical Decisions,(Mc.Graw Hill, New York,1970)

[12] K.B.Laskey and L.Martignon, " Bayesian Learning of Markov Fields Models"submitted to Neural Networks

[13] W.B.Poland, and R.Schachter,"Three approaches to probability model selection", Uncertainty in artificial intelligence:Proceedings of the tenth conference eds. R.Lopez de Mantara, and D.Poole (eds.)(Morgan Kaufman, San Mateo,1994)

[14] S.Grün, A.Aertsen, M.Abeles, G. Gerstein, and G.Palm,"Behavior– related neuron group activity in the cortex", Proc. 17th Ann. Meeting of the European neurosc. Ass.(Oxford University Press, ENA), (1994)

[15] G.Cooper and E.Herskovits,"A Bayesian method for the induction of probabilistic networks from data",Machine Learning 9 (1992),p.309–347

[16] D. Heckerman, D.Geiger, D.M.Chickering,"Learning Bayesian networks:The combination of knowledge and statistical data",Artificial Intelligence"(to appear).

Cognitive Modelling and Rule Extraction - Posters

Learning at subsymbolic and symbolic levels *

Alain GRUMBACH

E.N.S.T., 46 rue Barrault, 75634 Paris Cedex 13, FRANCE

grumbach@inf.enst.fr, phone : (33) 1 45 81 78 52, fax : (33) 1 45 81 31 19

1   Abstract

This papers deals with robot control based on sub-symbolic and symbolic knowledge, through a **cognitive** point of view, leading to an **artificial intelligence** model.

In a situation such as learning to play tennis, two knowledge forms are involved:
- implicit knowledge, e.g. sensori-motor associations; this knowledge is **subsymbolic**
- explicit knowledge, e.g. a teacher verbal advice, which makes use of **symbols**.

We suggest to consider robot learning in the same way : cooperation of sub-symbolic and symbolic knowledge.

We present a robot learning situation including two phases and corresponding models : - learning by example (subsymbolic), - learning by instruction (symbolic).

2   The problem

A tennis beginner learns different kinds of information :
- in the very beginning, the teacher shows him the right move and guides his/her arm; let us call this learning mode **learning by example** (LbE)
- when basic movements have been acquired, supervised learning goes on through verbal advice; let us call this phase **learning by instruction** (LbI)
- lastly, the beginner learns through his/her own experience; let us call it : **learning by doing** (LbD).

The aim of this work is to transfer this scenario to a robot learning situation. Let us consider a robot that moves in an area (the road) with obstacles and boundaries (see next figure). The robot must move without going out of the boundaries. It has visual sensors that allow it to see part of its front space. At each step it may choose among seven possible moves, corresponding to small speed variations. It may : - keep same speed - slightly increase speed - slightly decrease speed - slightly change speed direction toward left - slightly change speed direction toward right.

These choices are illustrated on next figure by little circles situated on the vertices of an hexagon, the center of which is the extremity of the current speed vector (M).

Legend : C : road, R : robot current position, r : previous robot position, V : robot vision area, M : 7 robot possible moves, m : center of M (vector rR = vector Rm)

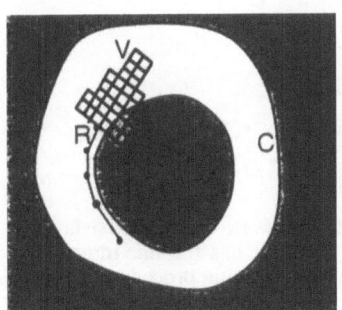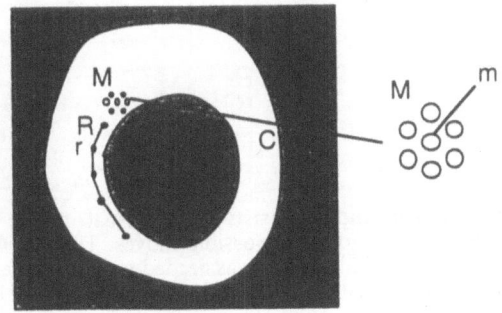

* This research was supported by French "Réseau Cogni-Centre" and "GDR 957"

This paper is published by courtesy of HERMES editor who published a previous French version in "Technique et Science Informatiques" vol. 12 n0 3/ 1993, p.347-369.

In this context, **learning by example** consists of the teacher showing the next move. **Learning by instruction** is exemplified by the teacher saying : "here is a strong left curve", or "if there is a strong left curve, you must slow down and turn left". **Learning by experience** would consist of using feedback to improve knowledge; for instance when the robot runs off the road, it must modify its knowledge so as to avoid such a failure in the future; this last learning ability is out of the scope of this paper.

The aim of this work is to model learning by example and by instruction using a representation that would be able to handle **both subsymbolic and symbolic** items. This leads us to connectionist models, which will be considered as a common framework for both kinds of information.

3 Related work

Our goal, i.e. learning through symbolic - subsymbolic interaction, has not been studied on a wide scale. We found few works that share part of our approach. The most relevant ones are :
- Nenov and Dyer [Nenov and Dyer 88] have designed the DETE system that makes correlation between visual and verbal information
- Hendler [Hendler 89] is working on hybrid systems that put stress on symbolic - subsymbolic interactions. He recently edited a special issue of Connection Science [Hendler 89] on that topic. He has designed architectures that allow cooperation of connectionist (inputs) and symbolic information. This approach has also been advocated by Feldman [Shastri and Feldman 84], and Shavlik and Towell [Towell and Shavlik 93] who created a framework based on a neural network rule-insertion algorithm (KBANN).
- Ritter, Martinez and Schulten [Ritter, Martinez and Schulten 89] and Sorouchyari [Sorouchyari 89] deal with robot sensori-motor coordination through neural network models (Kohonen's cognitive maps). Sorouchyari includes symbolic information creation, but does not actually use it for symbolic processing purposes.

4 Learning by example

While learning by example, sensorimotor information includes only visual input (picture) and chosen output (move). The teacher and learner both perceive visual information. The teacher computes this information and shows the correct move to the learner. The learner perceives both visual input (picture) and the teacher recommended corresponding output (move). To memorize this kind of situation involves making correlations between non symbolic information. An associative neural network is well suited.Let us call this neural network N_{p-a} .

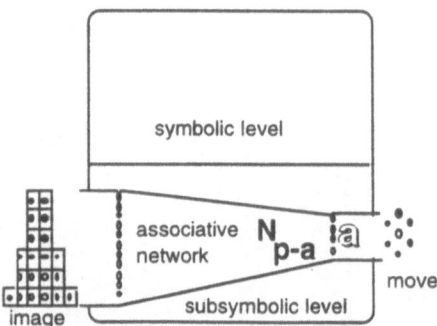

The network input consists of the visual image. The network output consists of 7 units corresponding to the 7 possible moves. The output unit that has the maximum activity is selected. No hidden layer was needed for this problem. The learning procedure is the classical delta-rule algorithm. The learning set consists of 13 examples corresponding to 13 typical situations.

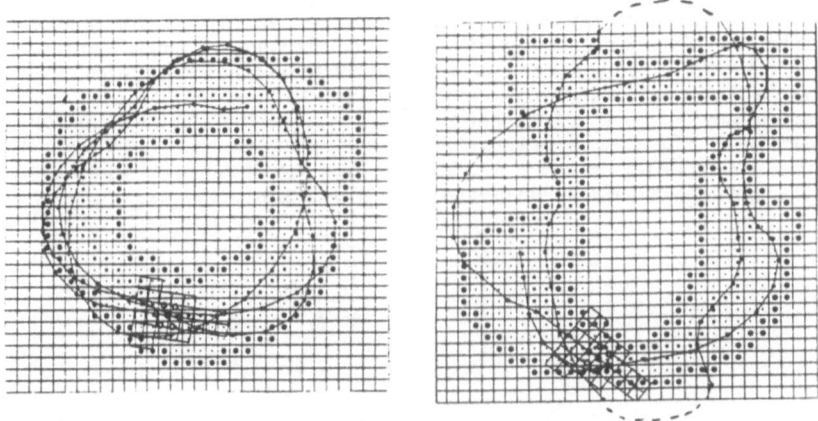

On this figure, we may see paths illustrating the skill of the robot after learning by example. Performance is tested on two roads, a smooth shaped one (left), and a more intricate one (right) where several failures occur.

5 Learning by instruction

Our point of view on symbolic knowledge use may be illustrated by the example of tennis playing learning. In the learning by example phase, the learner imitates the teacher's movement. When this basic behavioral knowledge becomes stabilized, it is complemented by learning by instruction through teacher's symbolic advice.

The teacher perceives the visual input (picture), computes this information, utters a symbolic phrase about it towards the learner.

Five typical phrases may appear :

1 - "here the road bends to the left "
2 - "here you are slowing down"
3 - "here you must slow down"
4 - "in the case of a sharp bend, do this"
5 - "if the road bends sharply to the left, you must slow down and turn left"

The learner perceives both the visual input (picture) and the symbolic phrase and memorizes the association. Let us consider some previous examples:

- "here the road bends to the left" : "here" refers to current situation (perception p); learning consists of creating the "left bend" symbol (symbol Sp) by memorizing the relation between current visual pattern and "left bend" label;
- "if the road bends sharply to the left, you must slow down and turn left" : learning consists of memorizing relation (symbolic rule) between "sharp left bend" symbol and "slow down and turn left" symbol.

This learning by instruction capability can be represented within the previous architecture:

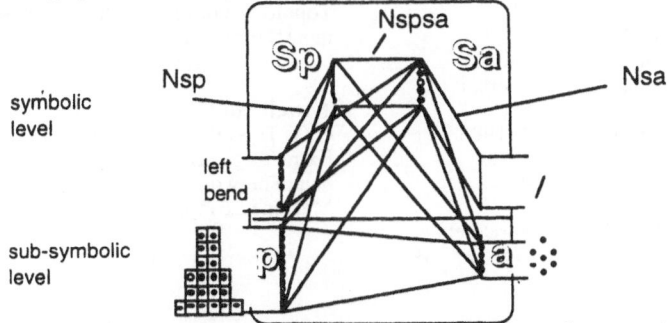

Sp : symbol corresponding to visual perception, Sa: symbol corresponding to action.

This model includes nine trapezoids; all of them are modelled by associative networks. The lower one is the Np-a network previously introduced. The Nsp network memorizes association between a picture and the correspondant symbol. The upper rectangle Nspsa, which associates the Sp and Sa symbols, consists of a set of condition-action rules, which may be considered as a degenerate associative network.

Once learning is complete, knowledge use consists of making input information (p) propagate through the full network structure (where a path exists). On next figure, we see two paths : p-> a (through Np-a network), p -> Sp -> Sa -> a. Activities coming from different paths are added up. The output unit with the maximal activity is selected.

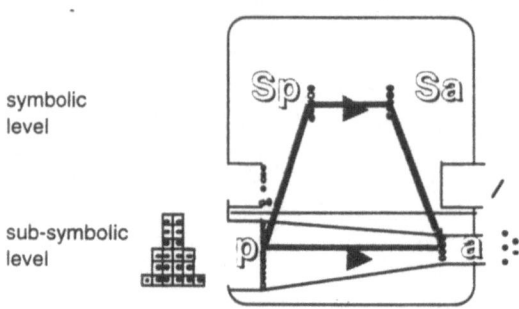

symbolic level

sub-symbolic level

6 Discussion

Let us emphasize the main features of this work.

Thanks to the proposed architecture, the system has learned two kinds of knowledge : sensorimotor (Npa) and symbolic (Nspsa), in a common framework based on a unic kind of representation (associative network). The symbols are grounded into the environment through perceptions or actions (Nsp and Nsa), and these symbols have been acquired through a learning process.

Acknowledgements :

I would like to thank students which have participated to this work : Sophie MIDENET, Mohamed SIALA, Nizar LAYANI, Bertrand GAVAUDAN, Hervé AZOULAY, Wolfgang SPERLING, and my colleagues Eric BONABEAU and Jean-Louis DESSALLES for their helpful comments.

BIBLIOGRAPHY

Grumbach 94
A. Grumbach
Cognition Artificielle. Du réflexe ... à la réflexion, Addison Wesley 1994

Hendler 89
J. Hendler
On the need for hybrid systems
Connection Science, 1(3), 1989

Nenov and Dyer 88
DETE system : Connectionists / Symbolic model of visual verbal association
ICNN, Vol. II, pp. 17-24, 1988

Shastri & Feldman 84
L. Shastri, J.A. Feldman
Semantic networks and neural nets
University of Rochester, TR 131 , 1984

Sorouchyari 89
E. Sorouchyari
Mobile robot navigation: a neural network approach
Journées Electronique EPFL, Lausanne 1989

Ritter, Martinez and Schulten 89
H. Ritter, T. Martinez, K. Schulten
Topology conserving maps for learning visuo-motor coordination
Neural networks, vol. 2, p. 159-168, 1989

Towell and Shavlik 93
G.G. Towell, J.W. Shavlik
Extracting refined rules from knowledge-based neural networks
Machine Learning, 1993

THE CONSTRUCTION OF EVALUATIVE MAPS: AFFECTIVE COMPUTATIONS IN THE AMYGDALA

Eamon P. Fulcher, Department of Psychology , Worcester College of H.E., Worcester, U.K. *email: e.fulcher@worc.ac.uk*

Abstract

A number of recent studies indicate that the amygdala is involved in affectively toned stimulus-reward associations, especially during classical conditioning. In previous publications a neural network model, EMNET (Evaluative Map NETwork) has been presented by this author which carries out stimulus-reinforcer computations by constructing an "evaluative map" of the environment [1,2,3]. EMNET, which is based on the idea that neurons may implement n-tuples [4], successfully captures a wide range of conditioning phenomena. Here, the strong similarities between EMNET and the amygdala are expounded, particularly the notion that the amygdala computes stimulus value. Interestingly, its simulated behaviour following "lesions" is interpreted as showing some resemblance to the pattern of behaviour observed in humans with severe amygdala damage.

The Amygdala

The amygdala is located in the rostromedial temporal lobe in mammals. It receives input from the thalamus, the visual and auditory association cortices, the olfactory bulb, and the hypothalamus, and projects to the hypothalamus, the forebrain, and various midbrain and brain stem nuclei. Hence, the amygdala receives low level sensory input and outputs to structures implicated in species-typical responses, including the autonomic nervous system (ANS).

It has been found to play a vital role in animal classical conditioning, especially the conditioned emotional response (CER). In this paradigm, a neutral stimulus, such as a brief high-pitched tone, is presented followed by an emotionally arousing stimulus (an 'unconditioned stimulus' or UCS), such as a foot shock (eliciting fear). After a handful of such trials, the animal displays an emotional reaction (the CER) to the neutral stimulus (which is termed the 'conditioned stimulus' or CS). So in the above example, the animal exhibits fear on hearing the tone. Affect is measured by recording those responses of the ANS associated with an emotional state (e.g. heart-rate). It is found, for example, that pairing a tone with a foot shock (which increases the heart-rate) results in the tone itself acquiring the ability to produce an increase in heart rate. One feature of CER, when compared to non-aversive reflexive conditioning, is that learning is more rapid. For example, fear conditioning may take as few as 20 trials, whereas conditioning of the nictitating membrane response may take up to 200 trials.

The role of the amygdala in conditioning has been derived from the observation that lesions to it have a disruptive effect on CER acquisition [5,6]. Neurons of the medial geniculate nucleus of the thalamus which project into the central nucleus of the amygdala have been observed to alter their response during CER learning [7]. Also, the firing of neurons in the amygdala correlate with the CS, [8]. Moreover, the conditioning of the heart rate response is disrupted during lesions to the amygdala [9]; and learning that certain tastes are noxious is also disrupted during lesioning [10,11]. Lesioning also affects the recognition of a superior aggressor [12], and the recognition of an attack situation [13]. A qualitative model compatible with the above data has been proposed in [14] in which this structure is viewed as a "stimulus value" feature detector. While a similar position is adopted in the model presented here, we consider such a process in computational detail.

The Amygdala as an Evaluative Mapper

Previous work that concern the more general, theoretical aspects of classical conditioning have opened the debate concerning what is learned during conditioning [15,16]. For example, the two-process view holds that learning involves two separate functional systems: these may be referred to as contingency mapping (learning that A leads to B) and evaluative mapping (determining stimulus value). The former is a more passive form of learning, while the latter is the learning of the importance of the stimulus for the organism's survival.

These two forms of learning have been modelled in previous work using EMNET, a modular neural network that simulates events in real time [1,3]. The model accounts for much of the data concerning the temporal aspects of classical conditioning, especially the inter-stimulus interval (ISI – the time between the onset of the CS and the onset of the UCS). Much of that work explored contingency mapping; here we examine further the neural structure, hypothesised to be the amygdala, that carries out the evaluative mapping.

Constraints on Modelling

The criteria that guide and constrain such modelling are given by the literature in the CER paradigms. These concern the temporal relationship between the inputs, the range of generalisation of learned responses, and the learning rates within structures of the model. For example, the data strongly indicates that the CS must precede the UCS and not vice versa, with an optimal ISI of 500 msecs. Also, the pattern of generalisation of a learned response is that of a normal distribution. In addition, the rate of response learning appears to be directly proportional to the output activity of the evaluative structure.

The Evaluative Map Network

In its simplest form (i.e., only considering a single evaluative dimension) the network consists of an Adaptive Combinatorial Unit (ACU) which is composed of a set of uniquely addressable memory locations (figure 1). The unit receives low grade sensory input and its function is to detect features in the input that have stimulus value (value with respect to some behavioural goal). Stored at each memory location j of the unit is some value v_j that reflects the learned evaluative strength of the input. Each activated v_j continually excites its own location integrator to some value x_j until input to location j is terminated. Due to internal feedback(to itself) x_j will remain active for some time after the termination of the input unique to location j. In essence then, x_j represents a 'memory trace' of the input. Since each signal path through the ACU is independent, more than one location integrator may be active simultaneously.

Indeed, this is necessary for Hebbian learning to occur in the ACU for stimuli that occur in close temporal proximity. An output integrator sums the activity of each location integrator, as well as feeding back to itself. Its activity z_j represents the output of the ACU. Modifications to the stored value v_j occur when at least two memory locations are simultaneously active and a transfer of value occurs through a Hebbian-delta rule. The equations that determine the activity patterns of location and output integrators are a set of difference equations similar to those explored in [17]. The main parameters of these equations are those that determine the attack, amplitude and decay of the location and output integrators. Given that the optimal ISI is 500 msecs, and that the minimum interval between two action potentials is assumed to be 1 msec, these parameters can be estimated.

In addition to the temporal learning outlined above, generalisation based on similarity is achieved by a spreading algorithm. The object here is to spread an updated v_j to its neighbouring memory locations (based on the Hamming metric, of distance H). Such spreading is determined by v_j, z_j, H, and a Gaussian function $g()$. Output of the unit is a scalar value in the range $\{-1,1\}$ which reflects the basic pleasant/unpleasant dimension.

"Lesioning" the Network

The modelling of neurobiological structures can be an interesting exercise in attempting to discover the sorts of architectures capable of achieving similar computational properties as the one under study. However, testing such models is not straightforward. Clearly, satisfying the design constraints provided by the available data (as is done here) is indicative of the model's usefulness. A second type of validation is to "lesion" components of the model since these should produce similar deficits in the model's behaviour as those which occur from real lesions or damage. In the following discussion, lesioning in EMNET is implemented by moving the values of a random sample of memory locations in the ACUs towards 0.

It is important to note that memory locations that have been modified through experience (i.e., CS evaluations) represent a prediction about the occurrence of a UCS (since they only acquire value through such temporal relationships). Hence lesioning the Evaluative Map network produces interesting results with regard particular stimulus sequences. If one regards an emotional reaction as resulting in part from an incorrect prediction, the immediate change to the "emotional response" of this network following lesions is that of inappropriate emotional reactions. Damage localised to CS locations means that any previously learned prediction is lost. This is interesting because it suggests that the emotional response may even be opposite in sign to one made before the lesion. For any system, human or animal, such a state is likely to be quite disruptive. Psychologically, it would imply that emotional reactions following damage to the amygdala become more sharp and inappropriate. Since the system is continually learning, then with increasing damage output of the unit tends towards 0. This may be interpreted as a change towards emotional "flatness".

These changes are remarkably similar to symptoms of dementia in which the amygdala is damaged by disease [18]. These are characterised In Alzheimer's by increases in agitation but with a gradual increase in passivity. Similarly, in Pick's disease there is an increase in apathy and depression, while in Huntingdon's chorea there are quite dramatic changes in emotion towards such states as euphoria, depression and paranoia. These symptoms sit well with the lesions to the Evaluative Mapper described above.

Conclusions

The amygdala is clearly an interesting structure and opens the door to modelling the early perceptual processes that result in emotional learning. Models such as EMNET described in this paper may be valuable in determining the computational properties of such structures.

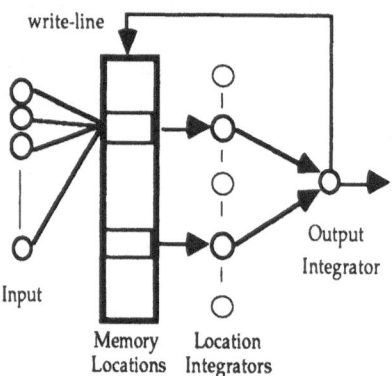

Figure 1. Elements of the Adaptive Combinatorial Unit.

References

[1] Fulcher, E.P. (1992) A distributed model of the representational states in classical conditioning. *Proceedings of the International Conference on Neural Networks, Brighton*, pp. 911-915.

[2] Fulcher, E.P. (1992) Unsupervised clustering with RAM discriminators. *International Journal of Neural Systems, 3 (1)*, 57-63.

[3] Fulcher, E.P. (In Press) Neural Networks and the Psychology of Learning. Chapman & Hall.

[4] Aleksander, I. (1989) The Logic of Connectionist Systems. In (&ed.) *Neural Computing Architectures*, London: Kogan Page.

[5] Davis, M., J.M.Hitchcock & J.B.Rosen (1987) Anxiety and the amygdala: Pharmacological and anatomical analysis of the fear-potentiated startle paradigm. In G.H.Bower (ed.) *The Psychology of Learning & Motivation*, 21. San Diego: Academic Press.

[6] Sakaguchi, A., J.E. LeDoux & D.J. Rein (1983) Medial geniculate but not auditory cortex lesions block emotional responses to auditory stimuli. *Society for Neuroscience Abstracts, 9*, 637.

[7] Weinberger, N.M. (1982) Sensory plasticity and learning: The magnocellular medial geniculate nucleus of the auditory system. In C.D.Woody (ed) *Conditioning: Representations of Involved Neural Functions*. New York: Plenum Press.

[8] Frysinger, R.C., B.S. Kapp, M.Gallagher, J.R.Haselton (1981) Amygdala central nucleus lesions: Effect on heart rate conditioning in the rabbit. *Physiology & Behaviour, 23*, 1109-1117.

[9] Kapp, B.S, R.C.Frysinger, M.Gallagher, & J.Haselton (1991) Amygdala central nucleus lesions: effects on heart rate conditioning in the rabbit, *Physiology & Behaviour, 23*, 1109-1117.

[10] Kemble, E.D. & J.A.Nagel (1973) Failure to form a learned taste aversion in rats with amygdaloid lesions. *Bulletin of the Psychonomic Society, 2*, 155-156.

[11] Nachman, M. & J.H.Asche (1974) Effects of basolateral amygdala lesion on neophobia, learned taste aversions, and sodium appetite in rats. *Journal of Comparative and Physiological Psychology, 87*, 622-643.

[12] Bolhuis, J.J., R.E. Fitzgerald, D.J.Dijk, & J.M.Koolhaas (1984) The corticomedial Amygdala and learning in an agonist situation in the rat. *Physiology and Behaviour, 32*, 575-579.

[13] Downer, J.D.C. (1962) Changes is visual gnostic function and emotional behaviour following unilateral temporal lobe damage in the "split-brain" monkey, *Nature, 191*, 50-51.

[14] LeDoux, J.E. (1992) Brain mechanisms of emotion and emotional learning, *Current Opinion in Neurobiology, 2*, 191-197.

[15] Levey, A.B. & I. Martin (1988) Classical conditioning in a cognitive era. *Biological Psychology, 27*, 153-166.

[16] Holland, P.C. (1990) Event representation in Pavlovian conditioning: image and action. *Cognition, 37*, 105-131.

[17] Reiss, M. & J.G.Taylor (1991) On temporal sequence storage. *IEEE International Conference on Artificial Neural Networks Proceedings, Bournemouth*, 129-133.

[18] Halgren, E. (1992) Emotional Neurophysiology of the Amygdala within the Context of Human Cognition. In Aggleton, J. (ed.), *The Amygdala: Neurobiological Aspects of Emotion, Memory and Mental Dysfunction*. New York: Wiley-Liss.

COMPUTATION, COGNITION, AND NEURAL NETWORKS

Frank van der Velde

Unit of Experimental and Theoretical Psychology
Wassenaarseweg 52, 2333 AK Leiden, vdvelde@rulfsw.leidenuniv.nl

Computation and Cognition. To understand computing with neural networks and how cognitive processes can be implemented in this way, it is imperative to distinguish between two kinds of machines and the forms of computation they produce: Finite-state automata (FA's), and machines such as the Turing machine (TM) or the pushdown automaton (PA). The difference concerns the use of (working) memory. In an FA, the memory is interwoven with the program. To increase processing capacity, new states have to be added to the machine, with new connections between these states and the old states. As a result, the structure of the machine changes and thus the function computed by the machine. In contrast, a TM or a PA is an FA connected to a memory that is external to the FA. The FA is the program of the machine. Because the FA (program) is separated from the memory, the processing capacity (memory) of the machine can be increased without changing the program, and thus without changing the function computed. To use an external memory, a program in a TM or PA has to produce representations (symbols) that can be stored in and retrieved from the memory. This shows why computation of this kind is symbol *manipulation*: The program manipulates symbols by the processes of storing and retrieving. With a repeated use of storing and retrieving, combinations (strings) of symbols can be stored in the memory, to be used for recognition or production.

The distinction between FA's and the TM or the PA coincides with the distinction between behaviorism and cognitive science [1]. The stimulus-response (S-R) models of behaviorism are FA's, whereas cognitive science is based on symbol manipulation. With behaviorism, however, one has "once again the nagging question of whether an analytical S-R account of complex behaviors is viable. The sheer number of individual associative bonds, and the problem of quantifying their individual and combined momentary status to predict behavior is mind-boggling" [2]. In behaviorism, information processing results from the reactivation of existing structures (the S-R combinations) that have previously been learned. The fundamental difficulty with this approach is the fact that many tasks of (complex) information processing consist of recognizing or producing familiar representations in new and unfamiliar combinations. In the case of natural language, for instance, familiar words or phrases can be recombined into novel sentences which have not been produced or recognized before. In other words, natural language is *generative*. This is also true for higher visual processing. We have the ability to find our way in a novel environment (e.g., an unknown city), because such an environment consist of new combinations of familiar visual structures. In fact "comprehension of a visual scene requires a kind of serial assembly operation similar in some respects to the integration of words in sentences" [3].

Symbol manipulation offers the possibility for cognition to be generative. Firstly, because symbol manipulation is more *productive*. Mathematically, TM's and PA's (with unlimited memory) can produce functions that cannot be produced by FA's. With limited memory capacity (as in any real system) a TM or PA is not more powerful in this sense, because FA's can be found that produce the same functions as these machines. However, a TM or a PA can produce a function on the basis of substantially less information, thus more productively, because it can store 'intermediary' results in the memory, which can be used later in the process [1].

Secondly, symbols allow *access* to information that is stored elsewhere. Hence, new information can be stored in such a way that it does not interfere with the information already available. For instance, human experts can have a knowledge base to the amount of 100.000 production rules or more. Such an amount of information is not obtained at once but in an incremental way, and the addition of new information does not destructively interfere with the information already available. With symbol manipulation, the processing system is able to store and get access to new information at new (distant) locations [4].

Because cognition depends on symbol manipulation, it seems that cognitive processing does not depend on the way it is implemented. However, cognition is related to the notion of ecological validity, which has two implications for the use of computing in cognitive science. Both implications reveal the importance of implementation in cognitive processing [1]. Firstly, cognitive processing consists of computational *processes* that can operate in real time, which likely means that they have to be implementable (at least in part) in a parallel distributed manner. Secondly, ecological validity involves a selection of representations (symbols). In computation theory, symbols can represent anything. Their meaning is not intrinsic to a computational system but is provided by the user. However, for a system to interact with the environment the meaning of symbols has to be intrinsic to the system, so that they represent (elements in) the environment within the system. This will be the case if the symbols are grounded in perception, as provided by an appropriate implementation [1].

Neural Networks. Computing with neural networks seems to be an appropriate way to meet the demands of implementation outlined above. After all, the neural networks of the brain have resulted from a long process of evolution in which the successful interaction with a hostile environment was crucial for survival. The form of processing that has evolved can be illustrated with the visual cortex. In primates, the visual cortex occupies about half of the neocortex. A similar magnitude holds for the human visual cortex. The visual cortex consist of about 30 distinct areas (modules), interconnected partly in parallel and partly in a hierarchical order. Many attributes of visual information (e.g. color, motion, shape, location) are processed in parallel in different areas. The set of modules is in particular divided over two anatomically distinct (but interconnected) parallel processing pathways.

The largest ('what') pathway deals with object identification, and the smaller ('where') pathway deals with location information. Thus, the visual information that enters the brain through the retina is reduplicated and distributed over separate modules and pathways that, in part, operate in parallel. In this way a computational *process* [1] has evolved that is highly successful in processing a two-dimensional continuous stream of visual input information.

The nature of the representations that have evolved can be illustrated with the cortical columns in the inferotemporal cortex (IT), which is the final module of the 'what' pathway involved in object identification. Cortical columns can be considered as the basic processing units in the cortex. Each column in the IT is about 0.4 mm wide. The cells in a particular column respond selectively to a set of related and relatively complex visual features. Cells in adjacent columns respond to different sets of features [5].

The selectivity to visual features in the IT can change as a result of persistent changes in the visual environment. Hence, they can be learned. Furthermore, the activity in a column can be sustained after the disappearance of the stimulus. This suggests that the cells in a column act as a cell assembly, which is a set of neurons with a relatively high probability of interconnection. Activation in an assembly can be retained because the neurons activate one another in turn. Thus, the IT columns show the existence in the cortex of representations in the form of persistent activities, or 'attractors' [6], in local cell assemblies. In case of the IT, the representations are moderately complex and abstract visual features, that are formed after a prolonged period of learning. They correspond to features that are commonly present in visual displays. A particular visual display will consist of a particular selection and ordening of these features. The features can be seen as a visual 'alphabet' used for coding visual scenes, not unlike the way in which words are used as elements in the processing of a sentence. The resemblance goes perhaps even further, because the IT is probably involved in language comprehension as well [3].

Representation as found in the IT is paradigmatic for representation in the cortex [6]. Thus, after a period of learning, representations are distributed over the cortex, each at its own *fixed* site (as the columns in the IT cortex). This ensures grounding in perception, because after learning connections will exist between the retina and a representation (its cell assembly), so that the representation (i.e., the attractor in the cell assembly) will be activated when the corresponding object is present in the visual field. Similarly, a representation can produce the appropriate action because connections will exist between the cell assembly (column) and the neurons that initiate the action. It is essential that the representations are at a fixed site in the cortex, because otherwise the pattern of connections described above has lost its meaning. In fact, the meaning of a representation is given by this pattern of connections [1, 7].

However, the existence of representations (or symbols) at a fixed site seems to contradict the possibility of symbol manipulation, in which symbols (apparently) are copied and transported from one site to another [4]. This contradiction has been identified as the largest problem in the unification of cognitive science and neuroscience [7, p207].

Recently, I have formulation a solution to this problem [1, 8] by showing how a TM or a PA, and thus symbol manipulation, can be implemented with neural networks. Current network models (feedforward nets, Hopfield nets, recurrent nets) do not have the ability of symbol manipulation [1, 9], because they do not possess the distinction between program and memory and therefore cannot distinguish between different tokens of the same symbol [8]. The networks used in the implementation of the TM and the PA [1, 8] do possess the necessary distinction between program and memory. Operation in the networks relies on the three basic operations of neurons: Excitation, inhibition and association (conditioning). The networks are of necessity modular (as is the cortex). They use representations (symbols) in the form of attractors in cell assemblies. The symbols can be copied ('tokened') and stored in memory. Nevertheless, they are grounded in perception, because their assemblies are at a fixed site. The memory (tape or stack) again consists of cell assemblies, which represent the positions in the memory. Symbols are tokened and stored when they are (temporarily) associated with position representations. A dynamical process ensures that symbols can be stored and retrieved in this way. I will discuss the use of these networks in the production of complex cognitive behavior [1, 8]. Indeed, "it would be hard to figure out how language [cognition] could function without some global rule-like operations, however implemented. The advantages of a neural network with this ability are immensely practical, for it would allow for far greater computational flexibility and generalisation than the current, simple networks can give themselves" [9].

References

[1] Van der Velde, F. (subm). Integrating connectionism and symbol manipulation: The importance of implementation. *Behavioral and Brain Sciences.*

[2] Amsel, A., & Rashotte, M. E. (1984). *Mechanisms of adaptive behavior: Clark L. Hull's theoretical papers, with commentary.* N.Y. Columbia UP

[3] Sereno, M. I. (1991). Language and the primate brain. *Proceedings 13th annual conference of the cognitive science society.* Hillsdale: Erlbaum.

[4] Newell, A. (1990). *Unified theories of cognition.* Cambridge, MA: Harvard University Press.

[5] Tanaka, K. (1992). Inferotemporal cortex and higher visual functions. *Current Opinion in Neurobiology,* 2, 502-505.

[6] Amit, D. J. (in press). The Hebbian paradigm reintegrated: Local reverberations as internal representations. *Behavioral and Brain Sciences.*

[7] Dennett, D. C. (1991). *Consciousness explained.* London: Penguin.

[8] Van der Velde, F. (1995). Symbol manipulation with neural networks: Production of a context-free language using a modifiable working memory. *Connection Science* 7, (3) (forthcoming).

[9] Anderson, J. A. (1994). Associative computation. In: d'Ydewalle, Eelen & Bertelson (Eds.), *International perspectives on psychological science, Vol 2.* (pp. 95-118), Hillsdale: Erlbaum.

AUTOMATIC SPEECH RECOGNITION SYSTEMS AND MODELS OF HUMAN WORD RECOGNITION: A COMPARATIVE ANALYSIS

Peter Wittenburg, David van Kuijk, Kay Behnke
Max-Planck-Institute for Psycholinguistics
Postbus 310, 6500 AH Nijmegen

1. Introduction

On a recently held workshop titled "Methods and Models of Word Recognition" [1] psycholinguists and speech engineers discussed their different views on spoken word recognition. In this paper we would like to continue this discussion by analyzing the similarities and differences between NN-based psycholinguistic models such as TRACE [2], Shortlist [3], and RAW [4] on the one hand and HMM-based automatic speech recognition systems on the other hand.

Psycholinguistic models try to incorporate results gathered from experiments with human subjects. They often use simplified mock input to avoid the additional complexities which real speech imposes. But the use of mock input might lead to wrong conclusions. The recently developed RAW model is the first model both working with real speech and claiming psychological relevance. Current HMM-based speech recognition systems neglect most evidence about human word recognition, nevertheless they currently achieve the best recognition scores. This performance is achieved by applying statistical knowledge extracted from large speech corpora. On corpora with spontaneous speech such as the Switch-Board corpus the achieved recognition results (the best HMM recognizers score 50%) are far from satisfactory and point to inherent limitations of this approach.

In this comaprative study we can just focus on two major topics: (1) What are the units of processing and (2) what are the temporal aspects of processing?

2. Units of Processing

In psycholinguistic theories of spoken word recognition various explicit concepts like phonetic features, phonemes, syllables, or words play a role. They are associated with characteristics like decision making and categorization. Psycholinguists still discuss the nature of the representations stored in the mental lexicon. They found some evidence in normal word recognition tasks that humans categorize the incoming speech stream into phoneme representations which are then used to access the mental lexicon. This is the reason that TRACE uses mock input which covers the confusion between phonemes and groups the bottom-up activations into phoneme activations. Shortlist is based on the same ideas, except that the feature layer currently is not implemented. These representations have lost much detail about the original signal. Following this representation the utterances /shipping/ and /ship inquiry/ will yield exactly the same activation patterns up to the point where information about the second syllable of /inquiry/ will come up.

In the recently developed RAW model which uses real speech as input the differences between the acoustic waveforms associated with the two words mentioned above will lead to small but distinguishable differences in the activation patterns already when the first syllable of /inquiry/ is processed. Differences in the acoustic pattern would lead to slightly damped activation patterns. RAW uses at the phonemic level a data reduction algorithm defined by the mapping of the 16-dimensional input to a 2-dimensional Kohonen map of a limited number of neurons, i.e. currently there is no categorization implemented as psycholinguistic theory would propose.

Speech engineers do not make a separation into lexical and pre-lexical processing. HMM-based recognizers are seen as decoders searching for the overall message with the highest a posteriori likelihood where the so-called Phoneme-Like-Units (PLUs) are nothing else than implicit instances in this computation. These PLUs which model some phonemic knowledge are defined as small transition graphs where each node represents a sub-state. Statistical pattern identification on speech corpora associates mixture density distributions with each such state. During processing each node which is part of an accepted path adds a local likelihood to the overall likelihood calculation. However, in modern HMM systems it is not really clear in phonetic sense what the sub-states or the PLUs really represent. Speech engineers consequently ignore any form of explicit notion of phonemic knowledge and any form of categorization. Such unconstrained formulations lead to the best recognition results up till now. However, a HMM cannot explain some human capabilities such as to recognize phonemes when nonwords are presented.

Syllables until now play almost no role in both worlds. Speech engineers argue that all important information about syllables is implicitly used, since it is part of the knowledge about words. Recently information about syllables was introduced into the Shortlist model to implement the metrical segmentation strategy guiding lexical access [5]. While Shortlist models this by an artificial boost of the activation, RAW relies on a separate flow of information extracted from the speech signal. At the word level speech engineers still like to see nodes just as instances in calculating the overall likelihood of the complete message. For psycholinguists the mental lexicon plays a central role where special mechanisms are applied. Here RAW looks for a new approach in so far that it incorporates attentional mechanisms to guide the building of lexical representations. These allow us to build up discriminative representations and to implement one-shot learning of new concepts which is of great importance. Statistical integration is only done for very similar signals. This is in contrast to HMMs where all forms have to be specified in advance and where the parameter identification is done on a large corpus by statistical means.

3. Temporal Aspects of Processing
Dynamics of the speech signal
The speech signal is a dynamic signal, i.e. the processing of static feature vectors cannot lead to satisfying results. We assume that the auditory processing makes use of the dynamics of the different phones in their acoustic/phonetic contexts. Plosives for example can only be recognized as a specific sequence of articulatory movements. Traditional psycholinguistic models like TRACE and Shortlist do not care about these aspects, since they use abstract phoneme representations. In RAW we intend to use at

first dynamic Kohonen map methods such as Temporal Kohonen Map [6] or Self-Organizing Feature Maps for Sequences [7] and later even methods simulating what is known from infants. A recently finished study shows the potential of dynamic maps [8]. In HMM mostly difference spectra are used to incorporate dynamics. Although it was shown that these improve the speech recognition rates considerably, it is not seen as a satisfying solution. The achieved window lengths are often not enough to include the relevant dynamic aspects. Modelling the phonemes with several PLUs is intended to cover the different production phases, however, the dynamics are still mapped onto static states. A solution can be found in hybrid HMM systems where dynamic supervised neural nets are used to calculate the observation probabilities.

Role of Competition

It has been shown that during word recognition the number of competitors for each moment in time influences the recognition process in humans. Words with many competitors cannot be detected as fast as words with only a few competitors. Psycholinguistic models simulate this by introducing lateral inhibition. Mathematically this leads to a delay in the activation curve of the finally winning unit and to contrast sharpenings. The same effect can be achieved by choosing another decision criterion which determines when a word is recognized. However, only competition can explain the effect that words which are embedded late in the input are not activated. Speech engineers relying on the decoder principle based on Bayesian statistics cannot accept any form of competition. Therefore, they apply more or less complex decision rules.

Temporal Order of Processing

All models and systems mentioned so far suffer from an exhaustive search space, i.e. for each incoming speech vector a lexical access will be started. In TRACE and Shortlist this means that for each incoming feature vector the complete network has to be copied to be able to distinguish the various paths. Shortlist improves this by reducing the size of the network to be copied. Both RAW and HMM include internal mechanisms to represent the temporal order. HMM do this by defining the transition probabilities and RAW incorporats the so-called gate signals which define temporal order in a stochastic way. HMM traces have to be started and stored per incoming speech vector to finally be able to determine the optimal one.

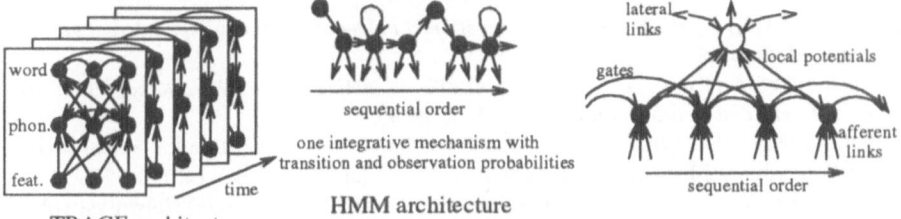

TRACE architecture   HMM architecture

A consequence of these architectural characteristics is that the processing in HMM and RAW is strictly left-to-right. Due to the stochastic definition of the gate signals in RAW it can cope with certain errors made by the speaker or distortions. The behavior of HMM in this respect is defined by the states and the transition probabilities between them. This makes it harder for HMM to deal with such phenomena. In TRACE and Shortlist all inputs are simply summed up at the price of copying the network which is not a plausible solution. After having built in the usage of prosodic cues in RAW, the behavior would become different in that RAW then would drastically reduce the number of word candidates receiving bottom-up information.

4. Conclusions

We tried to explain that psycholinguistically based models are relying on priciples of decision making and categorization. Therefore, it is assumed that using phonemes as input in simulations is adequate to study the interesting phenomena until now. Assuming such a strong abstraction from the details of the signal can lead to interpretation errors which limit the usability of such models. In contrast, all parameter identification in HMMs is based on statistical methods only. No decision or categorization should be carried out. RAW assumes a position which is in between these two extremes: most of the information is available and carried on, but it will normally not be used. At the layer of phonetic processing an attraction mechanism is assumed in line with the "Perceptual Magnetic Effect" found by P. Kuhl [9].

In our opinion it is extremely important to include prosodic information in a psycholinguistic model to limit the search space and to control the development of lexical representations by attentional mechanisms. The latter enables us to implement one-shot learning and speaker adaptation and, therefore, may let us bypass the problems of HMM resulting from running statistics over a whole corpus containing many variations. The stringent application of the HMM method which is seen as a big advantage by speech engineers creates problems for the psycholinguists to adopt these principles when simulating human behavior. In psycholinguistic models each linguistic level is associated with explicit concepts and specific mechanisms. Nevertheless, the knowledge from speech engineers will have to be used to implement models working with real speech as input. Speech recognition systems, on the other hand, will have to apply novel methods adopted from psycholinguistics to be able to cope with spontaneous speech.

References
[1] Workshop on "Models and Methods of Spoken Word Recognition". Unpublished Manuscripts. January 1995. Max-Planck-Institute for Psycholinguistics, Nijmegen
[2] McClelland, J., Elman, J. (1986). The TRACE model of speech perception. Cognitive Psychology, 18, 1-86
[3] Norris, D. (1992). Shortlist: A hybrid connectionist model of continuous speech recognition. Unpublished Manuscript.
[4] Wittenburg, P., van Kuijk, D. , Dijkstra, T. (1995). Modeling Human Word Recognition with Sequences of Artificial Neurons. Submitted to ICANN 95
[5] Cutler, A., Norris, D. (1988). The role of strong syllables in segmentation for lexical access. Journal of Experimental Psychology: Human Perception and Performance 14, 113-121
[6] Chappel, G.J., Taylor, J.G. (1993). The Temporal Kohonen Map. Neural Networks, vol. 6, 441-445.
[7] Wittenburg, P., van Harmelen, H. (1994). Self-Organizining Feature Map for Sequences. 39. International Scientific Colloquium, Univ. Ilmenau.
[8] Wittenburg, P., Neggers, H., Behnke, K., van Harmelen, H. (1995). Can Dynamic Self-Oragnizing Feature Maps Improve Automatic Speech Recognition? Submitted to ICANN 1995.
[9] Kuhl, P.K. (1991). Human adults and human infants show a perceptual magnet effect for the prototypes of speech categories, monkeys do not. Perception & Psychophysics 50 (2), 93-107.

Language Acquisition and the Necessity of a New Neural Network Approach

Kay Behnke & Peter Wittenburg

Max Planck Institute for Psycholinguistics
Postbus 310, 6500 AH Nijmegen (The Netherlands)
e-mail: behnke@mpi.nl

1   Introduction

Each theory related to the acquisition of a language has to take into account the fact that children are not biased toward learning a particular language, but have the potential to learn any language they are exposed to. Therefore, the cognitive abilities which help in acquiring a language must be of universal character.

While most of the research in this area is related to the question how children acquire adult's grammar, by investigating children's linguistic abilities at various stages of development, our concern is the ability of infants to segment the incoming speech stream into meaningful units. According to Jusczyk, there are at least four cognitive functions necessary for the recognition of fluent speech [2, 3]. This includes the discrimination of utterances of different words, the categorization of different utterances of the same word, the detection of relevant information for the discrimination and categorization processes, and the storage of sound patterns in memory for recognition of words and recovery of meaning.

In this connection it is of great importance how infants acquire the sound structure of their language environment. An efficient encoding of the input enables the infant to abstract from the huge amount of information in the speech signal and to focus attention on those properties which are relevant to distinguish among words in the native language.

According to our theoretical model the infant builds memory representations of the ambient language input which constitute a mapping of the distributional properties of the language–dependent speech sounds into a phonetic map (for further details, see [1]). The development of these representations, or phonetic categories, is basically the result of a self–organizing process. In this article we will present a neural network approach which is able to account for the developmental process of the acquisition of phonetic categories.

2   Demands on a neural network approach

The application and development of a neural network approach in cognitive science is worthless without a theoretical model of the process one is planning to simulate. The reason for this is that simulation results only have meaningfulness relative to an underlying theory. In this sense, the neural network serves as an alternative explanatory metaphor with the advantage that the behavior of the network can be tested under conditions which are not possible in real life. And, under conditions which were experimentally not tested so far and lead to further predictions with respect to the theoretical model as well as to further evaluation of the network model.

2.1 Assumptions about the acquisition process

It is not only the simulation results which determines the appropriateness of the network model with respect to the theoretical model, but in some sense also its underlying structure. That does not mean that a reproduction of the brain structure is necessary to get interpretable results from simulations, but we want to emphasize that in cognitive science a neural network should not be regarded as a *black box*. We call this the *functional appropriateness* of a neural network approach. In this sense, assumptions made in the theoretical model will also influence the network model.

With respect to the acquisition process our main assumption is that infants build memory representations of the ambient language input into a phonetic map in an initially self–organizing process. There is no "teacher" available which directs the mapping process. But on the other side, this directly poses the question, how infants are able to represent the characteristics of the ambient language without discriminative processes taking into account the intrinsic complex structure of speech?

A further assumption of the model is that the nature of the input is not static but increases in resolution during development. We assume that infants are not able to perform all details contained in the speech signal from the beginning on but that perception is constrained by an energy filter. Only speech signals above an energy threshold pass the filter which implicitly limits the variability in the incoming signal. However, the threshold value slowly decreases during development till a maximum level of detail. Because of the fact that vowels have the highest energy values within a speech signal, a conclusion from this assumption is that vowels are represented in the phonetic map very early during development. This is in accordance with psycholinguistic results (see e.g. [4]).

The incremental increase in resolution of the input signal has further implications for the learning process. First, based on the reduced resolution of the input we can assume that speech sounds coming from the same phonetic category are in general more similar to each other than speech sounds coming from different categories. Thus, the aim is to map the categories into areas in the map with strong correlated afferent weight vectors. Second, the underlying algorithm has to be able to deal with "new" phonetic categories in the input space based on the increase of information in the input during development. The new categories have to be mapped onto the network map without influencing the structure of already developed representations in the map. This is only possible if the initial dynamics of the learning algorithm are maintained during simulation, so that the mechanism how clusters of units develop in the map is independent of the point in time of the simulation.

3   The dynamic network architecture

The architecture of the neural network consists of a two–dimensional map of processing units in which each unit receives the same input vector at a simulation step. The input to each unit (i, j) is weighted by an afferent weight vector aff_{ij} which is initially located at a random point in the input space.

At a simulation step several internal values are computed for each unit (i, j) which influence the adaptation of the afferent weight vector: (1) a single activity value $singleAct_{ij}$ which describes the correlation between the current input vector and the current afferent weight vector aff_{ij}, (2) an average activity value $averageAct_{ij}$, averaging the single activity values over the direct neighborhood of unit (i, j), (3) a single cluster quality value $singleCQ_{ij}$ which describes the correlation between the current afferent weight vector aff_{ij} to its direct neighbors, and (4) an average cluster quality value $averageCQ_{ij}$, averaging the single cluster quality values over the direct neighborhood of unit (i, j).

The general idea behind the learning rule is the following:
At every simulation step the afferent weight vector of each unit is slightly adapted in a *random* direction independently of the current input vector. That means that the units in the map would "walk" within the input space when no input vectors were presented. The amount of adaptation is dependent on the current cluster quality values which indicate whether the unit is member of a cluster or not. The higher the cluster quality values, the higher the possibility that the unit is member of a cluster and therefore the smaller the adaptation in the random direction.

The presentation of an input vector leads to an adaptation of the afferent weight vectors according to a general Hebbian learning rule in the direction of the input vector. In our model, this adaptation is dependent on two factors: the average activity around the current unit and its average cluster quality. That means that only units which (1) are sensitive to the current input vector *and* (2) which build a cluster are adapted to the direction of the input vector.

The consequence of the learning rule is, that based on the random "walking" of the units in the input space, clusters of units with similar afferent weight vectors develop during simulation, but that only these will "survive" which are sensitive for one of the input categories.

In addition to the adaptation rule, there is a restriction which avoids that afferent weight vectors would concentrate in the statistical center of an input category. We define for each unit a unique receptor field around its current afferent weight vector. If an afferent weight vector would point into the receptor field of a neighboring unit after adaptation, the adaptation is not performed. Depending on the size of the neighborhood in the map, this constraint prevents the concentration of afferent weight vectors in a very small region within the input space and defines at the same time the resolution of the mapping process. The following simulation example illustrates that in more detail.

4   Simulation example

With the intention to model the high autocorrelation of subsequent speech vectors and the variability of a phoneme in different context we used sequences of vectors from predefined areas in the input space as input vectors (see 1(a)).

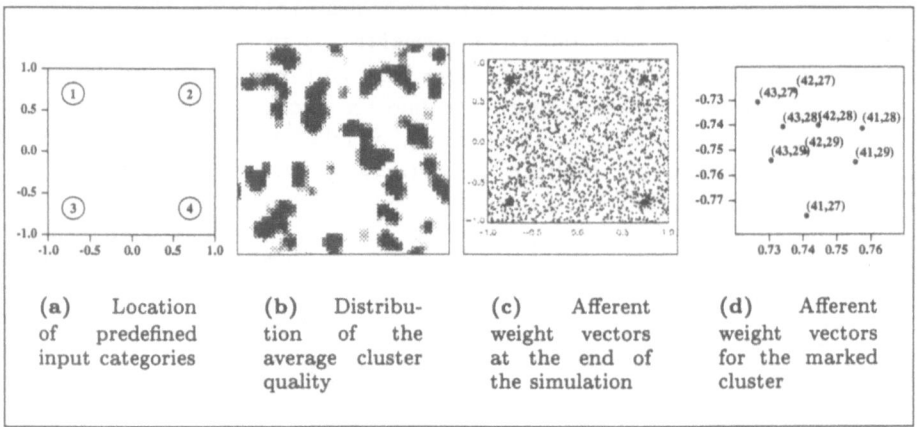

(a) Location of predefined input categories

(b) Distribution of the average cluster quality

(c) Afferent weight vectors at the end of the simulation

(d) Afferent weight vectors for the marked cluster

Figure 1: Result of a simulation

In figure 1(b) the distribution of the average cluster quality in the map at the end of a simulation is shown. Starting with randomly distributed afferent weight vectors, regions in the map developed containing units with strong correlated afferent weight vectors (indicated by black areas in the figure). The global distribution of the afferent weight vectors in figure 1(c) indicates that these units form clusters which are sensitive for the predefined input areas. However, most of the units are still unbounded to any cluster which makes it possible that new clusters based on changing input distribution can be formed. In figure 1(d) the afferent weight vectors of units which form a cluster are shown. As one can see, the vectors form a (partial) topological mapping.

We are currently investigating the behavior of the algorithm with real speech as input. The results will give us more information about the robustness of the algorithm.

References

[1] K. Behnke and P. Wittenburg. A neural network approach for the acquisition of phonetic categories. Submitted to ICANN' 95.

[2] P. W. Jusczyk. From general to language–specific capacities: the WRAPSA model of how speech perception develops. *Journal of Phonetics*, 21:3–28, 1993.

[3] P. W. Jusczyk. Infant speech perception and the development of the mental lexicon. In J. C. Goodman and H. C. Nusbaum (eds.), *The Development of Speech Perception*, pp. 227–270. MIT Press, Cambridge, Mass., 1994.

[4] J. F. Werker. Cross–language speech perception: Developmental change does not involve loss. In J. C. Goodman and H. C. Nusbaum (eds.), *The Development of Speech Perception*, pp. 93–120. MIT Press, Cambridge, Mass., 1994.

A Neural Model of Visual Attention

Piërre van de Laar

Tom Heskes

Stan Gielen

Department of Medical Physics and Biophysics, University of Nijmegen

RWCP* Novel Functions SNN† Laboratory

Nijmegen, The Netherlands

Abstract

We propose a biologically plausible neural model of selective covert visual attention. We show that this model is able to learn focussing on object-specific features. It has similar learning characteristics as humans in the learning and unlearning paradigm of Shiffrin and Schneider [8].

1 Introduction

In many pattern recognition and navigation tasks humans outperform computers. One explanation for this fact is that the software algorithm has to process all available information at a time, whereas a human quickly focusses its attention on the most important information, without paying too much attention to irrelevant information. The aim of this study is to model this process of information selection. In this article we roughly sketch a model for selective covert visual attention. Inspired by the human visual and attentional systems, we have made a biologically plausible model that might be able to explain some of the results of psychological experiments on attention, and especially on the learning of attention. Our model aims not to "appeal to some intelligent force or agent in explanations of attention phenomena" [3] but rather to be "a mechanism for attention selection" [1].

2 Selective Covert Attention

Johnston and Dark gave the following definition of selective attention: "Selective attention refers to the differential processing of simultaneous sources of information. In nature these sources are internal (memory and knowledge) as well as external (environmental objects and events)" [3]. As Johnston and Dark, we will follow "the vast majority of contemporary investigators and consider only external sources." Covert attention means that no overtly visible action is made to select between the different sources of information.

In our model we will consider both the bottom-up and top-down aspects of attention. In bottom-up attention the stimulus causes neural activity at the

---

*Real World Computing Partnership

†Foundation for Neural Networks

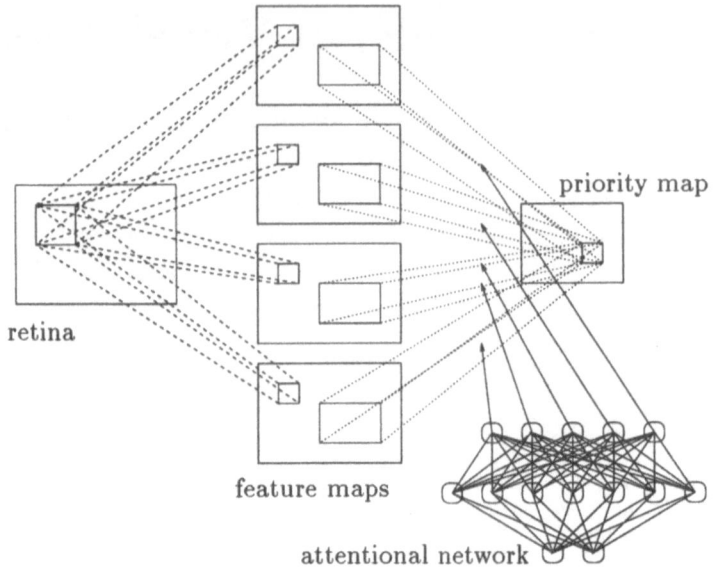

Figure 1: The architecture of the artificial visual system. From left to right: the retina, the feature maps and the priority map. The information going from feature maps to priority map is influenced by the attentional network, which receives a task-depending input.

different processing levels, which may affect the focus of attention. In top-down attention, the individual is internally biased towards particular stimuli, depending on the task or on the interpretation of the task to be performed. For example, an oval shape as the memory-set item in a distractor set of letters has been shown to give rise to a faster response for subjects who think the oval shape is a number (zero) than for those who start searching for a letter (o) [2].

3   A Neural Network Model

Our model, depicted in figure 1, receives only stationary images. These images are viewed by a "retina", which has a homogeneous distribution of receptors. The receptors pass their information on to the feature maps. Neurons in a feature map are topologically ordered, meaning that neighboring neurons have neighboring receptive fields. Translation-invariant pattern recognition is incorporated by applying weight sharing [5]. Each feature map has its own optimal stimulus which could be hardwired or learned, for example a color or a bar of a particular orientation.

The attentional network receives task-dependent input and controls the top-down aspects of attention by affecting the flow of information of the visual system from the lower levels to higher levels. In our model, the attentional network is a multi-layered perceptron, the output of which is used to gate the information flow from the different feature maps to the priority map. The priority map determines which part of the visual field contains the most inter-

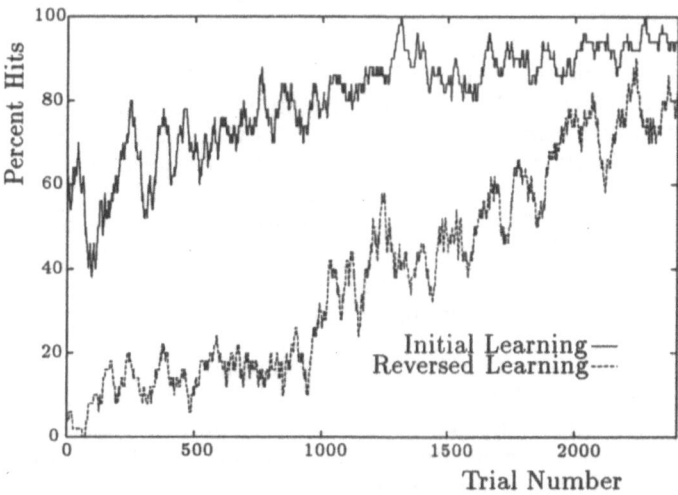

Figure 2: The reversal experiment: performance of detecting a target among distractors as a function of time before and after reversal. A hit occurs when our model focusses its attention on the target. After the initial 2400 learning steps, the target and distractor sets are interchanged and the reversed learning performance is measured over another 2400 learning steps.

esting information. Our attentional network is trained by backpropagation to enhance the relevant and to suppress the irrelevant information for the current task, which is specified by its input (see also [4]).

4   Comparison with Psychological Data

To test the selective attentional properties of the modeled visual system[1] we have compared its performance with the psychological data gathered in the learning and unlearning paradigm of Shiffrin and Schneider [8]. Qualitatively the performance of human subjects can be described as follows. Subjects who have to detect a target selected from a fixed set of targets among distractors selected from a fixed set of distractors, show a performance improving over time. However, when the target and distractor sets are reversed, the performance of these "trained" subjects drops far below the initial untrained level. Of course, continual training again leads to a better performance, but it takes a longer period of time to achieve the same accuracy as before reversal. Figure 2, which should be compared with figure 3 on page 132 of [8], illustrates that our neural-network model yields similar learning characteristics.

---

[1] Inputs are the characters {GMFP} and {CNHD}, which can be placed in either the left or the right visual field. The feature maps encode closure, bars of particular orientation in particular parts of the visual field and symmetries. The attentional network, which has noisy output neurons, was trained to yield maximal activation in the part of the visual field corresponding to the target position. In both situations, the task is the same: to detect the target among a set of distractors.

5 Discussion

The activity of the priority map can be used as input for control units in so-called shifter models (see e.g. [6]). These shifter models transfer the information inside the locus of attention to the higher levels of the visual system.

Our model only focusses its attention based on shape information. Attention towards a particular spatial position requires an extra module. This module may either influence the flow of information from the feature maps to the priority map or it can directly affect the priority map. Thus "two different attentional systems serve as sources of activation for [...] form and for location, although both might enter and amplify activity within the visual system at the same site" [7].

Our main goal remains to study the learning of attention. There is a vast amount of literature on all kinds of quantitative and qualitative psychological experiments, e.g. [2, 8, 9], that can be simulated with our neural-network model. In this paper we have given an illustrative example. In the future we will continue to test and improve our model in order to comply with the available psychological data without giving in to its biological plausibility.

References

[1] C. Bundesen. A theory of visual attention. *Psychological Review*, 97(4):523–547, 1990.

[2] J. Johnides and H. Gleitman. A conceptual category effect in visual search: O as letter or as digit. *Perception & Psychophysics*, 12(6):457–460, 1972.

[3] W. A. Johnston and V. J. Dark. Selective attention. *Annual Review of Psychology*, 37:43–75, 1986.

[4] J. K. Kruschke. ALCOVE: An exemplar-based connectionist model of category learning. *Psychological Review*, 99(1):22–44, 1992.

[5] Y. LeCun, B. Boser, J. S. Denker, D. Henderson, R. E. Howard, W. Hubbard, and L. D. Jackel. Backpropagation applied to handwritten zip code recognition. *Neural Computation*, 1:541–551, 1990.

[6] B. A. Olshausen, C. H. Anderson, and D. C. Van Essen. A neurobiological model of visual attention and invariant pattern recognition based on dynamic routing of information. *The Journal of Neuroscience*, 13(11):4700–4719, 1993.

[7] M. I. Posner and S. Dehaene. Attentional networks. *Trends in Neurosciences*, 17(2):75–79, 1994.

[8] R. M. Shiffrin and W. Schneider. Controlled and automatic human information processing: II. perceptual learning, automatic attending and a general theory. *Psychological Review*, 84(2):127–190, 1977.

[9] A. Treisman. Features and objects in visual processing. *Scientific American*, 255(5):106–115, 1986.

THE COMBINATION OF KNOWLEDGE IN FUZZY COGNITIVE MAPS

Paulo Camargo Silva

University of Erlangen-Nuernberg
Am Weichselgarten 9
D-91058 Erlangen
Tel.: (+49 9131) 85-9913
Fax : (+49 9131) 85-9905
camargo@immd8.informatik.uni-erlangen.de
Germany

Abstract

In this article we introduce matrices of combination for FCMs that have edges with weights in interval [0, 1] and FCMs that have edges with weights given by linguistics fuzzy quantifiers. We use a modification of operators of multi-agent modal logic for knowledge and belief and operators of fuzzy sets theory.

1 MULTI-EXPERT SYSTEMS

We consider a group of experts as a multi-agent system. For this we introduce modifications of two operators of multi-agent modal logic of knowledge and belief. It is possible, to base these operators on operators of fuzzy sets theory. In this way we obtain new forms of combined FCMs Kosko [1992a], Kosko [1992b] which are able in practical applications.

1.1 OPERATORS

The knowledge operators that we use in this part refer to a modification of modal operators of Halpern and Moses [1992] and Friedman and Halpern [1994]. However we avoid to present any theoretical development here. We consider that a FCM is the set of propositions (for fixed a) such as

$$C_i \overset{D_i}{\underset{a}{\tilde{v_i}}} C_j \tag{1}$$

where C_i, C_j are concepts and $\overset{D_i}{\underset{a}{\tilde{v_i}}}$ denotes the causal relation between concepts, v_i is the signal of relation (positive or negative) and D_i is the weight ($\{0, 1\}$, $[0, 1]$, linguistic fuzzy quantifiers). For example we can say that:

$$\text{Bad weather} \quad \overset{ALWAYS}{\underset{+}{\longrightarrow}} \quad \text{Freeway congestion} \tag{2}$$

where "Bad weather" is the concept C_i, "Freeway congestion" is the concept C_j, and $\overset{ALWAYS}{\underset{+}{\longrightarrow}}$ is $\overset{D_i}{\underset{a}{\tilde{v_i}}}$.

We define VAL as the value of (1):

$$VAL(C_i \overset{D_i}{\underset{a}{\tilde{v_i}}} C_j) = v_i.D_i \tag{3}$$

In this article we define an operator as a function of a set of propositions from type (1) to a proposition of same type. In other word:

$$S_G(p \overset{D_i}{\underset{a}{\tilde{v_i}}} q, p \overset{D_i}{\underset{b}{\tilde{v_i}}} q, \ldots, p \overset{D_i}{\underset{n}{\tilde{v_i}}} q) = p \overset{\overset{D_i}{\tilde{s_G^f}}}{v_i} q \tag{4}$$

and:

$$E_G^g(p \overset{D_i}{\underset{a}{\tilde{v_i}}} q, p \overset{D_i}{\underset{b}{\tilde{v_i}}} q, \ldots, p \overset{D_i}{\underset{n}{\tilde{v_i}}} q) = p \overset{\overset{D_i}{\tilde{s_G^f}}}{v_i} q \tag{5}$$

we consider $a, b, \ldots, n$ the experts of group, and $p \overset{D_i}{\underset{a}{\tilde{v_i}}} q$ is the causal relation between p and q in FCM of expert a, etc. Normally the combination of FCMs is made with FCMs that describe the same situation. The set of concepts has to be identical for the experts. This means that if the expert 2 introduces in his analysis a concept that the expert 1 does not use, this concept it to be added into the to FCM of expert1 and it is has zero causal connectivity.

The value of operators metioned above is given by:

(1) For the operator S_G^f:

$$VAL(S_G^f(p \overset{D_i}{\underset{a}{\tilde{v_i}}} q)) = \begin{cases} 0 & : \quad \forall_i.v_i = 0 \\ f(V_1) & : \quad \exists_i.v_i \neq 0 \end{cases} \tag{6}$$

where:

$$V_1 = \{v_i/v_i \neq 0\} \tag{7}$$

The function $f(V_1)$ can be defined in several forms. The FCMs that we have introduced above can have three different domains as ($\{0,1\}, [0,1]$, and linguistic fuzzy quantifiers). The combination function f has to be chosen relative the weight domain. For example, for the weight domain of linguistic fuzzy quantifiers we define $f(V_1)$ by:

$$f_{LUB}(V_1) = LUB(V_1) \tag{8}$$

where LUB denotes the "*least upper bound*" of (V_1). The operator S_G^f with a function $f_{LUB}(V_1)$ can be used to combine FCMs with linguistic fuzzy quantifiers, since these quantifiers are in a poset.

For weight domain [0, 1] we define $f(V_1)$ by:

$$f_{max}(V_1) = max(V_1) \tag{9}$$

where $max(V_1)$ denotes the maximun of set V_1. The operator S_G^f with a function $f_{max}(V_1)$ can be used to combine FCMs with weights in sets $\{0,1\}$ and $[0,1]$.

(2) For the operator E_G^g:

$$VAL(E_G^g((p\ \overset{D_i}{\underset{a}{\tilde{v_i}}}\ q)) = \begin{cases} 0 & : \ \exists_i.v_i = 0 \\ g(V_2) & : \ \forall_i.v_i \neq 0 \end{cases} \tag{10}$$

where:

$$V_2 = \{v_i/v_i \neq 0\} \tag{11}$$

Again, like in the case of f, the function $g(V_2)$ can be defined in several ways for the various weight domains. So that the first structure that we define for $g(V_2)$ is given by:

$$g_{GLB}(V_2) = GLB(V_2) \tag{12}$$

where GLB denotes the "*greatest lower bound*" of (V_2). The operator E_G^g with a function $g_{GLB}(V_2)$ can be used to combine FCMs with linguistic fuzzy quantifiers, since these quantifiers are in a poset.

The second structure that we define for $g(V_2)$ is given by:

$$g_{min}(V_2) = min(V_2) \tag{13}$$

where $min(V_2)$ denotes the minimun of set V_2. The operator E_G^g with a function $g_{min}(V_2)$ can be used to combine FCMs with weights in sets $\{0,1\}$ and $[0,1]$.

2   CONCLUSION

A FCM is a field of research including fuzzy sets theory and neural networks theory. There are much applications of FCMs. However the researches about FCMs suggest a combinations of FCMs that is possible to be made only with the most simple type of FCM, which have weights with values $\{0,1\}$. These FCMs are, in fact, crisp. They do not present weights in interval $[0, 1]$, and do not present weights with linguistic fuzzy quantifiers. There is not researches that combine complex FCMs,. In this article we present news forms of combination for FCMs with weights in interval $[0, 1]$, and weights given by linguistic fuzzy quantifiers.

3   REFERENCE

FRIEDMAN, N. & HALPERN, J.[1994] - A knowledge-Based Framework For Belief Change Part I: Foundations, Proceedings of Theoretical Aspects of Reasoning about Knowledge, Morgan Kaufman.

HALPERN, J., & MOSES, Y. [1992] - A Guide to Completeness and Complexity for for Modal Logic of Knowledge and Belief, Artificial Intelligence Vol.54, No. 3, April 319-379.

KOSKO, B.[1992] - Fuzzy Associative Memory Systems, in Kandel, A. (eds.), Fuzzy Expert Systems, CRC Press, London.

KOSKO, B. [1992] - Neural Networks and Fuzzy Systems: A Dynamical Systems Approach to Machine Intelligence, Prentice Hall, New York.

Robotics and Vision - Orals

A visually guided robot and a neural network join to grasp slanted objects

P. van der Smagt, A. Dev, and F. C. A. Groen

RWCP*Novel Functions SNN†Laboratory

Faculty of Mathematics and Computer Science, University of Amsterdam

Kruislaan 403, NL-1098 SJ Amsterdam

email: {smagt, anuj, groen}@fwi.uva.nl

Abstract

In this paper we introduce a method for model-free monocular visual guidance of a robot arm. The robot arm, with a single camera in its end-effector, should be positioned above a target, with a changing pan and tilt, which is placed against a textured background. It is shown that a trajectory can be planned in visual space by using components of the optic flow, and this trajectory can be translated to joint torques by a self-learning neural network. No model of the robot, camera, or environment is used. The method reaches a high grasping accuracy after only a few trials.

1 Introduction

Robot systems mostly depend on static visual information, taking into account only the *positions* of points of interest. On the other hand, living organisms use continuous feedback from their eyes in order to interact with their dynamically changing environment. In the meantime, there is sensory activity due to egomotion which must be taken care of.

In this paper it is investigated how visual feedback can be incorporated in a monocular robot arm control system, in order to position the arm directly above an object resulting in a state of rest, e.g., to grasp an object. It will be shown that by using time derivatives of the visual data, criteria can be developed which specify a trajectory which ends in a rest state (i.e., zero velocity and higher derivatives) at the end point. These criteria will be the visual setpoints along the followed trajectory. Thus it is possible that the eye-in-hand robot arm exactly stops on an observed object by use of optic flow, without having absolute visual depth information.

The use of optic flow for robot navigation is no novelty. Cipolla and Blake [1], for example, describe a system for estimating time-to-contact from the flow-field, and use this for obstacle avoidance of a robot arm with known kinematics. Sharma [2] uses time-to-contact derived directly from the optic flow as a basis for mobile robot trajectory planning. Vernon and Tistarelli [3] use visual velocity vectors to estimate depth in a complex visual scene, but again for a robot with known kinematics. In this paper, a method is presented which does not have any knowledge of the robot at the kinematic level; instead, a self-learning

---

*Real World Computing Partnership
†Dutch Foundation for Neural Networks

neural controller learns the camera–robot mapping directly from the behavior of the robot, without any prior knowledge.

In section 2 we will present the necessary theory for the deceleration task. Then, section 3 discusses how the required visual data can be obtained, even when the object is slanted. Section 4 describes the structure and implementation of the controller, and results are given in 5. A discussion follows in section 6.

2   The trajectory of an eye-in-hand system

The task of the neural controller is to generate robot joint accelerations $\ddot{\theta}$ which make the robot manipulator follow a trajectory leading to the goal state in which the end-effector (c.q., the camera) is placed on the object, and the robot arm is in rest. This trajectory will be expressed in the available visual domain, i.e., visual criteria $\xi(t)$ will be determined to which the trajectory of the observed object will have to adhere. Here we assume that no model of the object is available.

The world positions of the object and robot end-effector are given by x_o and $x_r(t)$, respectively, and we define the distance $d(t) \equiv x_r(t) - x_o$. We will use $d(t)$ to determine the trajectory that must be followed in the x, y, and z direction by the manipulator to reach the target. The goal state can be defined as reaching the state where $d(t \geq \tau) = 0$, in which τ is called the **time-to-contact**. This relation can also be interpreted as: all time derivatives of $d(t)$ must be zero at $t = \tau$:

$$\forall k \geq 0 : \quad d^{(k)}(\tau) = 0. \tag{1}$$

We assume $d(t) = ((d_x(t), d_y(t), d_z(t))$ and write $d(t)$ for either one of $d_x(t)$, $d_y(t)$, or $d_z(t)$.

By using closed loop control, the time axis is partitioned into intervals i, $i + 1$, $i + 2$, ... In each of these intervals we define a time-frame $t[i]$ with corresponding trajectory parameters $a_k[i]$. Then $d[i](t[i])$ is the $d(t)$ with respect to frame i. This choice is clear when we use a Taylor polynomial for $d(t)$, i.e.,

$$d[i](t[i]) = \sum_{j=0}^{n} a_j[i] t[i]^j + \varepsilon. \tag{2}$$

In figure 1 a trajectory is shown which is realized by feedback control. In this figure, the deceleration is computed for each time interval i, $i + 1$, ..., leading to a continuously adapted $d(t)$ and $\dot{d}(t)$.

Substituting (2) into (1), we will find a solution for the parameters of d as

$$\frac{a_n[i]^{n-k-1} a_k[i]}{a_{n-1}[i]^{n-k}} = \frac{1}{n^{n-k}} \binom{n}{k}, \quad 0 \leq k < n. \tag{3}$$

This form, however, has various disadvantages as a basis for feedback control. First, the time over which the motion is performed (i.e., the desired time-to-contact τ_d) cannot be controlled. Secondly, there are high-order terms a_n, a_{n-1}, etc., which must be available; in a real system such orders for $n \geq 2$ are difficult

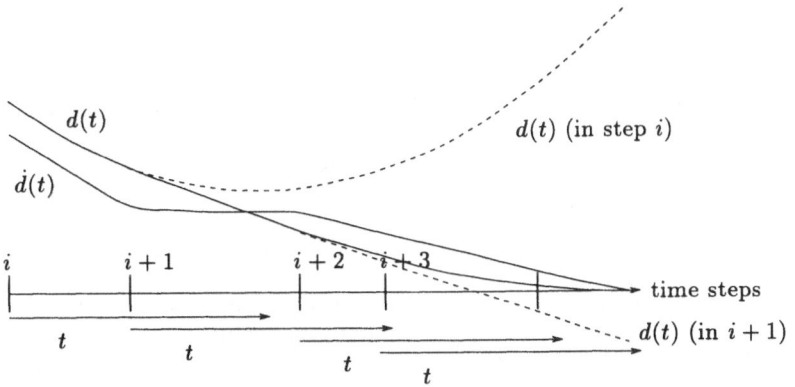

Figure 1: Exemplar trajectory followed by a feedback-controlled manipulator in a time scale. Both $d(t)$ and $\dot{d}(t)$ are are scaled to fit in the figure, and their signs are not shown. From i to $i+1$, etc., new measurements lead to a newly estimated deceleration, and thus the motion profile changes at $i+1$. The dashed lines give the predicted $d(t)$ which would have been followed if the deceleration would not have been changed.

to obtain with reasonable accuracy. Third, there is a total of $n-1$ equations that must be satisfied, i.e., the system is very complex.

To solve the first problem we introduce the τ_d as follows. Since $d(t)$ is approximated by a polynomial of order n, the nth derivative of the approximation of $d(t)$ must be constant in time; a_n is constant. Therefore, the $n-1^{\text{st}}$ must be linear in time. Consequently, the time to bring the $n-1^{\text{st}}$ derivative to 0 is equal to the quotient of the two: $(\tau - t[i]) = -a_{n-1}[i]/(na_n[i])$. By enforcing the desired time-to-contact and substitution into (3) we obtain

$$\frac{a_k[i]}{a_{n-1}[i]} = (-(\tau_d - t[i]))^{n-k-1} \frac{1}{n} \binom{n}{k}, \quad 0 \le k \le n \qquad (4)$$

where we write $(\tau_d - t[i])$ for $\tau_d[i]$. Satisfying these constraints leads to the desired trajectory. However, the constraints are all related and a simplification is in order. This leads to the following theorem.

Theorem 1 *Each of the components $d(t)$ of $\mathbf{d}(t)$ is described by a polynomial $d(t) = a_0 + a_1 t + \ldots + a_n t^n$. Then the stopping criterion $\forall k, 0 \le k < n :$ $d^{(k)}(\tau_d) = 0$ where τ_d is the desired 'time-to-contact', leads to the following constraint on the parameters of the successive intervals i of the trajectory:*

$$\frac{a_0[i]}{a_1[i]} = -\frac{(\tau_d - t[i])}{n}, \quad \forall i : 0 \le i < \nu, \qquad (5)$$

where $\nu \ge n$ and $(\tau_d - t[\nu]) \ge 0$.

The proof of this theorem is given in [4]. We will refer to (5) as the **time-dependent constraints**.

Theorem 1 shows that the trajectory of any n-order system can be described by a single constraint on the ratio of position and velocity of that system. The n

simultaneous criteria (4) which have to be satisfied at some time are replaced by satisfying a *single* constraint (5) during n subsequent steps. The 'simultaneous conditions' are 'laid out in time.'

Note the important advantage with respect to the time-independent constraints: the trajectory can be expressed in only the setpoint $(\tau_d - t[i])$ and $a_0[i]$ and $a_1[i]$; higher-order $a_k[i]$'s are not required.

3 Visual measurement of the stopping criteria

Our task is to learn the relationship between visual measurements and robot motion in order to satisfy the time-dependent constraints. In the above theory these constraints are expressed in parameters a_k. However, knowledge of these parameters requires knowledge of $\mathbf{d}(t)$, which we do not have.

Instead, the single hand-held camera can measure the position of the object projected on the camera's CCD, as well as its area that is projected on the CCD. We call these quantities $\xi_x(t), \xi_y(t)$, and $\xi_a(t)$. They are defined as

$$\xi_x(t) = f\frac{d_x(t)}{d_z(t)}, \quad \xi_y(t) = f\frac{d_y(t)}{d_z(t)}, \quad \xi_a(t) = \int_{\mathfrak{k}\in\mathcal{O}} da \quad (6)$$

where $\mathcal{O}$ is the object. These equations derive from a pinhole model of the camera. We will describe two visual measurements according to assumptions made on the orientation of the object surface the camera is looking at: 1. the surface is parallel to the image plane; and 2. no restriction on the object orientation.

3.1 Object surface parallel to image plane

The measured area of an object under the camera is now a simple function of the object distance and Cartesian area A: $\xi_a(t) = f^2 A/d_z(t)^2$. By multiplying ξ_x and ξ_y by $\xi_z = 1/\sqrt{\xi_a(t)}$, all quantities have similar relationships to their Cartesian counterparts, such that

$$\xi'_x(t) = \frac{d_x(t)}{\sqrt{A}}, \quad \xi'_y(t) = \frac{d_y(t)}{\sqrt{A}}, \quad \xi_z(t) = \frac{d_z(t)}{f\sqrt{A}}. \quad (7)$$

These $\xi'_x(t)$, $\xi'_y(t)$, and $\xi_z(t)$ are again fitted by polynomials:

$$\xi(t) = \sum_{i=0}^{n} v_i t^i + o(t^n) \quad (8)$$

where $\xi(t)$ is either $\xi'_x(t)$, $\xi'_y(t)$, or $\xi_z(t)$. Similarly, the v_i indicate the x, y, or z parameters. When writing c for c_x, c_y, or c_z, we conclude that

$$v_i = ca_i. \quad (9)$$

This means that the time dependent constraint can be written as

$$\frac{v_0[i]}{v_1[i]} = -\frac{(\tau_d - t[i])}{n}. \quad (10)$$

3.2 Non restricted object orientation

However, underlying the above relationship is the assumption that *the object surface normal is parallel to the viewing direction*, reflected in $\xi_z(t) = d_z(t)/\sqrt{A}$. If we relax the restriction on the direction of the object plane, the relative distance to the object cannot be measured from the object area alone [1]; the slant and tilt of the object plane are needed to compute the relative distance $\xi_z(t)$. We therefore will use another visual cue: the induced optic flow. From the optic flow, ideally a projection of the 3D motion field onto the 2D image plane, we want to compute the orientation ϕ of a planar surface on which the object lies or the planar surface of the top of the object. It can be shown [5] that every planar surface induces a second order optic flow $\mathbf{v} = (u, v)$ and can be written as

$$u(\xi) = \sum_{i+j\leq 2} u_{ij}\xi_x^i\xi_y^j, \quad v(\xi) = \sum_{i+j\leq 2} v_{ij}\xi_x^i\xi_y^j. \tag{11}$$

From the second order optic flow expansion coefficients $(u_{ij}, v_{ij})|_{\xi=0}$ the relative velocities of the camera, and the normal of the object plane can be computed [6] and are given by:

$$\begin{array}{lllll} u_{00} = -t_x & u_{10} = -t_z - \phi_x t_x & u_{01} = -\phi_y t_x & & u_{20} = \phi_x t_z \\ v_{00} = -t_y & v_{10} = -\phi_x t_y & u_{01} = -t_z - \phi_y t_y & v_{02} = \phi_y t_z \end{array} \tag{12}$$

with $t_i = \dot{d}_i(t)/d_z(t)$, where i indicates x, y, or z. The slant ϕ can be derived from these equations. The effect on the parameters v_i is that c is no longer a constant as before, but depends on *slant* and $d_z(t)$:

$$v_i = c(\phi, d_z)a_i. \tag{13}$$

4 Controlling the manipulator

We have thus developed a method for determining constraints in visual domain which lead to the desired goal state of rest in visual domain. In this section we will show how a controller for a robot arm can be constructed which uses the above theory. The controller consists of two parts: (1) a system which generates trajectories in visual domain; (2) a neural network which translates the trajectory in visual domain to joint motor commands.

4.1 A trajectory in visual domain

In the deceleration method described above, the visual acceleration of the camera $v_2[i]$ is determined from its position $v_0[i]$ and velocity $v_1[i]$ relative to the object, and desired time-to-contact τ_d. So we can say that $v_2[i]$ is the *desired* visual acceleration necessary to satisfy the time-dependent deceleration criterion; hence we prefer to write $v_{2d}[i]$. During deceleration, this $v_{2d}[i]$ is applied to the system, which leads to a new $(v_0[i+1], v_1[i+1])$ in the next time-step.

Instead of considering $v_{2d}[i]$ as the control variable for the controller, it is also possible to express a setpoint in a desired position and velocity at $i+1$.

So we say that the next *setpoint* is a pair $(v_{0d}[i+1], v_{1d}[i+1])$, the elements of which are computed as

$$v_{0d}[i+1] = v_0[i] + v_1[i] + v_{2d}[i], \tag{14a}$$

$$v_{1d}[i+1] = v_1[i] + 2v_{2d}[i]. \tag{14b}$$

The $v_{2d}[i]$ can be found by applying (5) to (14):

$$v_{2d}[i] = -\frac{nv_0[i] + v_1[i](n + (\tau_d - t[i]) - 1)}{n + 2(\tau_d - t[i]) - 2}; \tag{15}$$

while the optimal n is obtained with $n = \lceil -(\tau_d - t[i])v_1[i]/v_0[i] \rceil$, assuming local quadratic polynomials for the trajectory $d(t[i])$ that is followed. Thus we end up with a system which has as inputs the measured $v_0[i]$, $v_1[i]$, and the remaining $(\tau_d - t[i])$, and as output the desired $v_{0d}[i+1]$ and $v_{1d}[i+1]$.

4.2 Translating the trajectory to joint domain

We have shown that a trajectory can be determined in *visual* setpoints $(v_{0d}[i+1], v_{1d}[i+1])$ so as to satisfy the time-dependent constraint. The final question we have to ask is the following: how can we compute joint accelerations $\ddot{\theta}$ that make the robot-camera system follow those visual setpoints? The trajectory in visual domain must be translated to a trajectory in joint domain.

In the case that the area A of the object and focal length f of the camera were known constants, then the relationship between v_i and a_i would be known. In that case the trajectory $(v_{0d}[i+1], v_{1d}[i+1])$ in visual domain can be calculated from the trajectory in visual domain, and with known kinematics of the robot this can be used to compute a trajectory in joint domain. Of course, in our case the relationship between v_i and a_i is *not* known and it even varies with the size of the object. However, the mapping

$$v_0[i], v_1[i], v_{0d}[i+1], v_{1d}[i+1], \phi_x, \phi_y, \theta[i], \dot{\theta}[i], \ddot{\theta}[i] \rightarrow \ddot{\theta}[i+1] \tag{16}$$

is a function, i.e., does not have multiple solutions, since the end-effector velocity can be measured (in joint space), such that the visual measurements are disambiguated.

4.3 Controller structure

The resulting controller consists of the following parts:

1. the *vision/robot front-end* uses the measured $\xi(t)$ and $\theta(t)$ to determine the joint positions $\theta[i]$ and their derivatives $\dot{\theta}[i]$ and $\ddot{\theta}[i]$, as well as the visual trajectory parameters $v_j[i]$. Furthermore, the ϕ_x and ϕ_y are estimated;

2. these visual trajectory parameters are input to the *criterionator* which applies equations (15) and (14) to calculate the next visual setpoint $(v_{0d}[i+1], v_{1d}[i+1])$;

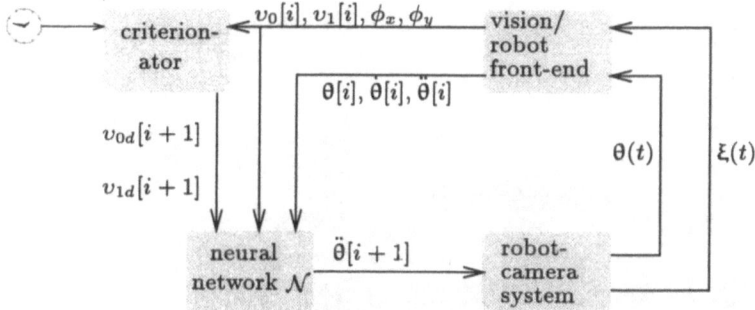

Figure 2: The structure of the time-to-contact control loop.

3. the current visual measurement $(v_0[i], v_1[i])$, the visual setpoint, and $\theta[i]$, $\dot{\theta}[i]$, and $\ddot{\theta}[i]$ are input to the neural network:

$$\mathcal{N}\left(v_0[i], v_1[i], v_{0d}[i+1], v_{1d}[i+1], \phi_x, \phi_y, \theta[i], \dot{\theta}[i], \ddot{\theta}[i]\right) = \ddot{\theta}[i+1]$$

4. the neural network generates a joint acceleration $\ddot{\theta}[i+1]$ which is sent to the robot:

$$\mathcal{R}\left(v_0[i], v_1[i], \theta[i], \dot{\theta}[i], \ddot{\theta}[i+1]\right) = \left(v_0[i+1], v_1[i+1], \theta[i+1], \dot{\theta}[i+1]\right).$$

Figure 2 shows the components of the time-to-contact controller in the control loop.

The neural network. The neural network that is used consists of a feed-forward neural network with six hidden units [4]. The neural network is implemented on two computers. One computer is used to control the robot. The learning samples it constructs are sent to a second computer, which does the actual learning. Weight matrices are periodically sent back to the controlling computer. The learning algorithm used is a conjugate gradient optimisation.

Learning samples can be simply generated from the measured visual and robot positions, leading to the following samples:

$$\left(v_0[i], v_1[i], v_0[i+1], v_1[i+1], \phi_x, \phi_y, \theta[i], \dot{\theta}[i], \ddot{\theta}[i]\right) \Rightarrow \ddot{\theta}[i+1]. \qquad (17)$$

5   Results

In order to measure the success of the method while applied to a simulated PUMA-like anthropomorphic robot arm with an upper and lower arm length of 50cm and 51cm, respectively, we measure the $\mathbf{d}(t)$, $\dot{\mathbf{d}}(t)$, and $\ddot{\mathbf{d}}(t)$ during the trajectory; with the simulator, these data are available. A correct deceleration leads to a $\mathbf{d}(t) = \dot{\mathbf{d}}(t) = 0$ when $\tau_d = 0$, i.e., at the end of the trajectory. The results of a run with the simulated robot are shown in figure 3. This graph shows the distance between the end-effector and the target object at $\tau_d = 0$. The results show that after only a few trials the positional error averages below one centimeter, while the velocity is below 0.5cm per simulator time unit.

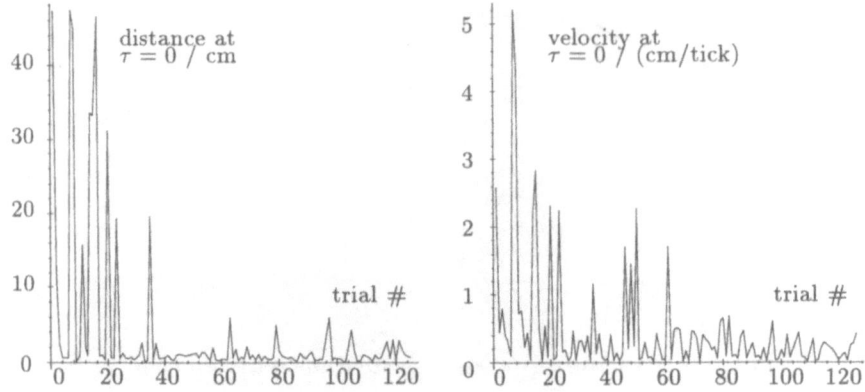

Figure 3: Distance and velocity at $\tau = 0$.

6 Discussion

A method has been proposed which can control the end-effector of a monocular robot in three-dimensional space by using optical flow feedback to obtain depth-information. The system does not use knowledge of the robot arm or camera, and does not require any calibration at all. Instead, from the visual feedback setpoints in visual domain are calculated, and these setpoints are translated to joint domain using an adaptive neural controller.

The choice of the controller input signals, by requiring only first derivative visual information, makes the system very viable for implementation on a real robot arm.

References

[1] R. Cipolla and A. Blake. Surface orientation and time to contact from image divergence and deformation. In G. Sandini, editor, *Computer Vision—ECCV '92*, pages 187–202. Springer-Verlag, 1992.

[2] R. Sharma. Active vision in robot navigation: Monitoring time-to-collision while tracking. In *Proceedings of the 1992 IEEE/RSJ International Conference on Intelligent Robots and Systems*, pages 2203–2208. IEEE, June 1992.

[3] D. Vernon and M. Tistarelli. Using camera motion to estimate range for robotic parts manipulation. *IEEE Transactions on Robotics and Automation*, 7(5):509–521, October 1990.

[4] Patrick van der Smagt. *Visual Robot Arm Guidance using Neural Networks*. PhD thesis, Dept of Computer Systems, University of Amsterdam, March 1995.

[5] J. J. Koenderink and A. J. van Doorn. Second-order optic flow. *Optical Society of America A*, 9(4), 1992.

[6] A. M. Waxman and S. Ullman. Surface structure and three-dimensional motion from image flow kinematics. *The International Journal of Robotics Research*, 4(3), 1985.

The Mobile Robot Rhino

J. Buhmann, W. Burgard, A.B. Cremers, D. Fox,
T. Hofmann, F. Schneider, J. Strikos and S. Thrun

Institut für Informatik III, Universität Bonn

Bonn, Germany

Abstract

RHINO was the University of Bonn's entry in the 1994 AAAI Robot Competition and Exhibition. RHINO is a mobile robot designed for indoor navigation
and manipulation tasks. The general scientific goal of the RHINO project is the
development and the analysis of autonomous and complex learning systems. This
paper briefly describes the major components of the RHINO control software as
they were exhibited at the competition. It also sketches the basic philosophy of the
RHINO architecture and discusses some of the lessons that we learned during the
competition.

1 General Overview

RHINO, shown in Figure 1, is a B21 mobile robot platform manufactured by Real
World Interface Inc. It is equipped with 24 sonar proximity sensors, a dual color
camera system mounted on a pan/tilt unit, and two on-board i486 computers. Sonar
information is obtained at a rate of 1.3 Hertz, and camera images are processed at a rate
of 0.7 Hertz. RHINO communicates with external computers (two SUN Sparcstations)
by a tetherless Ethernet link.

The RHINO project is generally concerned with the design of autonomous and
complex learning systems [8]. The AAAI competition ended an initial six-month
period of software design. Key features of RHINO's control software, as exhibited at
the competition, are as follows:

- **Autonomy.** RHINO operates completely autonomously. It has been operated
  repeatedly for durations as long as one hour in populated office environments
  without human intervention.

- **Learning.** To increase the flexibility of the software, learning mechanisms
  support the adaptation of the robot to its sensors and the environment. For
  example, neural network learning is employed to interpret sonar measurements.

- **Real-time operation.** To act continuously in real-time, any-time solutions [2]
  are employed wherever possible. Any-time algorithms are able to make decisions
  regardless of the time spent for computation. The more time that is available,
  however, the better the results.

- **Reactive control and deliberation.** RHINO's navigation system integrates a fast, reactive on-board obstacle avoidance routine with knowledge- and computation-intense map building and planning algorithms.

RHINO's software consists of a dozen different modules. The interface modules (a base/sonar sensor interface, a camera interface, and a speech interface) control the basic communication to and from the hardware components of the robot. On top of these, a fast obstacle avoidance routine analyzes sonar measurements to avoid collisions with obstacles and walls at a speed as high as 90 centimeters per second. Global metric and topological maps are constructed on-the-fly using a neural network-based approach combined with a database of maps showing typical rooms, doors and hallways. RHINO employs a dynamic programming planner to explore unknown terrain and to navigate to arbitrary target locations. It locates itself by continuously analyzing sonar information. In addition, a fast vision module segments images from two color cameras to find target objects and obstacles that block the path of the robot. RHINO's control flow is monitored by an integrated task planner and a central user interface.

The integration of a dozen different software modules, which all exhibit different timing and response characteristics, requires a flexible scheme for the flow and synchronization of information. The key principles for the design of RHINO's software are as follows:

- **Distributed control and communication.** Each module communicates with several other modules through Ethernet [3]. There is no single control unit, and communication is not centralized.

- **Asynchronous communication.** RHINO's software lacks a central clock. Each of the modules runs independently of the other modules. To resolve conflicts, certain modules (such as the on-board obstacle avoidance module) can take priority over other modules (such as the planner) in determining the robot's motion direction.

- **Software fault tolerance.** RHINO's software is designed to accommodate sudden failures of most of its software components. Almost all modules can be stopped and restarted at any time. Effective mechanisms ensure that restarted modules will immediately obtain the currently available global information.

The next sections present some of the key components of the RHINO approach in more detail: the obstacle avoidance module, the modules concerned with sensor interpretation and map building, the planner and explorer, and the visual routines. The article concludes with a discussion that highlights some of the lessons that were learned during the AAAI competition.

2   Fast Obstacle Avoidance

The obstacle avoidance runs on-board, independent of other software components such as the planner. Every 0.25 seconds, a new velocity and motion direction are chosen

Figure 1: The RHINO robot of the University of Bonn, Germany .

according to the most recent sonar measurements. To rapidly adapt to new situations, only the last three sonar sweeps are considered. RHINO can react immediately to changes in the environment and hard-to-see and moving obstacles such as humans.

The obstacle avoidance module controls both the velocity and the motion direction of the robot. At every instant in time, the velocity is determined such that no collision will occur within the next two seconds (2-sec rule). The motion direction is determined based on target points, which are generated by the planner (see below). To reach a given target, the robot can choose among different trajectories on which it will travel with different velocities. RHINO selects its motion direction by maximizing its translational velocity, denoted by v, while minimizing the angle to the target point, denoted by θ.

To determine v, a simplified model of the robot's environment is constructed. Proximity information, obtained from RHINO's sonar sensors, is used to construct a

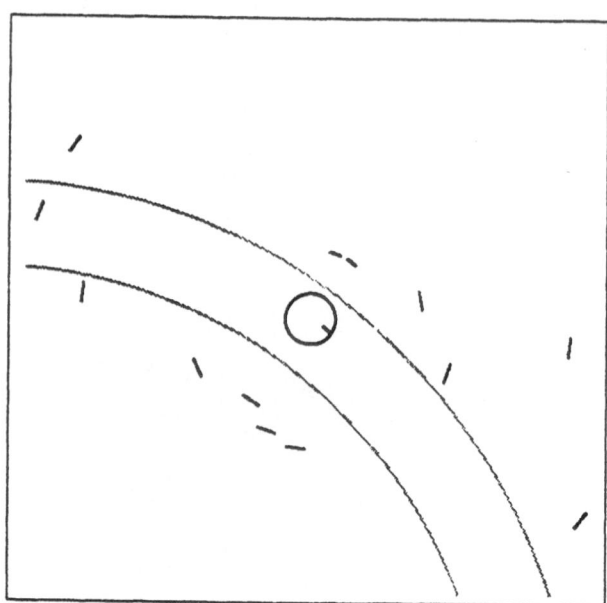

Figure 2: Obstacle line field. Each sonar reading is indicated by a line, centered around the robot. The trajectory, which is finally chosen by RHINO, is also shown.

two-dimensional obstacle line field. Every sonar reading is converted to a line in this field, as depicted in Figure 2. To avoid collisions with obstacles, the obstacle avoidance routine considers a variety of circular trajectories, one of which is shown in Figure 2. For each trajectory, the distance between the robot and the closest obstacle line along the projected trajectory is computed. This distance determines the translational velocity v, according to the 2-sec rule. The projected angle to the target point, θ, is calculated for the estimated robot position and orientation after 0.25 seconds. For both values v and θ a smoothed histogram is constructed. Because of the dynamic constraints, only a small number of trajectories are reachable within the next 0.25 seconds, and are consequently considered in the histogram. Finally, the trajectory that maximizes a weighted difference of v and θ is chosen. In order to increase the safety of the robot, a security distance of 10 centimeters is kept to surrounding objects. This security distance is increased to up to 30 centimeters, as the robot's velocity increases.

RHINO's obstacle avoidance approach is easily extendible to other sensors. For example, prior to the competition we successfully employed camera information to identify small obstacles on the floor, which block the path of the robot, as described below. Each visually detected obstacle is mapped into a few lines in the obstacle field, very much like the sonar information described above. However, visual information was not used by the obstacle avoidance routine during the AAAI competition, basically

(a)   (b)

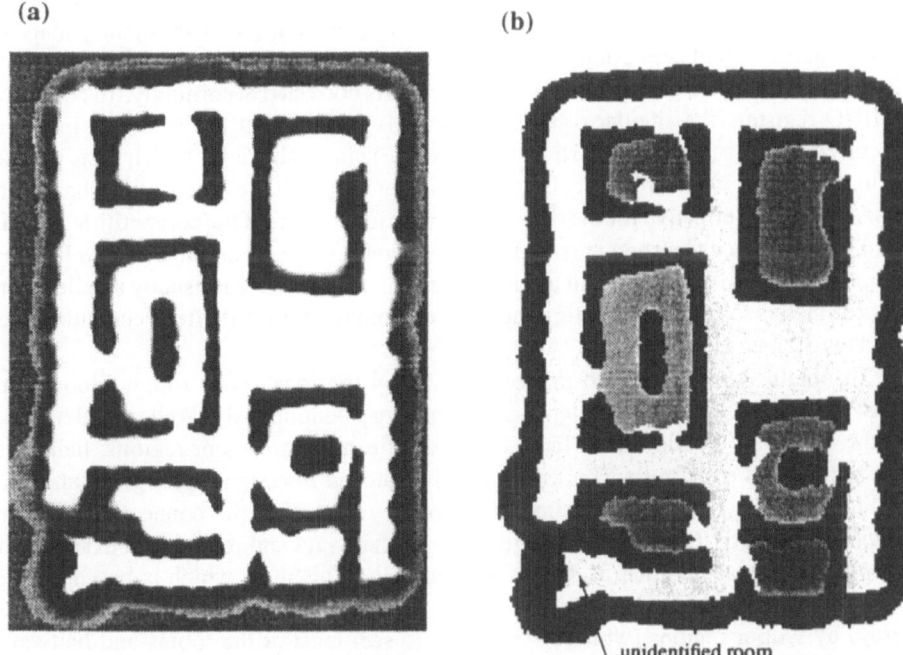

unidentified room

Figure 3: (a) This map of the competition ring was constructed from sonar measurements. Bright regions indicate free-space, and dark regions indicate walls and obstacles. Walls and obstacles are enlarged by a robot diameter. (b) Shown here is a topological analysis of the map. Obstacles are shown in black. As indicated by the different shading, the free-space is divided into 7 rooms/hallways (gray) by 9 door regions (white). The arrow points to an unidentified room in the competition ring.

because sonar information was fast and accurate enough in the competition ring.

3 Map Building and Position Control

RHINO's global navigation system builds and utilizes occupancy maps of the robot's environment. More specifically, when traveling through possibly unknown terrain, RHINO interprets its sonar readings to generate a two-dimensional, discrete probabilistic occupancy map. Sonar sensors are interpreted using an artificial neural network, which estimates the likelihood of occupancy of any point in a 3 meter-circle around the robot [7]. Multiple measurements are integrated using Bayesian inference [5]. Figure 3a shows a map which was constructed while we manually steered the robot through the competition arena. This map describes an area of approximately 30 × 20 meters. The hallways, rooms, large obstacles and doors can clearly be recognized.

To navigate based on global metric information, it is imperative that the robot be able to locate itself accurately in its map. RHINO is equipped with fairly accurate wheel encoders. However, even small angular errors in dead-reckoning can have devastating effects on the internal position estimation. In order to compensate for such error, the robot continuously matches its current sonar readings with its global occupancy map. If a mismatch is found between the occupancy map and the obstacles predicted based on the most recent sonar sweep, the internal position is corrected accordingly. In addition, RHINO registers the angular orientation of walls with respect to its current location to correct more accurately for rotational errors. This mechanism, which rests on the assumption that walls are typically perpendicular or parallel to each other, has been found to be very effective for the detection of rotational errors at the competition as well as in various office environments. If RHINO operates some 30 minutes with velocities of up to 90 centimeters per sec in unknown terrain, the total error is usually smaller than 30 centimeters. Without correcting the dead-reckoning, this error often accumulates as much as 30 meters.

To obtain topological information concerning the location of rooms, doors and hallways, RHINO analyses its metric occupancy map continuously. Walls are identified by thresholding. In addition, a large database of examples of door regions, hallways and rooms (and parts thereof) is continuously matched to assign topological labels to the unoccupied areas in the occupancy map. By analyzing the connectivity of the labeled map, RHINO is able to recognize doors, hallways and rooms. An example of a topologically labeled map is shown in Figure 3b. This map, which is based on the metrical map shown in Figure 3a, subdivides the terrain into 7 rooms and/or hallways (gray) by 9 door regions (white). As is easy to see, most of the rooms and hallways have been identified correctly. In the bottom left corner of that figure, however, a small room has not been identified. This is because due to sensor noise the occupancy map failed to capture a small wall—a problem that may particularly occur with very thin walls, such as those that were found at the competition.

The topological map-analyzer works continuously. At any point in time, it can be queried to output a topological map. However, the quality of the topological maps increases in time. The generation of the labels shown in Figure 3b requires approximately 15 minutes of processing time on a SUN Sparcstation 10. Note that the underlying database of topological examples consists of preselected prototypes based on occupancy maps which were constructed at the University of Bonn prior to the competition.

4 Planning and Exploration

The previous section presented RHINO's approach to mapping its environment. In this section we describe how occupancy maps are used when controlling the robot. RHINO's planner generates minimum-cost paths to arbitrary goal locations or, as described below, to unexplored regions. These paths are constantly refined and communicated to the obstacle avoidance routine, which then determines the final motion direction and velocity of the robot.

RHINO's main planning engine consists of a dynamic programming routine, which computes trajectories with minimum cost to a goal location [4]. The occupancy map

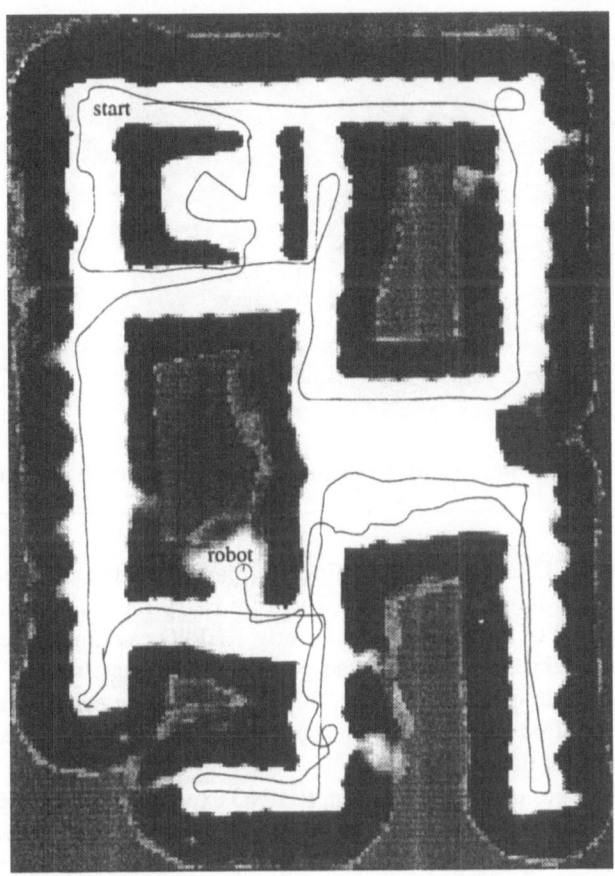

Figure 4: Occupancy map, constructed from scratch during 15 minutes of autonomous robot exploration. The robot's path, which starts at the upper left corner, is also shown.

is translated into a cost function, such that occupied territory results in high traversal cost, and free territory in low traversal cost. Dynamic programming propagates path information from the goal(s) to arbitrary locations in the map. Consequently, steepest descent results in a minimum cost path to the "cost-nearest" goal. Control can be generated at any time without any significant computation. However, deliberation time is traded for the quality of the resulting path.

Because occupancy maps are often too inaccurate to generate collision-free motion control, in dynamic environments, RHINO's planner commands only the rough motion direction, which is then finalized by the collision avoidance routine. Consequently, if unmodeled obstacles block the robot's path, the planner is faced with unexpected

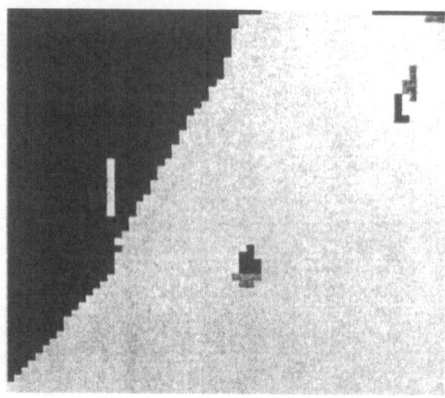

Figure 5: (a) Raw image and (b) segmented image in coarse resolution (9 segments).

robot actions. Dynamic programming pre-plans for arbitrary robot locations. This is because goal information is propagated for every location in the map, not just the current location of the robot. Consequently, RHINO can quickly react if it finds itself to be in an unexpected location, and generate appropriate motion directions without any additional computational effort. This rapid exception handling ability provides the necessary freedom for the collision avoidance routine to modify actions commanded by the planner at its own will.

In both stages of the competition, RHINO explored and mapped unknown terrain. RHINO's planning mechanism can easily be applied to generate explorative paths, lacking a specific goal point. If the set of goal positions is defined as the set of positions, for which no map information is available, RHINO moves straight to the unexplored. Figure 4 illustrates the path of some 15 minutes autonomous robot exploration in the competition ring. In this prototypical example, the main hallways have already been traversed, and RHINO continues to explore the unexplored rooms. RHINO's speed at the straight-line segments of the exploration path was generally between 50 and 90 centimeters per second. Further details on planning and exploration can be found in [7].

5 Vision and Object Recognition

The images from the color camera system are the input to a four-stage vision system, which solves two different tasks. First, it has to recognize important objects typically found in the environment (e.g., objects in an office environment). Second, it supplies valuable information for the robot navigation task by providing local occupancy maps to the map builder and obstacle locations relative to the robots position to the collision avoidance module. This second task, however, was not performed during the final runs at the competition. Here, map building and obstacle avoidance relied solely on sonar information, which turned out to be sufficiently reliable in the competition ring.

In the first stage of low-level processing, images are low-pass filtered and sub-sampled, to reduce the data transfer via the radio link and to pre-process the image for the next stage. This process is performed on one of the on-board i486 computers. Sampling in space (image size) and in time (frame rate) is done dynamically, dependent on the actual velocity of the robot. Thereby the transmission channel capacity is allocated in a task-driven way.

The second processing stage is done by an image segmentation algorithm which partitions the transmitted image into homogeneous, connected regions (*cf.* Fig. 5). Homogeneity is measured by a dissimilarity measure between neighboring image sites (pixels or blocks of pixels). For reasons of efficiency, the dissimilarity measure is restricted to a weighted squared sum of color and luminance differences between sites (with an additional threshold). Formally, the segmentation task can be described as a minimization problem of a cost function, which sums up the local inhomogeneities of all regions for a given partition. To achieve real-time performance without the need for special hardware, the segmentation is implemented by a fast region-merging scheme. The decision whether or not two neighboring regions should be merged depends on a comparison between the current costs and the costs after merging.

The third stage takes the segmented image as an input and seeks to identify and label certain elements of a typical indoor scene, e.g., the floor, walls and doors. Important further information for both navigation and object recognition can be derived: The distances and sizes of all objects or regions located on the floor are calculated based on knowledge of the position, viewing angle, etc. of the cameras. This distance and size information for walls and objects is incorporated into the occupancy map.

At the top of RHINO's vision processing architecture, a feature-based object recognizer detects objects of interest in the environment. The recognition module is able to learn from labeled examples of feature vectors, extracted from example images. For every type object, a Gaussian model (mean and covariance matrix) is estimated according to the maximum likelihood principle. Typical features are the normalized mean and the variance of object luminance, the mean and the variance of object color (hue and saturation), and geometrical features like the absolute size, width and height of the object, estimated based on the object location calculated in the third stage of the vision system. At the competition, we used between 30 and 50 training examples to model each of the 7 object classes. These example images showed different objects varying in object distance, lighting conditions and the choice of the class representative. The assignment to a class is done by minimizing the Mahalanobis distance to the class mean. If the likelihood is below a certain threshold, the candidate object is not accepted as a member of a known class and is considered unclassified.

Once a target object has been found, its location is communicated to the planner and other modules concerned with task and motion control.

6 Results and Discussion

This paper surveys the software architecture of the fully autonomous RHINO robot, as it was exhibited at the AAAI Robot Competition and Exhibition. RHINO is controlled by a dozen software modules which work and communicate asynchronously. Special emphasis is put on real-time operation, learning, and the integration of reactivity and global map knowledge. RHINO does not require prior knowledge on the locations of walls/obstacles, nor on the topology of its environment.

In the first stage of the competition ("office delivery"), RHINO had to move to a designated target location (see [6] for a detailed description of the competition). This stage consisted of three trials, two of which counted for the final score. Because we are specifically interested in navigation without prior information, we attempted to use the first trial for exploring and mapping the competition ring, and only the remaining two trials for the delivery task. However, although RHINO traveled fast, we learned that the arena could not be explored completely without prior information in the allotted time. Consequently, we had to "buy" a metric map for subsequent trials.

One of the major problems we encountered at the competition was RHINO's unreliable radio link. Unpredictable radio communication, possibly based on interference with other radio links, caused RHINO's on-board operating system (Linux, a PC-version of Unix) to suspend obstacle avoidance for periods of 10 seconds or more. In the second and third trial of stage one, RHINO suffered from severe communication failures and collided repeatedly with walls. Because of this, the first stage of the competition could not be completed, and RHINO was excluded from the finals of this stage.

The communication problem was fixed in the second stage of the competition by moving the radio link closer to the competition ring. In this stage ("cleaning-up an office"), RHINO was required to find and fetch objects like soda cans and paper wads, to pick them up, and to drop them in groups of three into a nearby trash bin. Since RHINO is currently not supplied with a manipulator, it indicated its intention to pick up and to drop objects by voice. RHINO used essentially the same exploration routines as in the first stage but at a reduced speed. In addition, the visual routines described above were employed for the identification of obstacles. At the competition, RHINO found most of the objects in the starting room and then continued to clean up the hallway. Here RHINO scored second, defeated only by a collaborating team of three robots, described in a different article in the same volume [1].

The AAAI competition ends an initial six-month period of software engineering. RHINO's software is generally applicable to autonomous navigation in indoor environments. In the future, RHINO shall operate 24 hours a day, interrupted only by battery charging. Our main scientific interest is the study and the design of autonomous, complex learning systems, which in the domain of robotics includes adaptive approaches to sensory processing and lifelong robot learning [8]. We are currently implementing various learning techniques that allow RHINO to adapt to new situations, and to acquire new skills necessary for achieving a broad variety of tasks.

Acknowledgment

Some of the low-level software (TCX [3], device drivers for the speech board and the cameras) were provided by Carnegie Mellon University, which is gratefully acknowledged. We also acknowledge the steady and helpful support by Real World Interface Inc. Travel to the competition would not have been possible without a generous travel grant by AAAI, and the invaluable assistance by Peter Lachart, Wolli Steiner and Peter Wallossek. One of the authors (T.H.) was partially supported by the Ministry of Science and Research of the state North Rhine-Westphalia.

References

[1] T. Balch, G. Boone, T. Collins, H. Forbes, D. MacKenzie, and J. C. Santamaria. Io, Ganymede and Callisto – a multiagent robot janitorial team. *AI Magazine*, Summer 1995, to appear.

[2] T. L. Dean and M. Boddy. An analysis of time-dependent planning. In *Proceeding of Seventh National Conference on Artificial Intelligence AAAI-92*, Menlo Park, CA, 1988. AAAI, AAAI Press/The MIT Press, pp. 49–54.

[3] C. Fedor. TCX. An interprocess communication system for building robotic architectures. programmer's guide to version 10.xx. Carnegie Mellon University, Pittsburgh, PA 15213, December 1993.

[4] R. A. Howard. *Dynamic Programming and Markov Processes*. MIT Press and Wiley, 1960.

[5] H. P. Moravec. Sensor fusion in certainty grids for mobile robots. *AI Magazine*, Summer 1988, pp. 61–74.

[6] R. Simmons. The 1994 AAAI robot competition and exhibition. *AI Magazine*, Summer 1995, to appear.

[7] S. Thrun. Exploration and model building in mobile robot domains. In *Proceedings of the ICNN-93*, IEEE Neural Network Council, San Francisco, CA, March 1993, pp. 175–180.

[8] S. Thrun. A lifelong learning perspective for mobile robot control. In *Proceedings of the IEEE/RSJ/GI International Conference on Intelligent Robots and Systems*, September 1994.

BACKGROUND INVARIANT FACE RECOGNITION

Rolf P. Würtz

Computing Science Dept., University of Groningen,
P.O. Box 800, 9700 AV Groningen, The Netherlands
Ph.: +31 50 636496, fax:+31 50 633800, email: rolf@cs.rug.nl

Abstract

As a contribution to handling the symbol grounding problem in AI an object recognition system is presented that is exemplified with human faces. It differs from earlier systems by a pyramidal representation and the ability to cope with structured background.

1 Introduction

Despite remarkable successes within purely symbolic contexts progress in Artificial Intelligence seems to be hampered severely by the symbol grounding problem, i.e. by the difficulties of extracting symbols to be manipulated from information obtained from the real world. A conditio sine qua non for a solution of this are very robust object recognition schemes. In this paper system for the recognition of human faces independently of small distortions and background is presented. The latter becomes problematic as soon as features are employed that extend beyond one pixel. This system is formulated here in a technical fashion, but at least the essential matching mechanisms can be formulated in neuronal dynamics using the framework of dynamic links. This as well as the details omitted here can be found in [2].

2 Representation of Images and Models

The processing of a retinal grey-level image in the primary visual cortex can be modeled by a wavelet transform based on complex-valued Gabor functions [1, 2]. The single wavelet is parameterized by its two-dimensional spatial frequency vector. The responses of all spatial frequencies of some fixed length form a *frequency level*, which assigns a small feature vector to all image points on an appropriate sampling grid.

The pyramidal arrangement has the advantage that all responses in the stored model which are influenced by the background can be discarded, but the object (which here is the face without hair) still represented well enough for recognition. See figure 1) for an illustration. The image to be analyzed consists of a full pyramid, a presegmentation is not required.

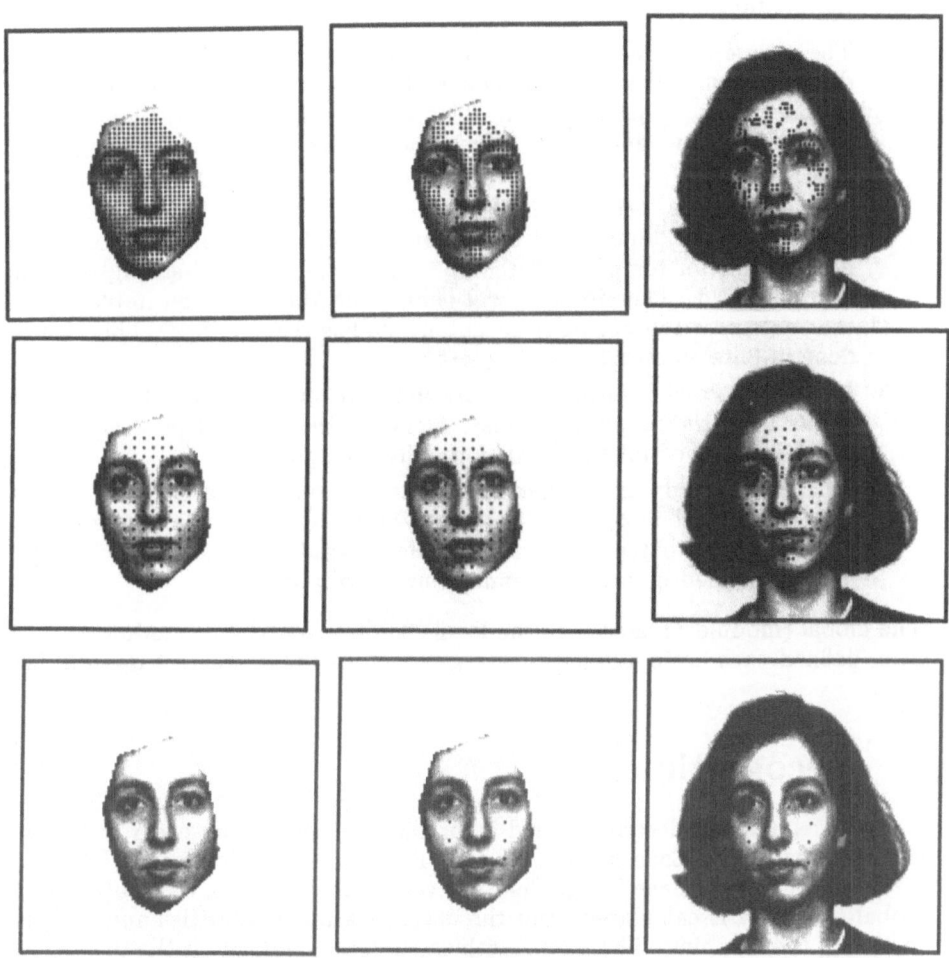

Figure 1: **Model representation and correspondences.** The leftmost column shows the representation of the stored models. Sampling density and minimal distance to the object boundary are dependent on the spatial frequency of the filter. The image is not presegmented and contains a full pyramid. The second and third column show the mappings created by the procedures from section 3. Phase matching leads to a mapping accuracy which is better than the grid resolution. As poor correspondences have been dropped, the remaining ones are fairly reliable at the lowest level already. Therefore, many persons can already be recognized on that level.

3 Matching

The pyramidal representations of image and model can be matched using the following modules:

1. The *coarse localization* of the counterpart of the model in the image is done by global template matching of the vectors of Gabor amplitudes on the lowest frequency level. This is not very expensive, because the resolution is low, and yields a first rough correspondence mapping.

2. Mappings acquired using only the amplitudes of the Gabor responses are not very precise, because the fine geometrical information resides in the phases. On the other hand, the phases or the full complex responses are not suitable for template matching because they depend strongly on the sampling grid. Therefore, a *local phase matching* has been implemented that enhances the accuracy of amplitude-based mappings. This can be done in parallel on all model points.

3. In order to cope with occlusion problems, it must be possible to *exclude points* from the mapping. This is done on the basis of poor similarity, which is also possible in parallel on all image points.

4. Finally, any mapping can be refined by local template matching with amplitudes from the next higher frequency level. For this, the model is split up into several patches that independently search for correspondences in an area defined by the coarse mapping already known.

The global (module 1) as well as the local template matching (module 4) have been defined such as to work with feature vectors and to neglect missing features.

4 Recognition

The procedures just described yield a correspondence map on every frequency level (see figure 1). Each of them can be used for recognition in the following way. An incoming image is matched to every model out of a database. Some global similarity is calculated from the actual feature similarities and the distortion of the mapping. The model with the highest similarity is the recognized one; the recognition is *significant* if the distance of the highest similarity to the distribution of all similarities exceeds a suitable threshold. This scheme has already been applied successfully in [1].

In the coarse-to-fine hierarchical matching procedure described above the notion of significance can been exploited to build a hierarchical recognition scheme: If a recognition is significant on one frequency levels, the higher ones need not be evaluated.

5 Results

In order to compare the background independence with the FACEREC algorithm [1] two model databases have been set up: **M1** with model segments that were rectangular and uniform in size and **M2** with model segments that were created by hand and excluded the hair from the face images. Thus the

Method	M1 ↔ I1		M1 ↔ I2		M2 ↔ I1		M2 ↔ I2	
	C	S	C	S	C	S	C	S
Hier. level 0	71	54	68	23	42	17	41	11
Hier. level 1	40	29	62	45	73	54	59	18
Hier. level 2	15	12	20	8	24	23	50	23
Hier. total	99	95	93	76	99	94	85	52
Level 2 only	94	89	86	69	94	93	78	49
FACEREC	95	93	92	81	19	1	14	1

Table 1: **Recognition Results.** This table shows the performance of the single hierarchy steps (levels 0,1,2), the total hierarchy, the highest level alone and FACEREC, an earlier system without mechanisms for background independence. Model base **M1** contains the faces without segmentation, **M2** the same ones with the hair removed. The **C** columns show the number of correct recognitions, the **S** the significantly correct ones. All numbers are percentages of the number of test images.

person's hair was treated as background, there face proper as object. Both model bases contained 83 persons looking straight into the camera.

For the experiments in table 1 two different image databases have been used: **I1** containing the same persons looking 15° to their side, which can be considered as a moderately difficult case compared to **I2** which contained **I1** and for each person one image with the head turned by 30° and one with a different expression. For each image database, the significance thresholds have been adjusted such that no false positive recognitions occurred. The important performance measure is thus the number of significant recognitions.

Table 1 shows that the total recognition performance of the hierarchical scheme is scheme is better than that of the highest frequency level alone. Thus the distinguishing features are distributed over the scales in a way which is not yet completely understood. The FACEREC system described in [1], which has no mechanisms for background independence, performs about as well as the hierarchical system with model database **M1**, but breaks down completely with **M2**, where the background dependence of the feature vectors becomes serious.

Acknowledgements: Part of this work has been funded by grants from the German research minister and the HCM program of the European community.

References

[1] Martin Lades, Jan C. Vorbrüggen, Joachim Buhmann, Jörg Lange, Christoph von der Malsburg, Rolf P. Würtz, and Wolfgang Konen. Distortion invariant object recognition in the dynamic link architecture. *IEEE Transactions on Computers*, 42(3):300–311, 1993.

[2] Rolf P. Würtz. *Multilayer Dynamic Link Networks for Establishing Image Point Correspondences and Visual Object Recognition*, volume 41 of *Reihe Physik*. Verlag Harri Deutsch, Thun, Frankfurt am Main, 1995.

Learning Structure from Motion: How to Represent Two-Valued Functions

Anuj Dev, Ben J.A. Kröse, Frans C.A. Groen

RWCP*Novel Functions SNN†Laboratory

Faculty of Mathematics and Computer Science, University of Amsterdam,
Kruislaan 403, NL-1098 SJ Amsterdam. email: anuj@fwi.uva.nl

1   Introduction

The optic flow is the vector field formed by the projection of the 3D-motion in the environment on the image plane of the observer. The optic flow vector $\mathbf{o}$ at image location $\mathbf{r}$ is thus a function of the scene depth $z_{\mathbf{r}}$ and the relative scene motion $\mathbf{m}_{\mathbf{r}} = (\mathbf{t}, \boldsymbol{\omega})_{\mathbf{r}}$ which can be written as

$$\mathbf{o}_{\mathbf{r}} = \mathcal{C}\left(\mathbf{r}, \mathbf{m}_{\mathbf{r}}, z_{\mathbf{r}}\right)$$

where $\mathcal{C}$ is some non-linear function which only depends on the camera mapping.

We are interested in a well known, but only partially solved, problem in computer vision: the extraction of scene structure and 3D-motion from the optic flow which is called *Structure from Motion*. Structure from motion thus boils down to finding $z_{\mathbf{r}}$ and $\mathbf{m}$ from a number of flow vectors $\mathbf{o}_{\mathbf{r}_1}, ..., \mathbf{o}_{\mathbf{r}_N}$. It is well known that the scene structure can only be recovered relative to the translational velocity of the scene. The relative depth is known as the time to contact $\tau_{\mathbf{r}}$. This means that for N ($N > 4$) flow vectors, five motion parameters $\hat{\mathbf{m}} = (\mathbf{t}/\|\mathbf{t}\|, \boldsymbol{\omega})$ and N relative scene depths can be computed from the optic flow.

To compute these, Bruss [3] uses a Least Mean Square approach to solve the desired parameters by minimizing the cost function E:

$$E = \sum_{r} \left(\tilde{\mathbf{o}}_{\mathbf{r}} - \mathcal{C}\left(\mathbf{r}, \hat{\mathbf{m}}, \tau_{\mathbf{r}}\right)\right)^2$$

where $\tilde{\mathbf{o}}_{\mathbf{r}}$ are the measured optic flow vectors. To constrain the optic flow, Adiv [1] and Negahdaripour [6] model the scene locally as a planar patch. Since a planar patch is defined by a three parameter vector $\boldsymbol{\phi}$, we can write the formed optic flow field as

$$\mathbf{o}_{\mathbf{r}} = \mathcal{C}\left(\mathbf{r}, \mathbf{m}, \boldsymbol{\phi} \cdot \mathbf{r}\right)$$

The least mean square approach described can again be applied. But since the number of estimated parameters is lower, more redundancy is introduced and the method is therefore less sensitive to a noisy optic flow field.

Both approaches described assume that the camera mapping, which is contained in $\mathcal{C}$ is known. In real world applications the calibration of the camera

---

*Real World Computing Partnership
†Dutch Foundation for Neural Networks

system is a tedious task and often impossible. We therefore propose not to calibrate the camera but to *learn* the structure from motion mapping.

In our approach we also model the scene locally as a planar patch as to introduce redundancy in the optic flow. If we model the corresponding optic flow field with a Taylor expansion, it can be shown that the Taylor expansion of the optic flow only contains non zero terms up to the second order [2]. This is only true for planar patches, a general rule is that the order l of the patch induces a $l + 1$ order of non zero coefficients The problem is thus to find the mapping W from the Taylor coefficients $(\alpha_1, \ldots, \alpha_8)$ to the unknown parameters of the planar patch: the orientation ϕ, and its motion $\hat{\mathbf{m}}$.

We learn the mapping W from a set of example data points $\mathbf{v}^i = (\boldsymbol{\alpha}, \{\phi, \hat{\mathbf{m}}\})^i$ which forms the training set. However, from the literature it is known that the mapping W is not a function in the sense that it is a $m \to 1$ mapping; for (almost) every set of optic flow vectors there exists a dual solution which is equally correct. This means that standard function approximators like Multi Layer Perceptrons (MLP) are unable to approximate W [1].

In this paper we propose a solution to learning multi valued functions [4]. We will represent the mapping as a manifold [2] in the product space of input and output. In this way the twofoldness of the output solutions is captured into the parameterization of the manifold. Once we have a parameterization of the manifold, we can search in this parameter space for a solution for the output given the input. In section 2. will generalize the notion of multi valued mappings and describe how such mappings can be learned as manifolds with an auto associative MLP. Section 3. shows how input output relations can be extracted from a learned mapping by preforming a local search in the parameter space. The experiments section will demonstrate this on the structure from motion problem.

2 Learning a manifolds parameterization

First we generalise the mapping by defining an input vector

$$\mathbf{x} = (\alpha_1, \ldots, \alpha_8)$$

in the input space $\mathcal{X}$ and an output vector

$$\mathbf{y} = (\phi, \hat{\mathbf{m}})$$

in the output space $\mathcal{Y}$. With this notation a data point $\mathbf{v}^i$ is a concatenation of an input vector $\mathbf{x}^i \in \mathcal{X}$ and output vector $\mathbf{y}^i \in \mathcal{Y}$. Geometrically, a mapping $\mathbf{x} \to \mathbf{y}$, $\mathcal{X} \to \mathcal{Y}$, can now be viewed as a manifold $\mathcal{M}$ which resides in the product space $\mathcal{V} = \mathcal{X} \times \mathcal{Y}$. The manifold will be denoted by its parameterization

$$\mathcal{M} : (\mathbf{F}(\mathbf{s})) = (\mathbf{F_x}(\mathbf{s}), \mathbf{F_y}(\mathbf{s}))$$

with

$$\mathbf{F}(\mathbf{s}) \in \mathcal{V} ; \quad \mathbf{F_x}(\mathbf{s}) \in \mathcal{X} ; \quad \mathbf{F_y}(\mathbf{s}) \in \mathcal{Y} ;$$

---

[1] Multi Layer Perceptrons converge to the expected output given the input

[2] We assume that the data points are drawn from a hypersurface, with the properties of a C^1 manifold, and contain additive noise.

the vector valued parameter functions and **s** the parameter vector which resides in the parameter space $\mathcal{S}$ with $dim\mathcal{S} \leq dim\mathcal{X}$. Figure 1 gives a graphical interpretation of the mappings. The first objective is to approximate the manifold from the given data points $\mathbf{v}^i$.

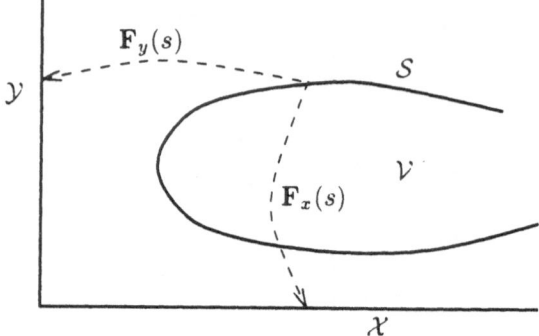

Figure 1: *The spaces $\mathcal{X}$ and $\mathcal{Y}$ form the product space $\mathcal{V}$. In $\mathcal{V}$ resides the manifold $\mathcal{M}$ which represents the mapping $\mathbf{x} \rightarrow \mathbf{y}$. The vector valued functions $\mathbf{F_x}$ and $\mathbf{F_y}$ map a point $\mathbf{s} \in \mathcal{S}$ to the spaces $\mathcal{X}$ and $\mathcal{Y}$.*

Since multi layer perceptrons MLP are general function approximators, we could learn the vector valued function $\mathbf{F(s)}$. The problem is that there are no learning examples available in the form of (s^i, v^i) pairs. But if we assume that there is a vector valued function [3] $\mathbf{F(s)}$ which is bijective onto $\mathcal{M}$, then there exists a vector valued function $\mathbf{G} = \mathbf{F}^{-1}$ which is also bijective so that

$$\mathbf{F(G(v))} = \mathbf{v} \quad \forall \mathbf{v} \in \mathcal{M}$$

The solution is now to learn both functions $\mathbf{F}$ and $\mathbf{G}$ within one MLP. We train the MLP to reconstruct its input vector $\mathbf{v}$ on the output layer and create a *bottleneck* by taking the number of hidden nodes the same as the dimension of the manifold $\mathcal{M}$. Note that this dimension is lower then $dim\mathcal{V}$, hence the *bottleneck*. In this way the hidden nodes span the parameter space $\mathcal{S}$, the mapping from the input layer to the hidden layer represents $\mathbf{G}$ and the mapping from the hidden layer to the output layer represents $\mathbf{F}$. These networks are known as auto associative MLP [5] and shown in figure 2.

3   Approximating mappings from the manifold

Given a parameterization of a manifold, the second objective is to give an approximation $\tilde{\mathbf{y}}$ of the mapping for a given $\tilde{\mathbf{x}}$. Since the mapping need not to be many-to-one, a function, there are several possible values of $\tilde{\mathbf{y}}$ "correct". These values $\mathbf{y}^i$ correspond to:

$$\tilde{\mathbf{y}}^i = \mathbf{F_y}(\mathbf{s}^i) \quad \forall \mathbf{s}^i | \mathbf{F_x}(\mathbf{s}^i) = \tilde{\mathbf{x}}$$

---

[3]Implicitly we say that there is no unique parameterization of a manifold.

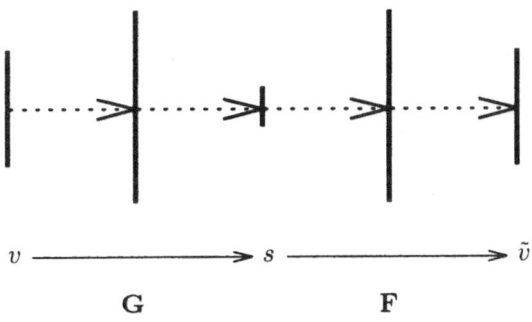

$v \longrightarrow s \longrightarrow \tilde{v}$

$\qquad$ **G** $\qquad$ **F**

Figure 2: MLP architecture which reconstructs its input. The dimensionality of the middle layer, dimS, is lower then the dimension of the input and output layer, dimV, and represents the parameter space. Note that we have drawn a five layer MLP which means that every (Borel continuous) vector valued functions **G** and **F** can be learned.

This means that there has to be an additional constraint on the desired mapping. For the gradient based method described next we will take the approximation which is closest, in Euclidean distance, in the input space: [4]

$$s^i = \min_i \left(\| \mathbf{F_X}(s^i) - \mathbf{F_X}(s^p) \| \right)$$

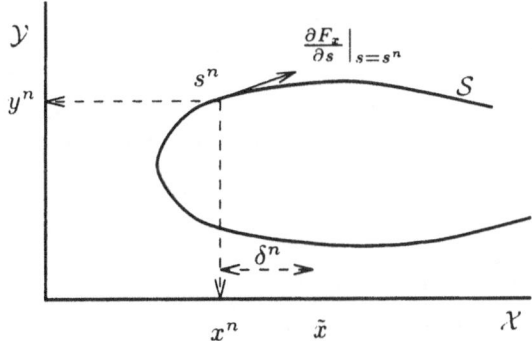

Figure 3: Approximating the output value for a given point $\tilde{x}$ in the input space. The distance of this point to the current position on the manifold is measured (δ). s is then changed so that this distance is minimized

The basic idea is to walk over the manifold from the starting point s^p toward the point where $\mathbf{F_X}(s) = \tilde{x}$. This is done by changing the parameter vector s so that the walking is in the direction of $\tilde{x}$. This means that we want to minimize the function E:

$$E = \| \tilde{x} - \mathbf{F_X}(s^p) \|^2 = (\delta^p)^2$$

---

[4] Actually s^p can be any point on the manifold but may be thought of as the previous state of the system that is being approximated by the AAMLP.

Any minimization scheme can be chosen. We use a gradient descent which (see figure 3) boils down to a well known iterative scheme:

$$\mathbf{s}_{n+1} = \mathbf{s}_n + \mu \delta^n \left. \frac{\partial \mathbf{F_x}}{\partial \mathbf{s}} \right|_{\mathbf{s}=\mathbf{s}^n}$$

where μ is a iterating constant. Note that $\partial \mathbf{F_x}/\partial \mathbf{s}$ is the Jacobian of the last layers of the MLP and can be calculated from them. [5]

4   Experiments

To recap the previous sections on learning and to lay out the method used for the experiments:

1. Find a parameterization of the mapping with a AAMLP given a dataset $\mathbf{v}$. We define the spanning error as the mean reconstruction error E_v:

$$E_v = \| \mathbf{v} - \mathbf{F}\left(\mathbf{G}\left(\mathbf{v}\right)\right) \| \quad \forall \mathbf{v} \in \mathcal{M}_{\text{test}}$$

where $\mathcal{M}_{\text{test}}$ is the set of test points on the manifold. E_v is a measure how well the manifold is approximated by the MLP. When learning the MLP, E_v is minimized using simple back propagation.

2. For a given point in the input space, find a solution to this mapping in the output space by minimising the input distance with a (local) search algorithm. The final error in the output space equals E_y:

$$E_y = \| \mathbf{y} - \mathbf{F_y}\left(\mathbf{s}^{opt}\right) \| \quad \forall \mathbf{v} \in \mathcal{M}_{\text{test}}$$

We have tested the approach with a perspective projection and unit focal length. The relation of the Taylor coefficients with the parameters of the planar patch is then known and given by [2]:

$$\alpha_1 = -\omega_y - t_x \,;\, \alpha_2 = t_z - o_x t_x$$

$$\alpha_3 = -\omega_z - \phi_y t_x \,;\, \alpha_4 = -\omega_y \phi_x t_z$$

$$\alpha_5 = -\omega_x - t_y \,;\, \alpha_6 = -\omega_z - \phi_x t_y$$

$$\alpha_7 = t_z - \phi_y t_y \,;\, \alpha_8 = -\omega_x \phi_y t_z$$

From this relation learning examples were generated by choosing random patch parameters (from the interval $[-1,1]$) and computing the Taylor coefficients $\alpha_1, \ldots, \alpha_8$. Since the interval is symmetric, each of the dual solutions is equal likely.

For clarity we tried to reconstruct the translation $(t_x/t_z, t_y/t_z)$ of the patch. To do so we used a $10-16-8-16-10$ network and trained it to reconstruct the 10 dimensional input consisting of $(\alpha_1, ..., \alpha_8, t_x/t_z, t_y/t_z)$. An spanning error

---

[5]Training the network with back propagation of the error and searching in the parameter space for a solution are almost identical. In the first case the error equals $\mathbf{v} - \check{\mathbf{v}}$ and the weights of the net are adjusted to minimize this error. In the second case the error equals $\mathbf{x} - \mathbf{F_x}(\mathbf{s}^p)$ and the parameter vector is changed to minimize this error.

$E_y = 0.02$ was found, which indicates that a good parameterization of the manifold was learned. After training we presented only the Taylor coefficients and initialized the middle layer, the parameter space, to a random value in the unit hypercube. The result of initializing the parameter layer with a random value and searching for a solution means that we can iterate to either of the dual solutions.

The plot gives the number of trials (vertical) in a bin against the iterations of the search in the parameter space with the corresponding error E_y in the recovered translation of the patch. As can be seen most of the trials converge very rapidly in less then 20 iterations to an error below 10%. It should be noted that we used a worst case initialization of the search vector, a random vector from a uniform distribution. If the trials were related, the patch translation didn't change much from one trial to the other, we could have initialised the search vector with the outcome of the previous search. This will reduce the number of iterations required to reach an error even further.

A comparison to the approaches described in the introduction is difficult since we also learn the camera mapping. If we only look at the number of iterations against the error, we can happily conclude that the errors are in the same order of magnitude for a given number of iterations.

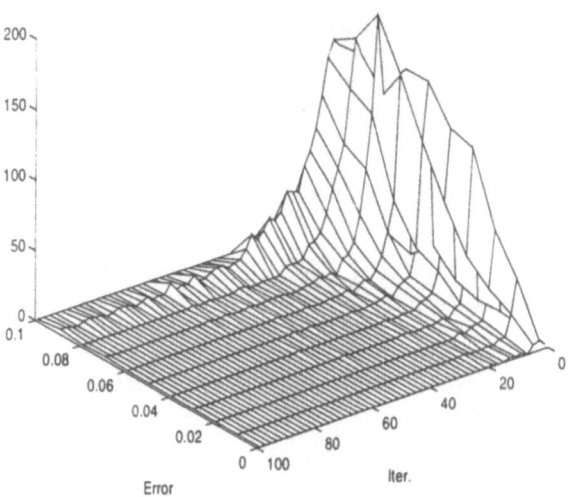

Figure 4: *Results of the gradient descent search in the parameter space, when trying to recover the translation of the patch from the Taylor coefficients of the optic flow field. There are 4096 trial performed. These trials are 'binned' where each bin contains the number of trials that fall within it. We have plotted the number of trials (vertical) against the coordinates of the bin: the end error (between 0 and 0.1) and the number of iterations (between 1 and 100).*

5 Conclusions

We have shown how to *learn* the solution to the structure from motion problem. Until now we have assumed that the learning samples are available for training the neural network. We are interested in the case where a camera is mounted on a mobile robot which can estimate the translational velocity of *moving* objects in the scene. In this case the motion of the *camera* can be estimated using a different type of sensors. In a *static* environment deliberate movements of the robot (camera) will now result in training samples for the network. Once the neural network is trained(in the static scene), patch parameters for a dynamic environment can be recovered.

The approach uses a two step method for approximating one (of several) solutions for a given optic flow field: iteratively learning the parameterization of the manifold and iteratively generating one solution. As can be seen from the experiments, an approximation of an output vector is usually close to a desired value within a few iterations, which means that once the manifold is learned the method is suited for real time applications.

References

[1] G. Adiv. Determining three-dimensional motion and structure from optic flow. *IEEE tansactions on PAMI*, 7(4):386–401, 1985.

[2] S. Ullman A.M. Waxman. Surface structure and 3-d motion from image flow: A kinematic approach. Technical Report CAR-TR-24, Center Automation Res., University of Maryland, 1983.

[3] B.K.P. Horn A.R. Bruss. Passive navigation. *Computer, Vision, Graphics and Image Processing*, 21(1):3–20, 1983.

[4] D. Demers. Learning global direct inverse kinematics. In *Neural Information Processing Systems*, 1991.

[5] M. Flemming G. Cottrell. Face recognition using unsupervised feature extraction. In *Proceedings of the International Neural Network Conference*, 1990.

[6] B.K.P. Horn S. Negahdaripuor. Direct passive navigation. *IEEE transactions on PAMI*, 9(1):168–176, 1987.

A natural object recognition system using Self-organizing Translation-Invariant Maps

Danny Roobaert & Marc M. Van Hulle

Laboratorium voor Neuro- en Psychofysiologie,

K.U.Leuven, Campus Gasthuisberg, Herestraat,

B-3000 Leuven, Belgium

Abstract

Translation-invariant representations are a key aspect of any object categorization system. In most neural realizations, these representations are either non-adaptive and often non-optimal, or adaptive and redundant. Especially in the case of natural images containing complex features, the neural code becomes highly inefficient. In this paper, we introduce an adaptive network model which overcomes a part of this redundancy by searching for translational invariance. The model implements a Self-organizing Translation-Invariant Map (STIM) using an unsupervised competitive learning rule. The latter employs generalized weight-sharing in combination with a symmetric, symmetry-breaking filter. A Kohonen-neighborhood update scheme ensures a proper covering of the feature space.

A number of STIMs are build as instantiations of a generic STIM module for natural object recognition. A local feature extracting STIM generates Gabor-like receptive fields, although with shapes which are better suited for the present application. In addition, more structured receptive fields are developed, enriching the neural representation. A second STIM endows the model with global translational invariance.

1 Problem outline

Natural object recognition is still a very difficult and underexplored domain in neural processing (and certainly in artificial intelligence). Due to the inherent complexity and variability of the natural environment, most of the existing object recognition models perform only well in an artificial environment. These systems have a hardwired architecture with fixed connections [4, 6, 8, 14, 16]. Using temporal neural dynamics to perform complex tasks, they are quite successful when crisp boundaries and geometrical line-structures are available. In this way, the construction of an edge-map as a first processing stage, as enunciated by David Marr, is generally accepted and considered trivial by the neuroscience community. However, in practice, edge-detection is not as trivial a task, as evidenced by the ambiguous performances of many advanced line- and edge detectors [1, 18].

Self-organizing adaptability, already from the earliest processing stages on, leads to a more flexible and less stage-dependent approach: as a result, feature detectors will develop with more specific forms. Especially for preprocessing purposes, this offers a way to surpass the rigidly-defined edge concept. Several authors have already proposed adaptive vision models. A well-known, albeit supervised model, is LeCun *et al.*'s handwritten numerals recognition system [11]. Unsupervised examples are *e.g.* Neven and Aertsen's model [13] and Fukushima's neocognitron [5]. However these authors did not build in an invariance principle at the *development* stage of the feature map, hence their maps contain many redundant features. In particular for processing natural images, when larger receptive fields give rise to much more degrees of freedom, in terms of synaptic weights, this leads to a very inefficient neural code.

In this paper a self-organizing map is presented which develops only representations that are translation-invariant. The main concept used is weight-sharing, as proposed by LeCun *et al.* [11] for parameter reduction. In contrast, we essentially use weight-sharing to generate translational invariance. The latter is enforced by a symmetric, symmetry-breaking filter and the adaptivity is induced by the unsupervised learning rule. Due to this invariance, the network is able to cope with more complex features –a requirement for processing natural images. The introduction of this map into a generic vision module, as part of a natural object recognition system, is straightforward and will be illustrated in this paper.

2 Self-organizing Translation-Invariant Map

In our STIM model, translation-invariance emerges in a bi-layered network as follows. The first layer, called F-layer, contains the shared feature maps while the second layer, called the I-layer, which receives converging inputs from the first layer, codes for invariance with fixed connections. The weight-sharing in the F-layer enables the translation-invariant coding of features.

Within the F- and I-layers, local Winner-Take-All (WTA) operations are performed in two hierarchical stages. Firstly, in the F-layer, WTA is performed to determine one winner per feature. Secondly, in the I-layer, a fixed number of the previous winners is retained. In the training phase, the final winner of these I-layer winners, determines the invariant feature to be updated. The secondary winners of the same set, further improve the output code. This population coding makes the response of the system not only more robust against errors but also, and more importantly, more sensitive to combinations and structural variants of the coded invariant features.

A symmetry-breaking filter, applied on the current F-input window, helps the self-organizing map to develop good representatives. This filter, a Gaussian, is symmetric and therefore introduces no *ad hoc* differentiation in the feature maps but only breaks the symmetry between identical, but shifted features.

The mathematical details will be given in a separate paper. Also in that paper a generalized Self-organizing Invariant Map (SIM) will be presented and discussed.

3 A natural object recognition model

Our model comprises the following four stages.

Stage 1: Contrast filter: Different image preprocessing strategies were explored. We have opted for a DOG-filter not only because of its biological relevance but also because of its clear computational advantage: DOG-filters only detect contrast differences in a fuzzy way. In addition, since they are simple, these filters can be made small without introducing a too large discretization error: hence, the decrease in spatial resolution by this stage is minimized. A single on-center and off-center DOG pair yields a complete, complementary half-wave rectified contrast map for stage 2.

Stage 2: Local feature extraction: An STIM develops a translation-invariant map of local features adapted to the current application *i.e.* it codes for features that are statistically relevant to the application. Since an input pair is received from the previous stage, two complementary maps develop.

Stage 3: Global translation-invariance: In this stage, a translation-invariant representation of the objects is developed, based on the local features extracted in the previous stage. Notice that an STIM is intended to be invariant for limited translational shift only and, in this way, it assumes an attentional system that more or less focuses the object.

Stage 4: Supervised learned perceptron: For user-purposes only, a single-layer perceptron is build in order to classify the outputs into object classes. The weights are formed by 'Learning Vector Quantization' (LVQ) [9].

4   Simulations

We have used 20 different pictures of 10 types of fruit as input images. These originals were randomly scaled (±10%), rotated (±10 deg), translated and noise added (10%) to obtain a set of 100 images. Half of them were used for training, the other half were used for testing. The images have a resolution of 100x100 pixels discretized into 256 grey-scales.

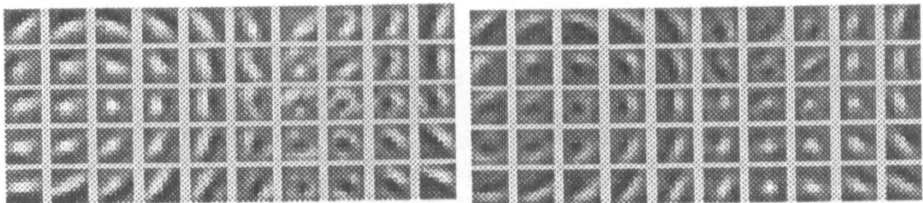

Figure 1: Stage 2 invariant feature map with 50 neurons. A typical map developed by a local STIM (left), and its complementary map (right). Notice the development of on- and off-center receptive fields, Gabor-like receptive fields, and in addition also more structured receptive fields providing an improved map of the environment and an improved neural code.

A typical example of the local STIM of stage 2 can be seen in Fig. 1. Apart from Gabor-like sets, also on-center and off-center cells emerge. Note that when the number of neurons is large enough, also curvature-selective, T- and H-junction-selective, and even more structured features develop. The latter not only improve the accuracy of the neural code, but also have a biological counterpart in the mammalian primary visual cortex [17].

Recognition rates varied between 92% and 96% correctness on the training set, while between 76% and 84% was obtained for the test set. However these results are only indicative, since dissimilarities in the sets can be heavily case-dependent.

5 Conclusion

Our STIM network has a large potential to outperform systems based on fixed filter techniques, since the features extracted are much more adapted to the environment. The popularity of the fixed filter approach derives from the many publications indicating the presence of *fixed* orientation sensitive cell-types in the primary visual cortex (*e.g.* [2, 7]). However, we disagree with such a simplistic view. There is considerable evidence for *adaptive* orientation-selective cells in the primary visual cortex [15]. In addition, more *structured* receptive fields such as the ones we have obtained, are not only computationally advantageous but have also a neurophysiological counterpart, as shown recently by Sun and Bonds [17]. Therefore, we oppose to Linsker's approach [12]. Starting from white noise, he obtains only orientation-selective cells as his most structured cell type. However, natural images are not like white noise and have much more complicated statistics [3]. In this way, the results of our simulations reflect this additional structure.

On a first glance, an STIM resembles a neocognitron [5]. However there are a number of crucial differences. Our architecture develops an invariant map from scratch and is aided by the symmetry breaker which guaranties that only one representative will win the competition. On the other hand, the neocognitron requires a pre-initialized map. Furthermore, our neighborhood update scheme yields a statistically better coverage of the input feature space and, in addition, generates a topographically-ordered map.

In summary, for our object recognition system, a hierarchical structure is used with the STIM module as a building block. Due to the high adaptivity of the STIM module, driven by a generic self-organizing learning mechanism, this module can be easily re-adapted to represent a variety of complex environments.

References

[1] Canny, J. (1986), *IEEE PAMI*, **8**, 679-697.

[2] Daugman, J.G. (1985), *J. Opt. Soc. Am.*, **2**, 1160-1169.

[3] Field, D.J. (1987), *J. Opt. Soc. Am. A*, **4**, No. 12, 2379-2394.

[4] Finkel, L.H., & Sajda, P. (1992), *Neural Comp.*, **4**, 901-921.

[5] Fukushima, K. (1980), *Biol. Cybern.*, **36**, 193-202.

[6] Grossberg, S., *et al.* (1989), *IEEE Trans. Biomed. Eng.*, **36**, 65-84.

[7] Hubel, T.H., & Wiesel, T.N. (1962), *J. Physiol.*, 106-154.

[8] Hummel, J.E., & Biederman, I. (1992), *Psych. Rev.*, **99**, 480-517.

[9] Kohonen, T. (1989), *Self-Organization and Associative Memory.* Berlin: Springer Verlag.

[10] Kohonen, T. (1993), *Proc. 1993 IEEE Int'l Conf. on Neural Networks, San Francisco*, 1147-1156.

[11] LeCun, Y., *et al.* (1989), *Neural Comp.*, **1**, 541-551.

[12] Linsker, R. (1988), *Computer*, March, 105-117.

[13] Neven, H., & Aertsen, A. (1992), *Biol. Cybern.*, **67**, 309-322.

[14] Olshausen, B.A., *et al.* (1993), *J. Neurosci.*, **13**, 4700-4719.

[15] Pei, X., *et al.* (1994), *J. Neurosci.*, **14**, 7130-7140.

[16] Seibert, M., & Waxman, A. (1989), *Neural Networks*, **2**, 9-27.

[17] Sun, M., & Bonds, A.B. (1994), *Visual Neurosci.*, **11**, 703-720.

[18] Zucker, S.W., *et al.* (1989), *Neural Comp.*, **1**, 68-81.

Affine Scale-Space for Discrete Pointsets

Ruud Geraets*   Alfons H. Salden†
Bart M. ter Haar Romeny
Max A. Viergever
Computer Vision Research Group, Utrecht University
Utrecht, The Netherlands

Abstract

In this paper we propose an affine invariant evolution scheme for discrete pointsets. A discrete parametrisation for pointsets in terms of vector distributions is introduced. Smoothing such a parametrisation with an affine invariant arclength-based Gaussian generates an affine invariant scale-space of curves. We demonstrate this for several pointsets. It is shown that self-intersections are preserved in the evolution process.

1   Introduction

In this paper we propose an affine invariant evolution process for discrete pointsets. We introduce a parametrisation for pointsets and a linear evolution scheme that enables us to turn a pointset into a time-dependent family of planar curves. The evolution of this family is invariant with respect to uni-modular or area-preserving affine transformations. For brevity we will denote in the sequel uni-modular affine transformations to affine transformations.

Note, that curve evolution schemes have been studied for several years in computer vision [1]. The input curves subjected to these schemes, however, are smooth.

2   Affine Invariant Evolution Equation for Pointsets

An ordering on a pointset allows the set to constitute a discrete curve. Connecting the first and last point of the set closes the discrete curve. We propose the following discrete arclength parametrisation for such discrete curves C. Let

$$C(d) = \sum_p \vec{X}_p \, \delta(d - d_p) \tag{1}$$

with the vectors $\vec{X}_p$ representing the points p of the pointsets and d the discrete arclength parameter. The values d_p denote the arclength values for which the points constituting the curve exist.

*ESPRIT Basic Research Action "Viewpoint Invariant Visual Acquisition" (VIVA), grant-nr. EC-BRA-6448
†Netherlands Organisation for Scientific Research (NWO), grant nr. 910-408-09-1.

To establish an affine invariant description of the pointset, we need a discrete version of the affine arclength and an affine description of the vectors X_p. To obtain a discrete affine arclength we use the well-known definition for the infinitesimal affine arclength ds_a

$$ds_a = |d\vec{x} \times d^2\vec{x}|^{1/3} \tag{2}$$

This implies for the discrete affine arclength parameter

$$\Delta s_a = |\Delta \vec{X} \times \Delta\Delta \vec{X}|^{1/3} \tag{3}$$

with

$$\Delta \vec{X} = \vec{X}_{i+1} - \vec{X}_i \quad \text{and} \quad \Delta\Delta \vec{X} = \vec{X}_{i+2} - \vec{X}_{i+1} \tag{4}$$

using the Euclidean coordinates of the $\vec{X}_i$.

The invariance of the vectors X_p is established in an affine invariant coordinate system generated as follows. We use two points of the set X_p to establish two basis vectors $\vec{E}_i$

$$\vec{E}_i = \vec{X}_i \quad i \in 1,2 \tag{5}$$

In this basis any other point in X_p is given by

$$\vec{X}_p = \alpha \vec{E}_1 + \beta \vec{E}_2 \tag{6}$$

with α and β affine invariant.

For the evolution equation we take

$$\frac{\partial C}{\partial t} = \frac{\partial^2 C}{\partial d^2} \; ; \; C(.,0) = C_0 \tag{7}$$

with t the scale-parameter (or time-parameter) and d the affine arclength parameter. With this choice for d, we establish an evolution equation invariant under the affine group action. Note that we take the parameter d to coincide with the discrete arclength parameter for $t = 0$ only, i.e. d does not depend on t. During the curve evolution the *frozen* arclength parameter d will turn into an arbitrary invariant parameter.

For the family of test functions needed to generate a C^∞ structure on this set of vector distributions we take the kernel (Green's function) of (7). Convolving (1) with the Green's function leads to an expression for a family of deforming planar curves

$$C(d;t) = \sum_p \vec{X}_p \frac{1}{\sqrt{4\pi t}} \exp(-\frac{(d - d_p)^2}{4t}) \tag{8}$$

3   Results

To study the consequences of our model, we demonstrate the affine evolution of the affinely related pointsets in figure 1. To indicate the ordering of the point set we connected the points. The evolution, however, will be based on the corner points only.

Figure 1: A star-shaped pointset (left image) and an unimodular affine related pointset (right image). The dashed lines indicate the ordering of the points. The evolution is based on the cornerpoints only.

The evolution of the curve is calculated at a number of time instances (see figure 2). Note that with increasing time the curve is shrinking towards an ellipse (for the case of smooth inputcurves see [1]).

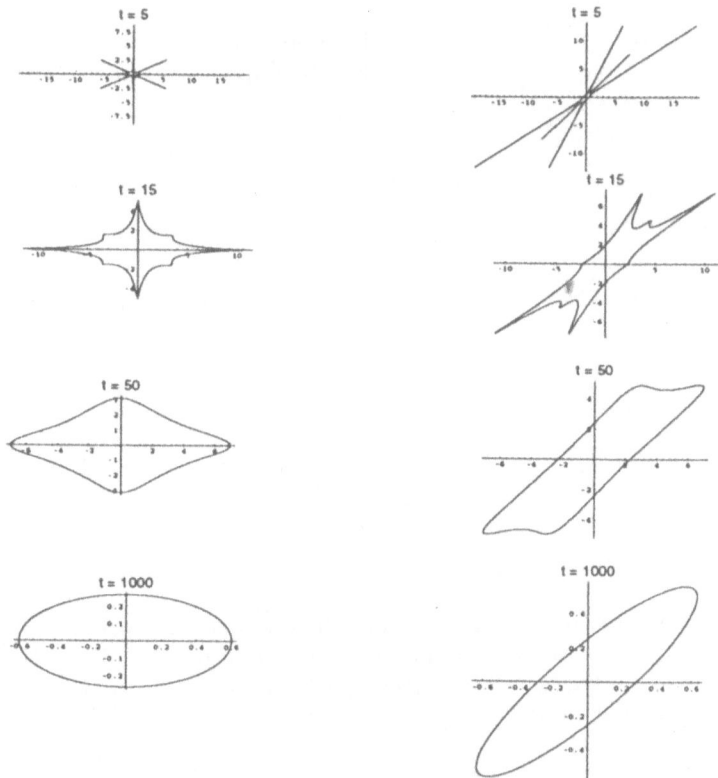

Figure 2: Affine invariant point set evolution. The left column arises from the left pointset from figure 1 and the right column arises from the right pointset from figure 1. Evolution times are $t = 5, 15, 50, 1000$. At all times the affine relation between the pointsets is preserved. Note that the pointsets evolve towards ellipses.

With the pointset parametrisation used in this paper, it is possible to evolve discrete curves with self-intersections. As an example, we took the lemniscate, see left image in figure 3. In the right image of figure 3 we created a discrete parametrisation for it. The ordering of the points is important here. We can find two alternative discrete parametrisations which are in fact the lemniscate and an "ellipse" with a severe constriction in the origin. Evolving the two objects in an affine invariant way results in the creation of objects which are just the lemniscate and an ellipse (figure 4).

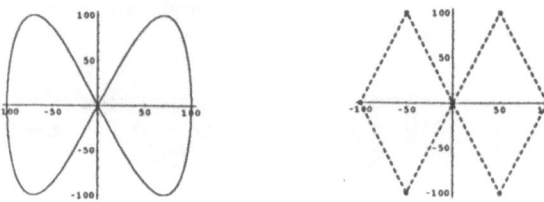

Figure 3: Lemniscate. The ordering of the points in the right image will turn the object into a discrete lemniscate or a discrete ellipse with a strongly constricted waist.

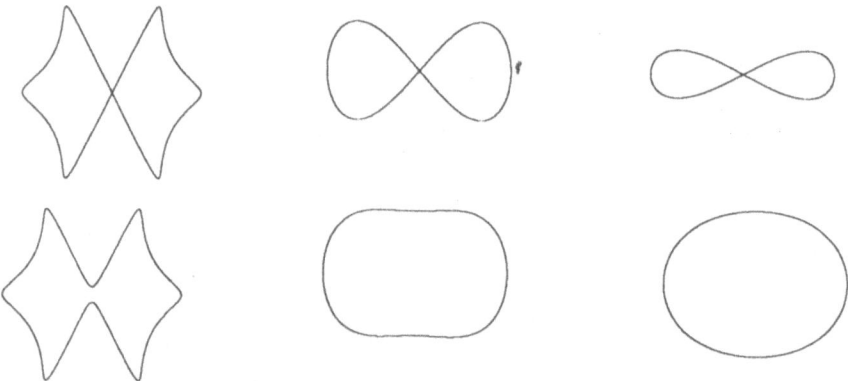

Figure 4: Evolution preserving topology. The upper sequence of images shows the evolution of the discrete lemniscate. The number of intersections is preserved. The lower sequence shows the evolution of the alternative point ordering, i.e. the discrete ellipse.

References

[1] Bart M. ter Haar Romeny, editor. *Geometry-Driven Diffusion in Computer Vision.* Computational Imaging and Vision. Kluwer Academic Publishers, Dordrecht, 1994. ISBN 0-7923-3087-0.

Robotics and Vision - Posters

Reports and Vision — Part V

Dual Processing of Visual Motion Reduces Smearing, Delay and Noise, but Yields the 'Wavy Edge' and 'Window-shift' Illusions

A.J. Noest (ajn@binf.biol.ruu.nl)
't Ven 41, NL-1115-HC Duivendrecht
The Netherlands

Abstract: Early stages of visual motion processing are modelled by two networks, each consisting of velocity-tuned cells. One network extracts the (possibly multi-valued) velocity-field, while the other extracts the patterns which move. The 'channel-coded' output from the velocity-net selects those co-local units in the pattern-net which are tuned to the local velocity, so preventing motion-smear and allowing compensation of target positions for unavoidable processing delays. Under special conditions, however, the model creates some artefacts which correspond well with known visual illusions.

1 Structure and 'normal' functions of the model

Psychophysical and neurophysiological data on visual motion processing in primates suggest that the underlying neural machinery consists of at least two concurrent subsystems. In my model, one network (V) represents the (semi-local) *velocities per se*, while the other (P) represents the *pattern(s)* in motion. This division of labour is reflected in the differences in local processing and lateral interactions within each network, as specified below.

The cells of both networks are tuned to positions $\vec{x}$ in the visual field X, as well as to velocities $\vec{v}$ in some range V. The $\vec{v}$-tuning of a unit corresponds to a distinct spatio-temporal *orientation* of its receptive field (RF). The structure of the retinal input to the V- and P-networks, the partial overlap of 'neigbour-ing' RF's, and the interactions within and between the two networks are all consistent with the notion that both networks form a smooth map of (part of) the 4-dimensional productspace $X \times V$.

The output of the V-net not only encodes the velocity-field, but it also 'gates' (enables) the P-units tuned to the 'correct' local $\vec{v}$. Normally, this selects a representation of the moving pattern which is free of the smearing which occurs in P-units tuned to other $\vec{v}$. Co-moving pattern integration for noise-reduction is also easily done using the selected P-units. Furthermore, a simple calibration of the *visual* position of each P-unit allows target positions to be compensated for processing-delays.

1.1 Extracting the velocity-field: the V-network

The task of the V-network is to measure, cluster, and segment the velocity field *per se*, roughly independent of the precise contrast-pattern. The V-units are thus what neurophysiologists call 'complex cells': they rectify the contrast-pattern, and compress its range of variation (when reasonably above the local noise). The actual motion measurement is done by spatio-temporal correlation, which is equivalent to computing the 'energy' in space-time oriented receptive fields (RF's). Each location is covered by RF's tuned to a range of $\vec{v}$-values.

At any location ($\vec{x}_0$,say), velocities are thus represented by a distribution $S(\vec{x}_0, \vec{v})$ of responses in the local $\vec{v}$-tuned units. A noiseless input with a well-defined local velocity produces a $S(\vec{x}_0, \vec{v})$ with a single, narrow peak in the $\vec{v}$-dimensions. Likewise, inputs with a smooth velocity-field produce a $S(\vec{x}, \vec{v})$ in which the local $\vec{v}$-peak position varies smoothly with $\vec{x}$. This $S(\vec{x}, \vec{v})$ representation has many advantages over the classical ('value-based') approach. For example, it allows multivalued ('transparent') fields to be handled without any problem (the distributions become multimodal in $\vec{v}$), and it also enables an easy and natural implementation of clustering and segmentation of realistic velocity-fields, which are often noisy, and only *piecewise* smooth [7].

V-Network interaction structure:

Simple (and naturrally occurring) lateral interactions within the V-system are sufficient for realizing the required clustering and segmentation tasks. Positive (excitatory) short-range coupling between units with roughly equal $\vec{x}$ and $\vec{v}$ (both along and across the $\vec{v}$-direction) do most of the clustering and noise-reduction, whereas negative (inhibitory) couplings across larger $\vec{x}$-distances transversal to $\vec{v}$ (and across V-space) improve the segmentation of the velocity-field. The range of the interactions is independent of the velocity parameter $\vec{v}$.

The adjacent sketch shows the spatial structure of the interaction kernel $C(\vec{x})$ between the V-units tuned to a single $\vec{v}$ (arrows). $C(\vec{x})$ has a 'mexican hat' shape transversal to $\vec{v}$ ($+$ = light, $-$ = dark), but a monotonic decay along $\vec{v}$.

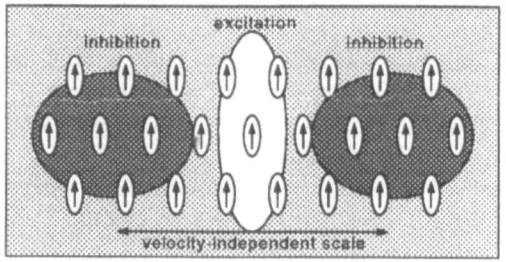

This kernel structure is similar to that in an earlier (qualitative) explanation [6] for the psychophysics of motion'assimilation' and 'contrast'. Analysis and simulations of the V-network within the present explicitly specified and wider-ranging model indeed reproduces these phenomena, as well as several others to be reported below. The 'mexican hat' shape of the interactions transversal to the $\vec{v}$-direction will appear in a crucial role with respect to the (Turing) instability in the V-net which gives rise to the 'wavy-edge' illusion (see section 2).

The output $S(\vec{x}, \vec{v})$ of the V-net is used to 'gate' (select) units in the P-net which have (nearly) the same $\vec{x}, \vec{v}$-indices. Normally, this serves to extract a representation of the moving pattern without motion-smear, and with compensation for delays, as discussed below.

1.2 Extracting patterns in motion: the P-system

The task of the P-system is to extract the contrast-structure which is (usually) 'dragged along' by the velocity-field. Thus, the P-units have a roughly linear contrast-response, but the RF's are again space-time oriented. Note that any pattern moving at velocity $\vec{v}$ is most faithfully represented by the class of

P-units tuned to $\vec{v}$, because these alone have RF's which are spatiotemporally 'aligned' with the X, t-locus 'swept out' by the moving pattern [1]. A collection of such $\vec{v}$-matched RF's (with various fine-structures) can then represent the local pattern in a way which is equivalent to the well-known situation for stationary patterns and purely spatial RF's [4]. Noise reduction by co-moving pattern integration is also easily done within the $\vec{v}$-matched set of P-units.

However, at each position $\vec{x}$, there are RF's tuned to a whole range V of velocities. Severe *motion-smear* would result if the local responses tuned to all $\vec{v} \in V$ would contribute to the P-system output, and realistic delays in the units would aggravate the problem further.

The solution is simple: Gate each local P-unit tuned to some $\vec{v}$ with the output of the co-local V-unit tuned to the same $\vec{v}$. Normally, this selects the least-smeared pattern representation, as discussed above. In addition, it allows delay-compensated positions to be assigned to moving targets.

The adjacent sketch shows the relation between the 'anatomical' positions of spacetime-oriented P-cells and their 'visual' positions, when compensated for signal delay. Because each cell is $\vec{v}$-tuned, only a single, fixed positional calibration per cell is required. Summing over the non-$\vec{v}$-matched RF's would evidently lead to motion-smear (aggravated by failed delay-compensation. *'Window-shift' illusion:*

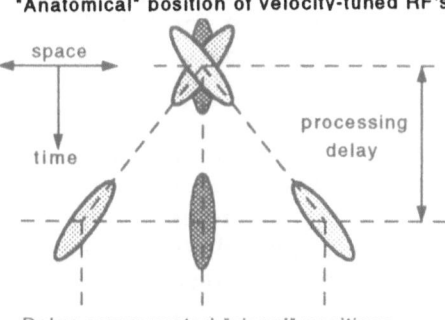

"Anatomical" position of velocity-tuned RF's

space

time

processing delay

Delay-compensated "visual" positions

This delay-compensation scheme can lead to paradoxical effects, even when the V-system functions normally. Take, for example, as visual input a moving grating, multiplied with a (preferably 'soft-edged') *stationary* window of several cycles wide. The V-network represents (mostly) the velocity of the moving grating, and gates the corresponding subset of P-cells, which represent the moving contrast pattern. The delay-compensation (30-60 msec) attributes a position to this pattern which is displaced along $\vec{v}$, away from the true position of the stationary window. Indeed, an illusory shift of the patch is seen [2] as soon as its 'contents' start to move. (The sharper the window's edge, the more it also produces signals encoding its true position, which reduce or compete with the shift.)

2 Turing-instability produces 'wavy-edge' illusion

A dramatic failure of the system can occur when a large-scale moving grating drives the V-network through an instability, leading to spatially periodic modulations on $S(\vec{x}, \vec{v})$. The normal selection of $\vec{v}$-matched P-cells by $S(\vec{x}, \vec{v})$ then fails near the minima of the modulation, so that a spatially periodic degree of 'lagging behind' and smearing appears on the grating edges. This artefact reproduces nicely what is known as the 'wavy-edge' illusion [5]. The instability in the V-system occurs as follows.

It proves sufficient to consider only those subsets of V-cells which are tuned to the edge velocity and situated on an $\vec{x}, t$-plane defined by one of the moving grating-edges. I thus drop $\vec{v}$ as an index, and keep only the spatial coordinate ξ along the edge. The dynamics of the 'soma potentials' $s(\xi)$ then is

$$\partial_t s(\xi) = -s(\xi) + \beta + C(\xi) * F[s(\xi)],$$

The V-cell output $S(\xi)$ is written as $F[s(\xi)]$ (with sigmoid $F[s]$), the bias plus input appears as β, and the convolution kernel $C(\xi)$, which models the lateral interactions, has the 'mexican hat' shape along ξ sketched earlier.

For small enough β, behaviour is 'normal', corresponding to the constant solution $s(\xi) = s_0 = \beta + \gamma_0 F(s_0)$, with $\gamma_0 = \int_{-\infty}^{+\infty} d\xi C(\xi) < 0$.

Introducing the perturbation $r = s - s_0$, and writing $F(s) = F(s_0) + f(r)$, one gets $\partial_t r(\xi) = -r(\xi) + C(\xi) * f(r(\xi))$. Expanding $f(r) = \sum_{n=1}^{\infty} \alpha_n r^n$, one has $\alpha_1 > 0, \alpha_3 < 0$, but α_2 could change sign, depending on β.

The Fourier transform of $C(\xi)$, denoted by $\gamma(\omega)$, should have a positive global maximum, say $\gamma_1 > 0$, at a finite frequency $\omega = \pm\Omega$. This is true for any realistic $C(\xi)$. The trivial solution $r(\xi) = 0$ loses stability when $\alpha_1 > 1/\gamma_1$, and we should then look for non-trivial solutions of the form $r(\xi) = r(\xi + 2\pi/\Omega)$.

In the generic case ($\alpha_2 \neq 0$), a second order Galerkin approach suffices. Thus, we truncate the expansion of the nonlinearity as $f(r) = \alpha_1 r + \alpha_2 r^2$, and write $r(\xi)$ accordingly as $r(\xi) = A_0 + A_1 \cos(\Omega\xi) + A_2 \cos(2\Omega\xi)$. The r-dynamics thus simplifies to 3 coupled ODE's for the A_n, and classification of the bifurcation amounts to reducing the A_n-flow to its normal form [3].

In terms of $\eta_n = \gamma_n \alpha_1 - 1$, $\kappa_n = \gamma_n \alpha_2$, with $\gamma_n = \gamma(n\Omega)$, the center manifold H becomes $A_m = -\kappa_m/(2\eta_m)A_1^2$; $m = 0, 2$, and projecting the A_n-flow within H onto A_1 yields the required normal form

$$\partial_t A_1 = \eta_1 A_1 - \alpha_2^2(\gamma_0/\eta_0 + \gamma_2/(2\eta_2))A_1^3.$$

Thus, the bifurcation is of the 'pitchfork'-type [3]. The terms γ_0/η_0 and $\gamma_2/(2\eta_2)$ in the A_1^3-coefficient are of opposite sign, but their sum is positive for all realistic parameter choices; this verifies the validity of the second-order analysis. Note that the unknown sign of α_2 drops out. The singular case $\alpha_2 = 0$ requires a higher-order analysis.

Acknowledgements: I thank Lex Toet for discussing and demonstrating the wavy-edge illusion, and SPIN/SNN for partial financial support.

References:

1. Burr,D.C., Ross,J. & Morrone,M.C. *Proc.Roy.Soc.London* B227 (1986), 249.
2. DeValois,R.L. & DeValois,K.K. *Vision Res.*31 (1991) 1619.
3. Guckenheimer,J. & Holmes,P. *'Nonlinear Oscillations, Dynamical Systems, and Bifurcations of Vector Fields'*. Springer, New York 1983.
4. Koenderink,J.J. & van Doorn,A.J. *Biol.Cybern.*55 (1987),367.
5. Marcus,J.T. & Toet,A. *Neurosci.Lett.*124 (1991), 239.
6. Nawrot,M. & Sekuler,R. *Vision.Res.*30 (1990) 1439.
7. Noest,A.J. in *'Shape in Picture'* (Y.-L. O *et al.*,Eds.), Springer, Berlin 1994.

THE DYNAMICS OF THE PERCEPTUAL ORGANIZATION IN APPARENT MOTION

M.A. Giese, G. Schöner[2], H.S. Hock[3]

[1] Institut für Neuroinformatik, Ruhr-Universität, D-44780, Bochum, Germany
[2] CNRS-LNC, 31, Chemin Joseph Aiguier, F-13402 Marseille Cedex 20, France
[3] Dept. of Psychology, Florida Atlantic University, FL 33431 Boca Raton, USA

The organization of feature elements to a complete visual percept has been considered as an interesting psychological problem since the time of Gestaltpsychology. Recently concepts from nonlinear dynamical theory have been proposed for the analysis of perceptual organization [3,4,7]. We present a simple neural network model for the perceptual organization processes in apparent motion. The model captures a large number of experimentally demonstrated dynamical properties of the perceptual organization process.

The model is based on the the *motion quartet*, an ambiguous psychophysical motion display. The display has been extensively analyzed in experimental literature and an extended empirical data-base is available. We believe that only the *quantitative* comparison of model-predictions with experimental results sufficiently restricts the class of possible model structures for biological phenomena. The motion quartet allows an exact quantitative parameterization because of its simple geometrical and temporal structure. This is important for a tight relationship between theory and experiment. Nevertheless the percept is based on cooperative phenomena, which we consider as functional basis of perceptual organization. We regard the motion quartet as prototype for the perceptual organization in motion perception.

The geometry of the motion quartet is determined by four dots placed in the corners of a rectangle with side-lengths H and V (cf. fig. 1a). Periodically, during the first half of a display cycle two dots in the diagonal edges of the rectangle are illuminated together (filled dots in figure 1b). In the second half the two other dots flash together (open circles). Choosing adequate time intervals for the frames and the repeat frequency results in an ambiguous percept. Subjects see alternatively anti-parallel horizontal or vertical motion. A mixed percept never occurs. By changing the aspect ratio $AR = V/H$ perception can be biased towards horizontal or vertical motion. Continuous variation of the aspect ratio reveals *hysteresis* between the horizontal and vertical percept. This reflects the stability of the underlying dynamical perceptual states [4]. Hysteresis is influenced in a characteristic way by *adaptation effects*. Vice versa the adaptation depends on the stability of the adapting percept [5]. Subjects experience spontaneous changes between the horizontal and the vertical percept when they observe the display for several frames . Experimental data on the

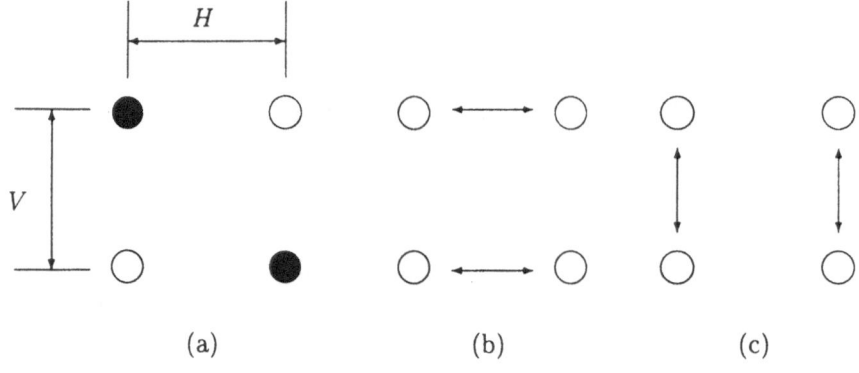

(a)   (b)   (c)

figure 1

statistics of the perceptual switches is available [6,7]. The model captures the
complex interdependence between adaptation and stability and also reproduces
the statistics of perceptual switches.

Our model seems biologically plausible and is formulated in a mathematical
language which allows an easy transfer of the results to technical problems. It
is based on a simple neural network structure described by Amari [1]. Each
neuron represents a possible elementary motion. The biological correlate of
each model neuron may be rather an ensemble of cells or mechanisms which
are involved in the representation of an elementary motion [2]. The neurons
have cooperative connections yielding to a competition between the horizontal
and the vertical motion percept. The interactions depend on the geometrical
relation of the interacting motion detectors. The single neuron dynamics has
the form:

$$\tau \dot{u}_i = -u_i + \sum_j W_{ij} \, f(u_j) + S_i - h_i(t) + \xi_t$$

The membrane potential u_i is dependent on the input signal S_i. The input is
positive when the elementary motion detector is activated by the stimulus. An
elementary motion is perceived when the output of the corresponding neuron
$y_i = f(u_i)$ is near one. f is a monotonous threshold function. The weight matrix (W_{ij}) specifies the cooperative interactions between the neurons. τ is the
time constant of the membrane potential. The white noise process ξ_t represents
stochastic fluctuations which lead to spontaneous perceptual transitions.

Adaptation effects are modeled by assuming dynamically varying thresholds
$h_i(t)$ of the neurons. The threshold dynamics can be written

$$\tau_{ad} \dot{h}_i(t) = -h_i(t) + a_i(t)$$

where the neuron activity variable a_i is a function of the membrane potential u_i. For $\tau_{ad} \gg \tau$ frequently activated neurons become less sensitive over

Hysteresis–loop and switching probability

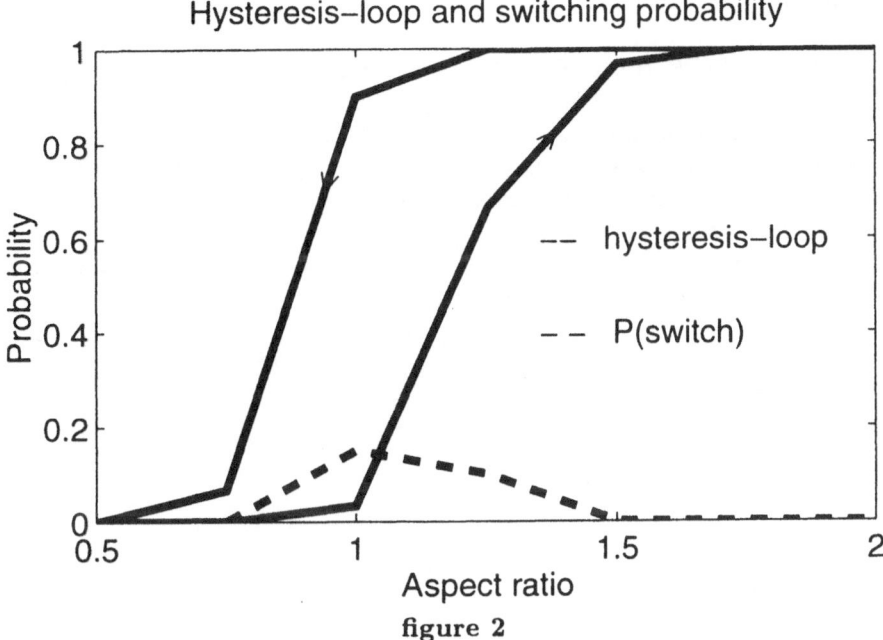

figure 2

time. We present the model in two versions. A simple version with only 2 neurons, representing vertical and horizontal perceived motion, allows an analytical mathematical analysis and reveals profound insights in the underlying dynamical and stochastic phenomena. Analysis is possible in the stationary limes as well as in the transient regime. In the complex version the model contains 8 neurons, one for each possible elementary motion in the display. We show how the simple model can be extended systematically to this complex neural structure.

In detail the model reproduces quantitatively the following experimentally observed effects [4,5,6]: 1) Perceptual bistability and spontaneous switching between horizontal and vertical percept; 2) hysteresis and the shape of the hysteresis-loop for continuous change of the aspect ratio; 3) statistics of the perceptual switches and switching-times; 4) dependence of hysteresis and switching-rate on the stimulus size and on the rate of change of the aspect ratio; 5) interdependence of percept stability and adaptation; Figure 2 presents as an example the hysteresis loop and switching rate as calculated from the model, which closely correspond to the psychophysical results from [4]. The model can be extended easily to explain further psychophysical effects and to more complex neural structures.

Supported by the Studienstiftung des deutschen Volkes.

Literature:

[1] Amari, S.: *IEEE Trans. Syst. Man Cyb.*, **SMC-2**(5), 643-657 (1972).

[2] Carmesin, H.O.: *Acta Phys. Slovaca*, **4**(5), 311-330 (1994).

[3] Ditzinger, T, Haken, H.: *Biol. Cyb.*, **61**, 279-287 (1989).

[4] Hock. H.S., Kelso, J.A.S., Schöner, G.: *J. Exp. Psych.*, **19**(1), 63-80 (1993).

[5] Hock, H.S., Schöner, G., Hochstein, S.: *in preparation.*

[6] Hock, H.S. *et al.:in preparation.*

[7] Kruse, P., Stadler, M.: In Haken, H., Koeppchen: *Synergetics of perception.* Springer, Berlin (1993).

Multiscale Image Segmentation based on a Receptive Field Model

Koen L. Vincken
André S.E. Koster
Max A. Viergever

RWCP[1] Novel Functions SNN[2] Laboratory
Computer Vision Research Group, Utrecht University
Utrecht, The Netherlands

Abstract

A multiscale method (the hyperstack) is proposed to segment multidimensional image data. Hyperstack segmentation is based on the scale space concept of front-end vision. A scale space is created by blurring the original image using linear Gaussian kernels of increasing width.

The segmentation algorithm then starts with the linking of voxels at adjacent levels in scale space, followed by a root selection to find the voxels that represent the segments in the original image. Results (obtained by downward projection from these roots) will be presented for segmentation of MRI brain data.

1 Introduction

Multiscale image processing derives from the idea that a visual system analyses a scene at multiple levels of resolution simultaneously. The scale space, a one-parameter family of blurred replicas of the input image, is based on the diffusion equation and was proposed by Witkin [4] and Koenderink [2] as the image representation for linear multiscale analysis. The blurring kernel corresponding to this diffusion concept is the Gaussian function, which thus is the mathematical equivalent of the receptive fields in biological vision systems.

A hyperstack uses the scale space to obtain a segmentation of the input image. By using this multiscale approach the global image information can effectively be included. In [3] we have shown that the hyperstack performs well on noisy images and is particularly strong in segmenting both small and large objects simultaneously.

The basic idea of the hyperstack is to define relations (*linkages*) between voxels in all pairs of adjacent scale space levels, such that the levels at larger scales—containing the global information—guide the collection of voxels in the original image at the smallest scale (the *ground level*).

---

[1] Real World Computing Partnership
[2] Dutch Foundation for Neural Network

2   The hyperstack process

The hyperstack segmentation method consists of four phases (see Fig. 1).

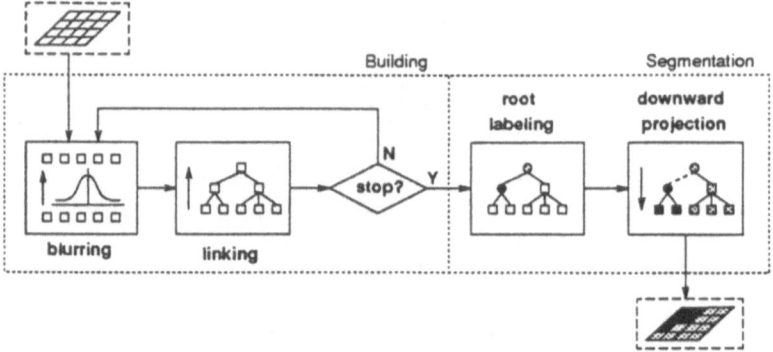

Fig. 1. *Schematic of the hyperstack image segmentation process.*

Blurring. The hyperstack starts by creating a scale space—which is a set of replicas of the original image—blurred by a Gaussian kernel of increasing width. According to the property of scale invariance [1] the blurring strategy has to follow an exponential sampling in scale space. Hence, the linear scale parameter $\tau = n \cdot \delta\tau$, $n \in \mathbb{N}$, is related to the absolute scale σ_n at level n by $\sigma_n = \varepsilon \cdot \exp(n \cdot \delta\tau)$, where ε is taken to be the smallest linear grid measure of the imaging device.

Linking. During the linking phase voxels in two adjacent levels are connected by so-called *parent-child* linkages. Linking is a bottom-up process in the sense that we start linking the children of level 0 (the ground level) to parents in level 1, then find parents in level 2 for the children in level 1, and so on. Since only parent voxels that have been linked to before are considered children in the next linking step, convergence of the scale-tree is assured.

Research to different statistical and heuristic components has shown that a general and robust linking scheme should use three components: (i) the traditional intensity difference component, (ii) the ground volume criterion, and (iii) the mean ground volume intensity component. (The ground volume of a voxel is defined as the number of ground voxels with a route to that voxel.)

For all three components it can be argued that a large value should be preferred for a child to link to that parent. For instance, if the mean ground volume intensity of the child closely resembles the corresponding field of the parent, then it is natural for the child to merge into that parent segment: they both represent (part of) a segment with that intensity value.

In the case of *probabilistic linking* child voxels are allowed to link to more than one parent. In particular, partial volume voxels will have more than one 'attractive' parent. The probability value assigned to each link is derived from the corresponding linkage strengths by straight normalization. Hence, the sum of the probabilities of all parent linkages of one child equals 1.

Root labeling. The roots—*i.e.*, voxels in the scale-tree that represent segments in the original image—are identified after all levels are connected through the linkages.

The two components that turned out to be robust root criteria were: (i) the mean ground volume intensity difference component, and (ii) the ground volume intensity variance component.

Downward projection. In the actual segmentation phase root values are projected downwards from every root to the connected voxels in the ground level. By using a unique value for each root it is guaranteed that the segments in the original image are distinguishable. This allows for quantitative validation of segmentation results. For visual inspection, the mean intensity of the pixels within each segment is a reasonable alternative.

3 Results

In Fig. 2 we have shown a two-dimensional probabilistic segmentation of a small part of a 2D MR brain image, based on the highest object probability of each ground voxel (*i.e.*, denoting the probability that each pixel belongs to the cerebellum). Probabilistic linking clearly outperforms its conventional counterpart in which a voxel can only be part of one object.

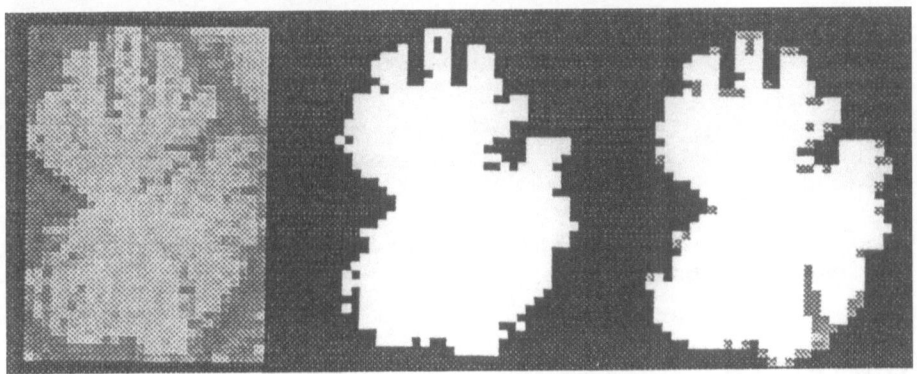

Fig. 2. *Segmentation of the cerebellum of an MR brain image. Original image (left), conventional hyperstack segmentation (middle) and a probabilistic result (right).*

In Fig. 3 the surplus value of probabilistic over conventional segmentations is shown for the ventricle system. The original image is an MR image of the brain containing 32 slices of 128^2 and a grey-level resolution of 12 bits. The conventional segmentation has several serious shortcomings.

In Fig. 4 a high-resolution 3D MR image has been rendered. Because the resolution is significantly higher (128 slices of 256^2) than the ventricle image, we did not compare the result to multi-parent linking. We have segmented this high resolution image (128 slices of 256^2) into three segments: brain, skull/skin, and background. The brain has been extracted and separated from the other parts by a simple pixel-connect operation.

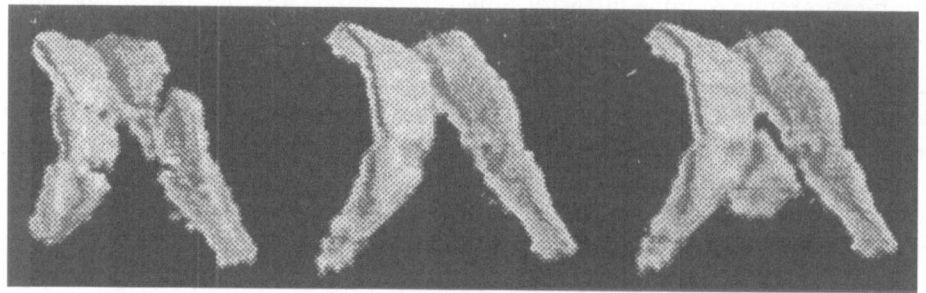

Fig. 3. *Renderings of a normal (left) and a probabilistic (middle) segmentation of the ventricle system in a 3D MR image of the brain. Adding objects is very simple (right).*

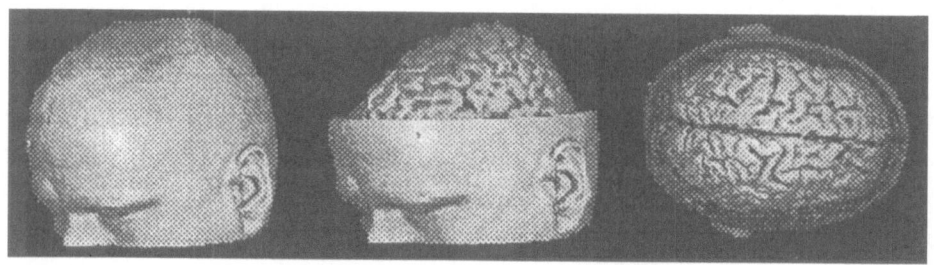

Fig. 4. *Renderings of a high-resolution 3D MR data set, based on single-parent linking.*

References

[1] L. M. J. Florack, B. M. Ter Haar Romeny, J. J. Koenderink, and M. A. Viergever. Scale-space and the differential structure of images. *Image and Vision Computing*, 10:376–388, 1992.

[2] J. J. Koenderink. The structure of images. *Biol. Cybern.*, 50:363–370, 1984.

[3] K. L. Vincken, A. S. E. Koster, and M. A. Viergever. Probabilistic segmentation of partial volume voxels. *Pattern Recognition Letters*, 15(5):477–484, 1994.

[4] A. P. Witkin. Scale space filtering. In *Proc. International Joint Conference on Artificial Intelligence*, pages 1019–1023. Karlsruhe, W. Germany, 1983.

A Neural Network Architecture for Scene Interpretation

Boehme, H.-J., Braumann, U.-D., Heinke, D., Gross, H.-M.

Department of Neuroinformatics, Technical University of Ilmenau
Ilmenau, Germany

1 Introduction and scenario

Due to the characteristics of the information entering intelligent biological systems interacting with a natural environment these systems must try to find the best solution of a decision situation by means of a permanent interaction between the incoming data stream and the internal decision making process.

Against this background a neural overall architecture was outlined and implemented to develop an emerging behavior–oriented object understanding on the basis of self–organization by interaction with its surroundings (analysis of complex visual scenes). The information stream entering the architecture is internally represented according to its characteristic in order to make the currently best possible decision based on this representation depending on the concrete sensory situation, using an online-learning-mechanism. This processing principle is evident for three reasons: First, to avoid too much possible multiple alternatives. Second, to do the most promising action for the best behavior respectively the most rapid adaptation to various external influences. And third, to enable and to do some correction of previously made wrong decisions [1].

The work is based on experimental findings in psychophysics, which show that a visual scene is processed in parallel and in sequence leading to the theory of selective visual attention ([3], [2]). Following the searchlight metaphor we consider a visual attention process as an internal scanning accomplished over the scene without eye movements building the frame for our hypothesis on behavior-oriented self-organization of object understanding.

The environment of our system consists of stationary, colored 2D-scenes. The scenes contain objects against structured backgrounds. The objects vary in both their arrangement and their appearance. Neither rotation nor scaling of the size of the objects has been done.

Concerning the learning process it is important to assume that objects appear as stable visual structures which are emerging by scanning various scenes and then lead to an increase of the effectiveness of the scanning behavior.

2 Overall architecture

Fig. 1 depicts the overall architecture of the model of the self–organization of an object understanding. The Early Processing stage computes features of every location of the input scene in the domains of luminance and color. By means of calculation of the difference between the local feature extents and the mean feature extent of the overall scene these features are combined to a saliency

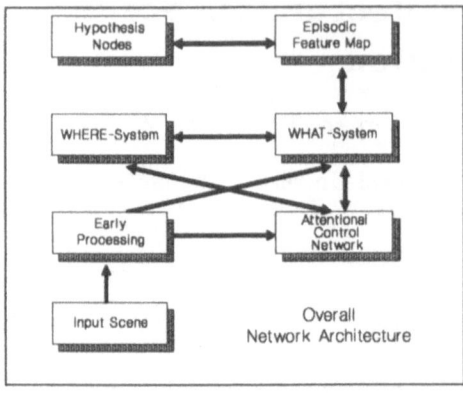

map of the image which is the base of the purely data–driven move of a window of attention over that image. At each stopping point of that window a complex feature (on the base of the same feature domains) is extracted for the taken detail. The Where- and What-system together with the Attentional Control Network form the first hierarchical level for knowledge representation and generation of scanning behavior. By means of interactions of Where- and What-system stable transitions between features appearing within the input

Figure1: *The overall architecture.*

scenes are detected, represented and directly used for the knowledge–based influencing of the scanning process, controled via the Attentional Control Network.

While at the first representation level hypotheses on the next most likely move can be generated and verified, the second representation level tries to take sequences of successfully verified single moves and keep them as candidates for objects or parts of objects.

Rather than dealing with details of the architecture we want to focus the attention to the overall behavior of our model.

3 Model behavior and simulation results

The most important substructure of the model where the interactions between the two representation levels take place is the Feature Transition Memory (FTM) of the What–system. This FTM contains clusters of units (nodes) which encode relative moves (moves of the window of attention) from one starting feature to various target features thereby using a location map of the Where-system. First let us consider the behavior of the first representation

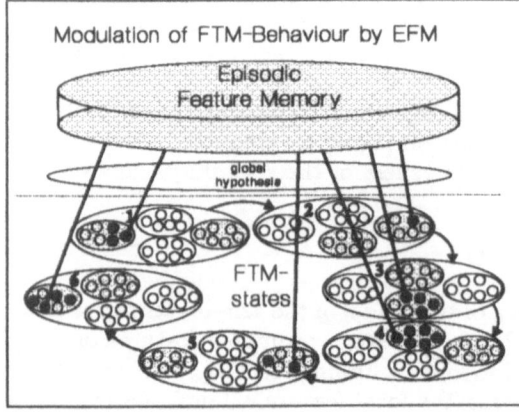

Figure2: *As an example six stages of the process of hypothesis making are shown. The nodes that would have been selected just on the basis of local hypotheses alone are marked gray. The black labelled nodes depict those which finally were selected by the fusion of local and global hypotheses in order to control the scanning of the image.*

level without interaction with the second one. Each single node of the FTM represents just one single feature transition. Via simple reinforcement learning the FTM tries to make hypotheses about the most likely next feature transition. By means of modulation of the scanning behavior the internal hypotheses are verified.

During this learning procedure the FTM becomes more and more successfull in forcing the window of attention to that positions where the expected feature matches with the actual one. As the most important result the model prefers to scan objects because of their spatial stability of feature relations. Figure 3 outlines the effect of this learning process. However, the local hypotheses

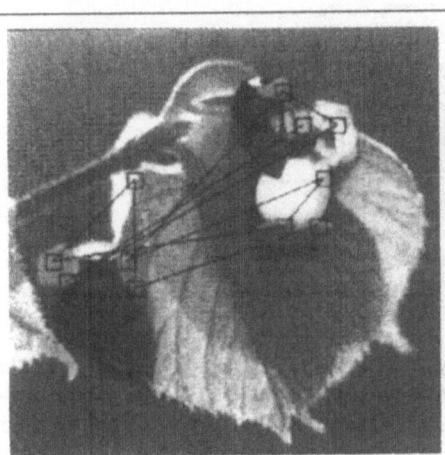

Figure3: *The left example shows a purely datadriven scanpath. The right simulation was made after one presentation of the scene gallery. Here red lines mark knowledge based moves. Moves which were tested but not really executed are ignored.*

which can be generated by this first level of representation one should rather consider as the preliminary stage for an implicit object understanding since more global relationships between features as candidates for objects or parts of objects cannot be represented yet. This is the crucial motivation for the development of the Episodic Feature Memory (EFMem) as the second level of the processing hierarchy. The EFMem involves the Episodic Feature Map (EF-Map) containing uniformly structured and functionally autonomous columns arrangend in clusters, too, and a Hypotheses Layer (HL). Inside the EFMap a self-organizing establishes assemblies of columns standing for sequences of successfully predicted feature transitions (moves of the window of attention). The nodes of the HL (hypothesis nodes) represent the different established column assemblies in the EFMap. The mapping between the column assemblies and the hypothesis nodes changes during the organization process. The activity of the hypothesis nodes shows to which extent the various assemblies have emerged in the EFMap and thus determines the strength of influence of the EFMem upon the FTM. This influence is affecting the nodes of the FTM indirectly via the corresponding columns of the EFMap. Therefore it is possible to support the modulation of the scanning behavior by means of knowledge about whole scan pathes prefering objects or parts of objects.

The cooperation of the FTM and EFMem is shown in Fig. 2. Therefore, several stages of the FTM following each other are depicted symbolically. By

 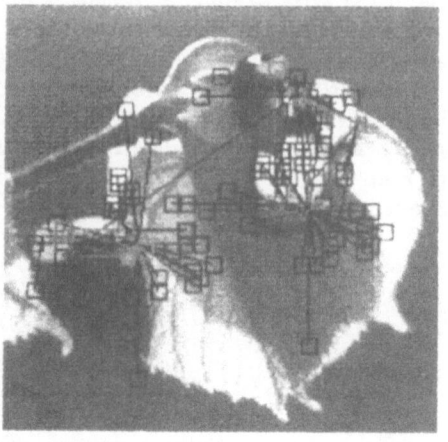

Figure4: *The left simulation was made without the effect of a global hypothesis whereas the right could benefit from that influence. See text for more details.*

means of the inclusion of local feature hypotheses in global hypotheses and the backward–directed projection of them into the FTM the overall model is able to include knowledge about even more complex stable relations into the control of the scan path.

The two simulation results depicted in Fig. 4 clarify how the backward–directed projection from EFMem onto the FTM has effect on the scanning of real scenes. The scan path in the left image was exclusively modulated through the available local hypotheses on features following each other immediately. In both images a different number of green squares mark the not really executed moves which were suggested by hypotheses but could not be verified, while the successfully (really) executed moves are marked red. In the right simulation example the scan path for the same image was modulated by hypotheses affected by the EFMem. It shows that compared to the previous example far less non–verifiable hypotheses arose. Faulty hypotheses cannot be completely suppressed with the current implementation stage because the complete gallery of scenes (100 images) has been presented to the system only once, a heavy overlapping within the EFMap–column–assemblies has to be handled, and the presented architecture is just an approach towards an episodic representation of feature sequences and subject to further refinement.

References

[1] Gross et al.: Proc. ICANN–94, Vol. 1, pp. 58–61, Springer 1994

[2] Treisman, A.M. et al.: Cognitive Psychology (12), pp.97–136, 1980

[3] Wolfe, J.M. et al.: AI and the Eye, pp.79–103, 1990

Neural dynamics parametrically controlled by image correlations organize robot navigation

Hartmut NEVEN[1] *and Gregor SCHÖNER*[2]
[1]Ruhr-Universität Bochum D-44780 Bochum, Germany
email: neven@neuroinformatik.ruhr-uni-bochum.de
[2]CNRS-LNC, F-13402 Marseille Cédex 20, France

1   Introduction

Behavior-based robot designs confront the problem how different elementary behaviors can be integrated. We address two aspects of this problem, the stabilisation of decisions based on changing behavioral requirements and the fusion of multiple sources of qualitative sensory information. These issues are studied in the context of a vision-guided mobile robot that is endowed with the ability to reach a goal while it avoids obstacles. Behavior is organized from the "inside" of the robot. Even in the absense of external stimuli the internal dynamics generate behavior. By exploiting image correlations the visual sensors provide coarse qualitative estimates of spatial relations. These estimates are immediately coupled into neural dynamics realized by neural fields that generate obstacle avoidance and homing behaviors of an autonomous mobile robot. The background of the neural field approach is theoretical work on the function of cerebral cortex ([1]). Neural fields represent a state space approach to estimation and control. Convergent information cooperates and divergent information competes in shaping the stable attractor states. The states of the dynamical systems are on the one hand instrumental for the control of behavior on the other hand they provide concise hypotheses for the interpretation of sensory input. The concept of dynamic state deals effectively with contradictory information and assures that only behaviorally relevant information is extracted from the input. The navigation scheme works succesfully in real-time with image resolutions as poor as 32^2 pixels.

2   The calculus of neural fields: Stimuli as forces

We employ a simplified version of the calculus of neural fields (see [5] for broader discussion). First continuous variables, x, are chosen to characterize the robot behavior. In terms of these behavioral variables the desired and undesired behaviors must be expressable as simple points or sets. Examples are such variables as robot position and robot velocity. The basic equation describes the evolution of the state x in the presence of a stimulus $s(x, t)$:

$$\dot{x} = \frac{1}{\tau_x} \int_{-\infty}^{\infty} (x' - x) \exp \left[-\frac{(x' - x)^2}{c^2} \right] s(x', t) dx' \qquad (1)$$

Multimodal input distributions lead to multistability and bifurcations. Hysteresis phenomena may result if the input distribution varies in time. These nonlinear effects play an important role in the technique. The prescription of how sensory and internal variables shape the stimulus $s(x, t)$ is a primary design task when applying neural fields. One direct method is to transform sensory or internal events into a value $x_0(t)$ of the behavioral variable and a coupling strength $\eta(t)$ of a stimulus peak, which might be a measure of confidence or importance of the information. The stimulus then takes the form

$$s(x, t) = \eta(t)\delta(x - x_0(t)) \qquad (2)$$

with the Dirac δ-function.

3 Making and stabilizing decisions: Heading to a goal while avoiding obstacles

As a cue for collision danger we exploit the radial optical flows that result from the rectilinear robot motion in the direction of the optical axis. These flow fields permit estimating time-to-contact ([4]), in our case separately for the left and the right camera. We employ a region-based correlational algorithm ([3]) to compute the optic flow.

To control motion we use the forward velocity v and the angular velocity ω as behavioral variables. Behavior is generated by integrating neural field dynamics of the form

$$\dot{\omega} = \frac{1}{\tau_\omega} \int_{-\infty}^{\infty} (\omega' - \omega) \exp \left[-\frac{(\omega' - \omega)^2}{c_\omega^2} \right] (s_{\omega,\text{goal}}(\omega', t) + s_{\omega,\text{obs}}(\omega', t)) d\omega' \qquad (3)$$

The contributions to the dynamics reflecting target acquistion (index goal) and obstacle avoidance (index obs) are added to form the total stimuli.

Space permits only the discussion of the angular velocity dynamics. Target acquisition is based on stabilizing the rotation rate required to reduce to zero the angle under which the goal is seen. The idea of how obstacles couple into the dynamics is this: Add two symmetric terms to the rotation rate dynamics, one specifying rotation to the left at maximal rate, the other rotation to the right at maximal rate. The strengths of these two contributions depend on time-to-contact as determined by the right camera and by the left camera, respectively. Whichever side has a smaller time-to-contact, strengthens the contribution for turning to the contra-lateral side, so that a tendency is created to turn away from the side with smaller time-to-contact ([2]). Over a wide range of time-to-contact differences between left and right the resultant dynamics of rotation rate is bistable. Hysteresis makes sure that as the form of the contributions change due to fluctuations and changing sensory information while driving, a decision to turn either left and right is maintained, until overwhelming evidence for the opposite decision has been acquired. The Figure shows on its left hand side the temporal evolution of the complete stimulus for the ω-dynamics.

4 Estimating the robot position relative to a home base by fusing visual and dead reckoning information

The task to be solved here is the generation of estimates of ego-position in a coordinate system centered at the home base that enable the system to generate homing behavior. The two cartesian coordinates, x and y and the heading direction, φ, determined as the angle between vehicle heading and x-axis, are therefore each considered a behavioral variable, governed by a neural field dynamics of the form

$$\dot{x} = \frac{1}{\tau_x} \int_{-\infty}^{\infty} (x' - x) \exp\left[-\frac{(x'-x)^2}{c_x^2}\right] (s_{x,\text{visual}}(x',t) + s_{x,\text{motion}}(x',t)) dx' \quad (4)$$

Two available sources of information are to be integrated through this dynamics. A contribution, $s_{.,\text{vision}}$, is based on a computed "optic flow", but now between a memorized image characterizing the home base and the current image. The second contribution, $s_{.,\text{motion}}$, is based on temporal integration of the movement commands generated by the target acquistion and obstacle avoidance dynamics (previous Section), that is, a simple form of dead reckoning.

In order to relate the correspondance vectors (dh, dv) directly with the motion parameters we have to make coarse approximations. The resulting estimates serve to specify the location of the contributions to the neural field dynamics of distance, x, and direction, φ, to the home base. The strength of these constrictions are simply the confidence values of the flow computation, so that in the absence of a reasonable match between memory view and current image the vision sense does not contribute to the position estimation dynamics. The equations are:

$$s_{x,\text{visual}}(x,t) = \sum_{(h,v)} corr((h,v)) \delta\left(x - \frac{X}{1 + \frac{v}{dv}}\right) \quad (5)$$

$$s_{\varphi,\text{visual}}(\varphi,t) = \sum_{(h,v)} corr((h,v)) \delta\left(\varphi - \left(\frac{1}{\frac{h}{f}(\frac{dh}{f} + \frac{h}{f}) + 1} - \frac{dh}{f}\right)\right) \quad (6)$$

These two formulas suffice to determine all three position coordinates if one takes reference to memory images taken in different viewing directions.

Although above expressions lead to spurious information when the assumptions underlying the approximations are violated, this does not matter in closed loop. This is because of the fact that the flow field computation yields lower confidence values whenever these assumptions are violated.

The second source of positional information are the copies of the movement commands sent to the vehicle. By parametric choice, the visual contribution dominates whenever the correlation algorithm yields high scores, while the dead reckoning dominates whenever the visual module generates low correlations between current and memory image. Thus the system decides autonomously to which degree it relies on visual infomation. The right hand side of the Figure shows the evolution of the relative strength of the two types of contributions. Note that the state of the neural field functions as a hypothesis regarding the sensory input.

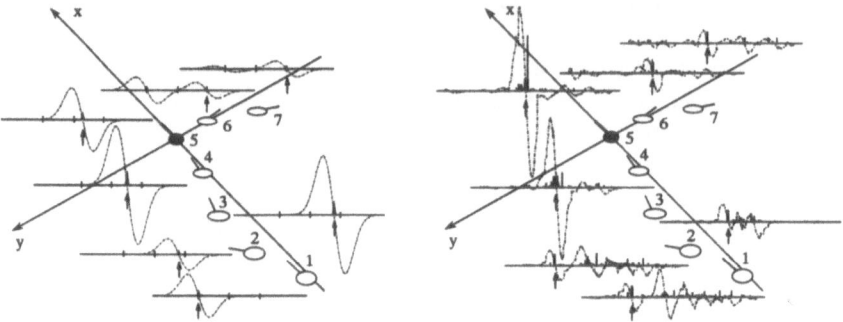

Figure 1: Right: The dynamics of rotation rate, ω illustrated at 7 points during a run. The solid line represents input $s(\omega, t)$ the dashed line the resultant dynamics $\dot{\omega}$. As long as the potential is monostable the neural fields implement essentially an ordinary control system. Nonlinear effects play a role when the potential becomes bistable. Left: The dynamics of position estimation (here, x). Near the home base (position 5) the visual entries (thin bars) are dominant over the entries from dead-reckoning (thick bar) leading to correct homing in spite of dead-reckoning errors.

Acknowlegdements

We thank A Aertsen, V Braitenberg, O Faugeras, C von der Malsburg, B Poucet, W von Seelen and T Smithers along with their collaborators for inspiring discussions.

References

[1] S Amari. Dynamics of pattern formation in lateral-inhibition type neural fields. *BiolCyb*, 27:77–87, 1977.

[2] V Braitenberg. *Vehicles. Experiments in Synthetic Psychology*. MIT Press, Cambridge, Mass., 1984.

[3] HH Bülthoff, JJ Little, and T Poggio. A parallel algorithm for real-time computation of optical flow. *Nature*, pages 337–549, 1989.

[4] DN Lee and DE Reddish. Plummeting gannets: A paradigm of ecological optics. *Nature*, 293:293–294, 1981.

[5] G Schöner and Ch Engels. Dynamic field architectures for autonomous systems. In *From Perception to Action*, pages 242–253, 1994.

Statistical Pattern Recognition - Orals

Statistical Ideas for Selecting Network Architectures

B. D. Ripley

Professor of Applied Statistics, University of Oxford

Oxford, UK

Abstract

Choosing the architecture of a neural network is one of the most important problems in making neural networks practically useful, but accounts of applications usually sweep these details under the carpet. How many hidden units are needed? Should weight decay be used, and if so how much? What type of output units should be chosen? And so on.

We address these issues within the framework of statistical theory for model choice, which provides a number of workable approximate answers.

This paper is principally concerned with architecture selection issues for feed-forward neural networks (also known as multi-layer perceptrons). Many of the same issues arise in selecting radial basis function networks, recurrent networks and more widely. These problems occur in a much wider context within statistics, and applied statisticians have been selecting and combining models for decades. Two recent discussions are [4, 5]. References [3, 20, 21, 22] discuss neural networks from a statistical perspective.

1 A smoothing example

Figure 1(a) shows a dataset which is a simulation of an example from [29] where it illustrates smoothing splines. Throughout we assume that the true noise variance, $\sigma^2 = (0.2)^2$, is known.

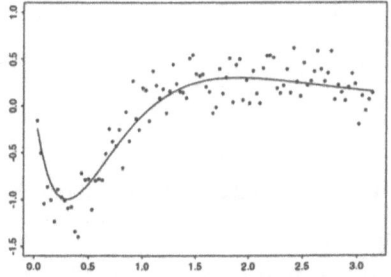 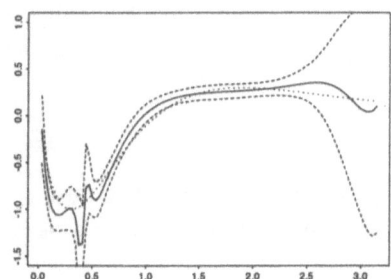

Figure 1: (a) 100 data points from a curve-fitting problem, with the true curve. (b) Curve fitted by 1–8–1 neural network, with ± 2 standard error bands and true curve (dotted).

Figure 1(b) shows a fitted curve from a neural network with one linear output unit, least-squares fitting and 8 hidden units. The limits shown are ± 2 standard errors, obtained from the theory of non-linear regression (based on local linearization: [25]) since this is a non-linear regression problem. Clearly there is over-fitting, and as Figure 2 shows, there are many local minima of the fitting function. (These *are* local minima; the Hessian was confirmed to be positive definite.) Using weight decay provides one way to combat over-fitting; least squares is replaced by minimizing

$$\sum_{i=1}^{n} [y_i - f(x_i; \mathbf{w})]^2 + \lambda \sum_{ij} w_{ij}^2 = S(\mathbf{w}) + \lambda \|\mathbf{w}\|^2, \tag{1}$$

possibly omitting the 'bias' weights or giving them a proportionally lower penalty (as here).

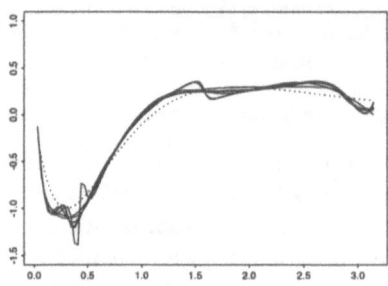

Figure 2: Curves from 1–8–1 neural networks fitted from different starting weights.

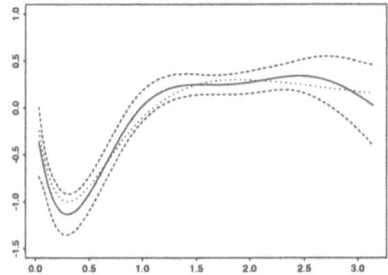 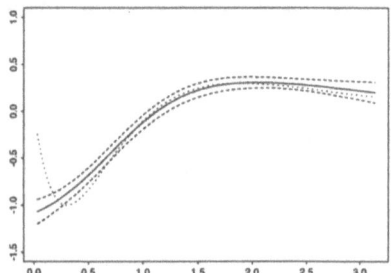

Figure 3: Curve fitted by 1–8–1 neural network, with ± 2 standard error bands and true curve. (a) $\lambda = 10^{-3}$ and (b) $\lambda = 0.1$.

Figure 3 shows the result of using weight decay. Without knowing the truth, we cannot be sure which part is correct. These weight decay figures were not picked from thin air, as a Bayesian argument ([22, 24]) gives some guidance, suggesting $(0.002–0.2)\sigma^2$ for a regression network. When weight decay is used, multiple minima can still occur, although none were found here.

The effect of weight decay is to reduce the variability of the fit, at the cost of bias, since the fitted curve will be smoother than the true curve. Model selection is choosing a point on the bias/variance compromise (in the terminology of [6]). Thus it is surprising that the standard errors in Figure 1(b) are tighter than those in Figure 3(a). In fact the former are optimistic: the local linearization breaks down too quickly, and in any case does not take into account the other local minima.

We now have two ways to control the smoothness of the fit, the number of hidden units and λ. The two interact, as for moderate values of λ the number of hidden units has very little effect on the fitted function. (We might do better to use a direct *regularization* penalty, such as the $\lambda \int (f'')^2$ used in spline smoothing, see [2].) How do we choose the degree of smoothness from the data alone?

2 Cross-validation, NIC and p_{eff}

Cross-validation ([27]) is perhaps the most widely used method to select architectural parameters, but the term is often mis-used in the neural networks field. In the language of pattern recognition ([24]), the n available examples may be divided into training, validation and test sets. The training set is used to train the network, that is to choose its parameters (weights). The performance on the validation set is used to select one of a competing family of networks (or a combination of them). Finally, the test set is used to measure the performance of the selected network(s). Note that the validation set must be distinct from the test set for the assessed performance to be valid.

Since examples are normally not abundant (and large validation and test sets are needed to measure performance well), cross-validation divides the available samples up and re-uses them. In its simplest form (V-fold CV), the examples are divided into V roughly comparable sets, and each is used in turn as a validation set with the remainder used as the training set. This allows every example to be used for validation. Leave-one-out cross-validation (LOOCV) takes $V = n$, so predicts each example from the network trained on the remaining examples. We can use cross-validation as a test-set method for one network, but if we wish to both validate and test we need a double layer of cross-validation.

Cross-validation can be very CPU consuming for neural networks (especially LOOCV). The presence of local minima causes difficulties; we can attempt to track local minima by starting at a solution to the full problem ([14, 15]), but removing one outlying point can change dramatically the structure of the local minima.

References [28, 12, 13, 16, 17, 18] take another approach, a generalization of Akaike's AIC criterion. The *deviance* is minus twice the log-likelihood; in a regression problem with known σ^2 this is the least-squares measure S divided by σ^2. (If σ^2 is *not* known, the deviance is $n \log(S/n)$.) Then

$$\text{AIC} = \text{deviance} + 2p$$

where p is the total number of parameters. This is only appropriate for maximum likelihood (here least-squares) fitting; when a regularizer such as weight decay is used, p

must be replaced by $p^* = \text{trace}[KJ^{-1}]$ where

$$J = -\text{E}\frac{\partial^2 g(X_i, \theta_0)}{\partial\theta\,\partial\theta^T} \qquad \text{and} \qquad K = \text{Var}\frac{\partial g(X_i, \theta_0)}{\partial\theta}.$$

Here the averages are done under the true distribution, θ_0 is the 'least false' parameter (the best possible fit to the true distribution) and g is the log density plus the regularization divided by n. (This follows from standard statistical theory; [7, 24]. Without regularization $p^* = p$.) These quantities can be estimated from the derivatives and the Hessian evaluated at the fit to the training set, replacing θ_0 by $\hat{\theta}$ and expectations by averages over the training set. Moody's p_{eff} is p^*, and this version of AIC is called NIC. The underlying idea of NIC/AIC is to estimate the deviance for a test set of size n, compensating for the fact that the weights were chosen to fit the training set.

Stone in [28] proved (but did not actually state) that for large training sets LOOCV and NIC are equivalent, and this seems to work well in practice. Thus NIC gives the effect of LOOCV at much less computational effort. However, the theory of NIC relies on a single well-defined minimum to the fitting function, and it can be unreliable when there are several local minima. Leave-one-out cross-validation will also often not leave a single local minimum, and makes order $1/n$ perturbations, which are too small to assess the fit accurately. In consequence, LOOCV has the reputation of being very variable, and 10-fold cross-validation is usually a better guide.

Size	λ	SSq	SSq$_{CV}$	p^*	NIC	Bayes
2	0	3.47	4.29	7	4.03	—
2	10^{-3}	3.57	4.78	5	3.98	3.30
3	0	4.12	4.81	10	4.92	—
3	10^{-3}	3.40	4.54	6	3.93	3.30
3	10^{-2}	3.61	4.91	6	4.11	3.72
3	10^{-1}	5.40	6.20	6	5.94	6.08
5	0	3.14	4.57	16	4.42	—
5	10^{-3}	3.21	4.51	17	4.59	3.38
5	10^{-2}	3.61	4.91	6	4.11	4.11
5	10^{-1}	5.38	6.18	6	5.92	5.98
8	0	2.89	5.35	25	4.89	—
8	10^{-3}	3.21	4.54	18	4.67	4.74
8	10^{-2}	3.58	4.94	7	4.16	4.63
8	10^{-1}	5.36	6.14	7	5.90	5.83

The table shows various quantities for a range of fits to the data of Figure 1(a). 'SSq' is the fitted sum of squares, and 'SSq$_{CV}$' is the 10-fold CV prediction of the fit to another dataset. (Clearly for 3 hidden units and $\lambda = 0$ the optimizer got stuck in a local minimum.) The table shows that NIC and SSq$_{CV}$ are usually similar, and that

p^* can be much less than p and is effectively controlled by λ. Figure 4 shows that there are only small differences between the models with small NIC (and small SSq_{CV}).

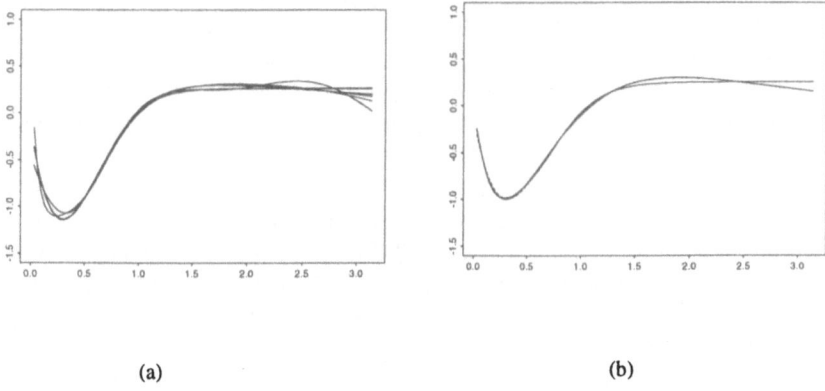

(a)   (b)

Figure 4: (a) The fitted curves for all the well-supported models. (b) Curves fitting these models to the true curve (with no added noise). Only the two hidden-unit model shows bias.

3   Bayesian model selection

In the Bayesian framework ([5]) it has been established for many years (at least since the 1948 edition of [8]) that the way to choose models is via the *Bayes factor*, (ratios of) the probability density of the data given the model. We have a prior density $p(\theta)$ over the parameters (weights) which must be integrable. Weight decay corresponds to an assumption of independent normal distributions for the weights (as is well-known in the context of ridge regression).

Finding the Bayes factor implies an integration over weight space which is normally impossible, but good approximations are known ([4, 5, 24]). In fitting we will maximize $E(\theta) = \log \ell(\theta; \text{data}) + \log p(\theta)$. One approximation is

$$\log P(\mathcal{T} \mid \text{model}) \approx E(\widehat{\theta}) + \frac{p}{2} \log 2\pi - \frac{1}{2} \log |H| \tag{2}$$

where H is the Hessian of E. This is easily computed from the fit.

For a regression problem, $\log \ell(\widehat{\theta}; \text{data})$ is essentially the sum of squares divided by $2\sigma^2$, so (2) was converted into a corrected sum of squares, shown in the table. The agreement with the other methods is good, although no results are available without weight decay.

In this approach there is no difficulty with multiple local minima, which are combined by integration over the weight space (although (2) will need to be summed about each local minimum).

Markov chain Monte Carlo methods ([1]) have been used to explore the model space and so compute Bayes factors, for example in [26, 11].

4 Model averaging

Bayesians have long known that the logical endpoint of their theory is to average the predictions of their models, using the Bayes factors as weights, and this has become computationally feasible recently ([26, 10, 4]). In particular, rather than choosing a hyperparameter such as λ, we should integrate over a proper distribution. This is feasible with a coarse grid for λ. It also insists that we average over the predictions from local minima, computing the weights from (2).

Model averaging has been discussed in non-Bayesian contexts, for example Wolpert's [30] *stacked generalization*. It seems to have been missed that this was part of Stone's first paper on cross-validation ([27], pp. 126–7). His suggestion was to take a convex combination of the predictions of several models, using cross-validation to choose the weights of the combination. Averaging (often unweighted) has often been proposed for neural networks (e.g. [9, 19]).

There is little argument that model averaging is a good idea for predictions (and neural networks have almost no explanatory purpose). The main argument against is computational; for many pattern recognition problems it is difficult to get one network to predict fast enough, let alone many. But if it is computationally feasible, it has been a long established practice in applied statistics to average over a few models. The Bayesian view averages over all models, even those which do not fit well, and this has been applied ([26]), but Madigan & Raftery in [10] advocate an "Occam's window" to exclude all but models with good fits, and to exclude models which subsume others that are included. Thus only the simplest models which fit well are included in their averaging process.

References

[1] Besag, J., Green, P., Higdon, D. and Mengersen, K. (1995) Bayesian computation and stochastic systems. *Statistical Science* 1995.

[2] Bishop, C. Improving the generalization properties of radial basis function neural networks. *Neural Computation* 1991; 3:579–588.

[3] Cheng, B, and Titterington, D. M. Neural networks: a review from a statistical perspective (with discussion). *Statistical Science* 1994; 9:2–54.

[4] Draper, D. Assessment and propagation of model uncertainty (with discussion). *Journal of the Royal Statistical Society series B* 1995; 57:45–97.

[5] Gelman, A., Carlin, J. B., Stern, H. S. and Rubin, D. B. *Bayesian Data Analysis.* Chapman & Hall, New York, 1995.

[6] Geman, S., Bienenstock, E. and Doursat, R. Neural networks and the bias/variance dilemma. *Neural Computation* 1992; 4:1–58.

[7] Huber, P. J. The behavior of maximum likelihood estimates under nonstandard conditions. In: Le Cam, L. M. and Neyman, J. (eds) *Proceedings of the Fifth Berkeley*

Symposium on Mathematical Statistics and Probability. University of California Press, Berkeley, 1967, 1:221–233

[8] Jeffreys, H. *Theory of Probability.* Third edition. Clarendon Press, Oxford, 1961.

[9] Lincoln, W. P. and Skrzypek, J. Synergy of clustering multiple backpropagation networks. In: Touretzky, D. S. (ed) *Advances in Neural Information Processing Systems 2.* Morgan Kaufmann, San Mateo, CA, 1990, pp. 650–657.

[10] Madigan, D. and Raftery, A. E. Model selection and accounting for model uncertainty in graphical models using Occam's window. *Journal of the American Statistical Association* 1994; 89:1535–1546.

[11] Madigan, D. and York, J. Bayesian graphical models for discrete data. Technical report 239, Department of Statistics, University of Washington, 1993.

[12] Moody, J. E. Note on generalization, regularization and architecture selection in nonlinear learning systems. In *First IEEE-SP Workshop on Neural Networks in Signal Processing.* IEEE Computer Society Press, 1991, pp. 1–10.

[13] Moody, J. E. The *effective* number of parameters: an analysis of generalization and regularization in nonlinear learning systems. In: Moody, J. E., Hanson, S. J. and Lippmann, R. P. (eds) *Advances in Neural Information Processing Systems 4.* Morgan Kaufmann, San Mateo, CA, 1992, pp. 847–854.

[14] Moody, J. and Utans, J. Principled architecture selection for neural networks: Application to corporate bond rating prediction. In: Moody, J. E., Hanson, S. J. and Lippmann, R. P. (eds) *Advances in Neural Information Processing Systems 4.* Morgan Kaufmann, San Mateo, CA, 1992, pp. 683–690.

[15] Moody, J. and Utans, J. Architecture selection strategies for neural networks: Application to corporate bond rating prediction. In: Refenes, A.-P. (ed) *Neural Networks in the Capital Markets.* Wiley, Chichester, 1995, pp. 277–300.

[16] Murata, N., Yoshizawa, S. and Amari, S. A criterion for determining the number of parameters in an artificial neural network model. In: Kohonen, T., Mäkisara, K., Simula, O. and Kangas, J. (eds) *Artificial Neural Networks.* North Holland, Amsterdam, 1991, pp. 9–14.

[17] Murata, N., Yoshizawa, S. and Amari, S. Learning curves, model selection and complexity of neural networks. In: Hanson, S. J., Cowan, J. D. and Giles, C. L. (eds) *Advances in Neural Information Processing Systems 5.* Morgan Kaufmann, San Mateo, CA, 1993, pp. 607–614.

[18] Murata, N., Yoshizawa, S. and Amari, S. Network information criterion determining the number of hidden units for artificial neural network models. *IEEE Transactions on Neural Networks* 1994; 5:865–872.

[19] Perrone, M. P. and Cooper, L. N. When networks disagree: Ensemble methods for hybrid neural networks. In: Mammone, R. J. (ed) *Artificial Neural Networks for Speech and Vision.* Chapman & Hall, London, 1993, pp. 126–142.

[20] Ripley, B. D. Statistical aspects of neural networks. In Barndorff-Nielsen, O. E., Jensen, J. L. and Kendall, W. S. (eds) *Networks and Chaos – Statistical and Probabilistic Aspects.* Chapman & Hall, London, 1993, pp. 40–123

[21] Ripley, B. D. Neural networks and related methods for classification (with discussion). *Journal of the Royal Statistical Society series B* 1994; 56:409–456

[22] Ripley, B. D. Neural networks and flexible regression and discrimination. In: Mardia, K. V. (ed) *Statistics and Images.* Carfax, Abingdon, 1994, pp. 39-57 (Advances in Applied Statistics 2).

[23] Ripley, B. D. Flexible non-linear approaches to classification. In: Cherkassky, V., Friedman, J. H. and Wechsler, H. (eds) *From Statistics to Neural Networks. Theory and Pattern Recognition Applications.* Springer, Berlin, 1994, pp. 105–126.

[24] Ripley, B. D. *Pattern Recognition and Neural Networks.* Cambridge University Press, Cambridge, 1995.

[25] Seber, G. A. F. and Wild, C. J. *Nonlinear Regression.* Wiley, New York, 1989.

[26] Stewart, L. Hierarchical Bayesian analysis using Monte Carlo integration: computing posterior distributions when there are many possible models. *The Statistician* 1987; 36:211–219.

[27] Stone, M. Cross-validatory choice and assessment of statistical predictions (with discussion). *Journal of the Royal Statistical Society series B* 1974; 36:111–147.

[28] Stone, M. An asymptotic equivalence of choice of model by cross-validation and Akaike's criterion. *Journal of the Royal Statistical Society series B* 1977; 39:44–47.

[29] Wahba, G. and Wold, S. A completely automatic French curve. *Communications in Statistics* 1975; 4:1–17.

[30] Wolpert, D. H. Stacked generalization. *Neural Networks* 1992; 5:241–259.

Developments in Probabilistic Modelling with Neural Networks—Ensemble Learning

David J.C. MacKay

Department of Physics, Cambridge University

Madingley Road, Cambridge CB3 0HE, United Kingdom

mackay@mrao.cam.ac.uk

Abstract

Ensemble learning by variational free energy minimization is a framework for statistical inference in which an ensemble of parameter vectors is optimized rather than a single parameter vector. The ensemble approximates the posterior probability distribution of the parameters.

In this paper I give a review of ensemble learning using a simple example.

1 Ensemble Learning by Free Energy Minimization

A new tool has recently been introduced into the field of neural networks. In traditional approaches to model fitting, a single parameter vector $\mathbf{w}$ is optimized by, say, maximum likelihood or penalized maximum likelihood; in the Bayesian interpretation, these optimized parameters are viewed as defining the mode of a posterior probability distribution $P(\mathbf{w}|D, \mathcal{H})$ (given data D and model assumptions $\mathcal{H}$).

The new concept introduced by Hinton and van Camp (1993) is to work in terms of an approximating *ensemble* $Q(\mathbf{w}; \theta)$, that is, a probability distribution over the parameters, and optimize the ensemble (by varying its own parameters θ) so that it approximates the posterior distribution of the parameters $P(\mathbf{w}|D, \mathcal{H})$ well. The objective function chosen to measure the quality of the approximation is a *variational free energy*,[1]

$$F(\theta) = - \int d^k \mathbf{w} \, Q(\mathbf{w}; \theta) \log \frac{P(D|\mathbf{w}, \mathcal{H}) P(\mathbf{w}|\mathcal{H})}{Q(\mathbf{w}; \theta)}. \tag{1}$$

The numerator $P(D|\mathbf{w}, \mathcal{H}) P(\mathbf{w}|\mathcal{H})$ is, within a multiplicative constant, equal to the posterior probability $P(\mathbf{w}|D, \mathcal{H}) = P(D|\mathbf{w}, \mathcal{H}) P(\mathbf{w}|\mathcal{H}) / P(D|\mathcal{H})$. So the free energy $F(\theta)$ can be viewed as the sum of $- \log P(D|\mathcal{H})$ and the Kullback-Leibler divergence between $Q(\mathbf{w}; \theta)$ and $P(\mathbf{w}|D, \mathcal{H})$. $F(\theta)$ is bounded below by $- \log P(D|\mathcal{H})$ and only attains this value for $Q(\mathbf{w}; \theta) = P(\mathbf{w}|D, \mathcal{H})$. For

---

[1] Variational free energy minimization is a well-established tool for the approximation of probability distributions in statistical physics (Feynman 1972). The free energy can also be described in terms of description lengths, as in Hinton and van Camp (1993).

certain models and certain approximating distributions, this free energy, and its derivatives with respect to the ensemble's parameters, can be evaluated.

Hinton and van Camp (1993) considered a regression network with one non-linear hidden layer and showed that a *separable* Gaussian approximating distribution $Q(\mathbf{w}; \theta)$ can be optimized with a deterministic algorithm.

Hinton and Zemel (1994) have applied the same approach to the optimization of an autoencoder. The hidden-to-output part of an autoencoder is viewed as defining a generative model employing latent variables that live in the hidden layer of the model. The optimization of such a generative model is challenging, requiring, for every given data example, an implicit or explicit computation of the posterior probability distribution P of the latent variables. Hinton and Zemel (1994) view the input-to-hidden 'recognition' part of the autoencoder as defining an approximating distribution Q for this distribution P. A single objective function F can then be defined for simultaneous optimization of the generative model and the recognition model. The Helmholtz machine (Dayan *et al.* 1995) is a further generalization of these ideas.

In a broader statistical context, Neal and Hinton (1993) have shown that it is possible to view the Expectation-Maximization (EM) algorithm in terms of a free energy minimization. The Bayesian (ML II) approach to the optimization of hyperparameters in a hierarchical model (reviewed in (MacKay 1992)) can also be derived as a free energy minimization (MacKay 1995a). The deterministic Boltzmann machine can be derived as a free energy approximation to the Boltzmann machine (Radford Neal, personal communication). And MacKay (1995b) has obtained an algorithm for decoding certain binary codes by variational free energy minimization.

2   Inferring a Gaussian distribution

For background reading on Bayesian methods, the textbook of Box and Tiao (1973) is recommended.

The popular one-dimensional Gaussian distribution is parameterized by a mean μ and a standard deviation σ:

$$P(x|\mu, \sigma) = \frac{1}{\sqrt{2\pi}\sigma} \exp\left(-\frac{(x-\mu)^2}{2\sigma^2}\right) \equiv \text{Normal}(x; \mu, \sigma^2). \tag{2}$$

Let us examine the inference of μ and σ given data points x_n, $n = 1 \ldots N$, assumed to be drawn independently from this distribution. When inferring these parameters, we must specify their prior distribution. This gives us the opportunity to include specific knowledge that we have about μ and σ (from independent experiments, or on theoretical grounds, for example). If we have no such knowledge, then we can construct an appropriate prior which embodies our supposed ignorance. In either case, it may be appropriate to consider *conjugate priors*; these are priors which have a functional form which integrates naturally with data measurements, making the inferences have an analytically convenient form. The conjugate prior for a mean μ is a Gaussian, $P(\mu|\mu_0, \sigma_\mu) = \text{Normal}(\mu; \mu_0, \sigma_\mu)$. In the limit $\mu_0 = 0, \sigma_\mu \to \infty$, we obtain the *noninformative prior* for a location parameter, the flat prior. This is 'noninformative' because it is *invariant* under the reparameterization $\mu' = \mu + c$. The prior $P(\mu) = \text{const.}$ is also an *improper* prior, that is, it is not normalizable.

The conjugate prior for a standard deviation σ is a gamma distribution, conveniently defined in terms of the inverse variance $\beta = 1/\sigma^2$:

$$P(\beta) = \Gamma(\beta; b_\beta, c_\beta) = \frac{1}{\Gamma(c_\beta)} \frac{\beta^{c_\beta - 1}}{b_\beta^{c_\beta}} \exp\left(-\frac{\beta}{b_\beta}\right), 0 \leq \beta < \infty \tag{3}$$

This is a simple peaked distribution with mean $b_\beta c_\beta$ and variance $b_\beta^2 c_\beta$. In the limit $b_\beta c_\beta = 1, c_\beta \to 0$, we obtain the noninformative prior for a scale parameter, the $1/\sigma$ prior. This is 'noninformative' because it is invariant under the reparameterization $\sigma' = c\sigma$. The $1/\sigma$ prior is less strange looking if we examine the resulting density over $\log \sigma$, or $\log \beta$, which is flat. This is the prior that expresses ignorance about σ by saying 'well, it could be 10, or it could be 1, or it could be 0.1, ...' Scale variables such as σ are usually best represented in terms of their logarithm. Again, this noninformative prior is improper.

In the following examples, I will use the improper priors for μ and σ.

2.1 Maximum likelihood and marginalization: σ_N and σ_{N-1}

The task of inferring the mean and standard deviation of a Gaussian distribution from N samples is a familiar one, though maybe not everyone understands the difference between the σ_N and σ_{N-1} buttons on their calculator. Let us recap the formulae, then derive them.

Given data $D = \{x_n\}_{n=1}^N$, an 'estimator' of μ is

$$\bar{x} \equiv \sum_{n=1}^N x_n / N, \tag{4}$$

and two estimators of σ are:

$$\sigma_N \equiv \sqrt{\frac{\sum_{n=1}^N (x_n - \bar{x})^2}{N}} \text{ and } \sigma_{N-1} \equiv \sqrt{\frac{\sum_{n=1}^N (x_n - \bar{x})^2}{N-1}} \tag{5}$$

There are two principal paradigms for statistics: sampling theory and Bayesian inference. In sampling theory (also known as 'frequentist' or orthodox statistics), one invents 'estimators' of quantities of interest and then chooses between those estimators using some criterion measuring their sampling properties; there is no clear principle for deciding which criterion to use to measure the performance of an estimator; nor, for most criteria, is there any systematic procedure for the construction of optimal estimators. In Bayesian inference, in contrast, once we have made explicit all our modelling assumptions, our inferences are mechanistic. Whatever question we wish to pose, the rules of probability theory give a unique answer which consistently takes into account all the given information. Human-designed estimators and confidence intervals have no role in Bayesian inference; human input only enters into the important tasks of designing the hypothesis space, and implementing inference in that space. The answers to our questions are probability distributions over the quantities of interest. We often find that the estimators of sampling theory emerge automatically as modes or means of these posterior distributions when we turn the handle of Bayesian inference.

In sampling theory, the estimators above can be motivated as follows. $\bar{x}$ is an unbiased estimator of μ which, out of all the possible unbiased estimators of μ, has smallest variance (where this variance is computed by averaging over an ensemble of fictitious experiments in which the data samples are assumed to come from an unknown Gaussian distribution). The estimator $(\bar{x}, \sigma_N)$ is the maximum likelihood estimator for (μ, σ). The estimator σ_N is *biased*, however: the expectation of σ_N, given σ, averaging over many imagined experiments, is not σ. This motivates the invention of σ_{N-1} which can be shown to be an unbiased estimator. Or to be precise, it is σ_{N-1}^2 which is an unbiased estimator of σ^2.

We now look at some Bayesian inferences for this problem, assuming non-informative priors for μ and σ. The emphasis is thus not on the priors, but rather on (a) the likelihood function, and (b) the concept of marginalization. The joint posterior probability of μ and σ is proportional to the likelihood function illustrated by a contour plot in figure 1a. The log likelihood is:

$$\log P(\{x_n\}_{n=1}^N | \mu, \sigma) = -N \log(\sqrt{2\pi}\sigma) - \sum_n (x_n - \mu)^2/(2\sigma^2), \qquad (6)$$

$$= -N \log(\sqrt{2\pi}\sigma) - [N(\mu - \bar{x})^2 + S]/(2\sigma^2), \qquad (7)$$

where $S \equiv \sum_n (x_n - \bar{x})^2$. Given the Gaussian model, the likelihood can be expressed in terms of the two functions of the data $\bar{x}$ and S, so these two quantities are known as 'sufficient statistics'. The posterior probability of μ and σ is, using the improper priors:

$$P(\mu, \sigma | \{x_n\}_{n=1}^N) = \frac{P(\{x_n\}_{n=1}^N | \mu, \sigma) P(\mu, \sigma)}{P(\{x_n\}_{n=1}^N)} \qquad (8)$$

$$= \frac{\frac{1}{(2\pi\sigma^2)^{N/2}} \exp\left(-\frac{N(\mu-\bar{x})^2 + S}{2\sigma^2}\right) \frac{1}{\sigma_\mu} \frac{1}{\sigma}}{P(\{x_n\}_{n=1}^N)} \qquad (9)$$

This function describes the answer to the question, 'given the data, and the noninformative priors, what might μ and σ be?' It may be of interest to find the parameter values that maximize the posterior probability (though it should emphasized that posterior probability maxima have no fundamental status in Bayesian inference). Differentiating the log likelihood with respect to μ and $\log \sigma$ we find the maximum likelihood solution: $\{\mu, \sigma\}_{\mathrm{ML}} = \left\{\bar{x}, \sigma_N = \sqrt{S/N}\right\}$.

There is more to the posterior distribution than just its mode. As can be seen in figure 1a, the likelihood has a skew peak. As we increase σ, the width of the conditional distribution of μ increases. And if we fix μ to a sequence of values moving away from the sample mean $\bar{x}$, we obtain a sequence of conditional distributions over σ whose maxima move to increasing values of σ.

The next question we might ask is 'given the data, and the noninformative prior on μ, and assuming a particular value of σ, what might μ be?'

The posterior probability of μ given σ is

$$P(\mu | \{x_n\}_{n=1}^N, \sigma) = \frac{P(\{x_n\}_{n=1}^N | \mu, \sigma) P(\mu)}{P(\{x_n\}_{n=1}^N | \sigma)} \qquad (10)$$

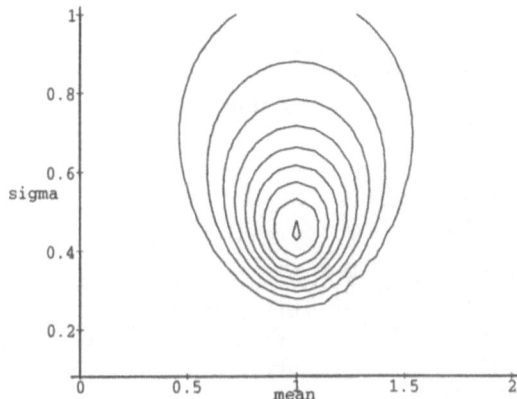

a)

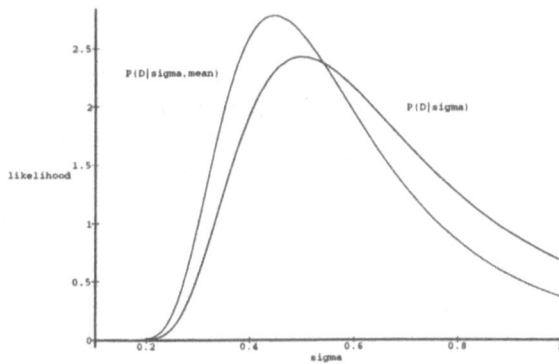

b)

Figure 1: **The likelihood function for the parameters of a Gaussian distribution.**
a) Contour plot of the log likelihood as a function of μ and σ. The data set of $N = 5$ points had mean $\bar{x} = 1.0$ and $S^2 = \sum(x - \bar{x})^2 = 1.0$. Notice that the maximum is skew in σ. The two estimators of standard deviation have values $\sigma_N = 0.45$ and $\sigma_{N-1} = 0.50$.
b) The two graphs show: the likelihood as a function of σ, with μ fixed to $\bar{x}$, i.e., $P(D|\mu = \bar{x}, \sigma)$ [this is a vertical section through the peak in (a)]; and the 'evidence' (marginalized likelihood) for σ, $P(D|\sigma)$, assuming a flat prior on μ (rescaled by an arbitrary constant). The evidence is obtained by projecting the probability mass in (a) onto the σ axis. The maximum of $P(D|\mu = \bar{x}, \sigma)$ is at σ_N. The maximum of $P(D|\sigma)$ is at σ_{N-1}.

$$\propto \quad \exp(-N(\mu - \bar{x})^2/(2\sigma^2)) \tag{11}$$

$$= \quad \text{Normal}(\mu; \bar{x}, \sigma^2/N). \tag{12}$$

We note the familiar $\sigma/\sqrt{N}$ scaling of the error bars on μ.

Let us now ask the question 'given the data, and the noninformative priors, what might σ be?' This question differs from the first one we asked in that we are now not interested in μ. This parameter must therefore be *marginalized* over. The posterior probability of σ is:

$$P(\sigma|\{x_n\}_{n=1}^N) = \frac{P(\{x_n\}_{n=1}^N|\sigma)P(\sigma)}{P(\{x_n\}_{n=1}^N)}. \tag{13}$$

The data-dependent term $P(\{x_n\}_{n=1}^N|\sigma)$ appeared earlier as the normalizing constant in equation (10); one name for this quantity is the 'evidence', or marginalized likelihood, for σ. We obtain the evidence for σ by integrating out μ; a noninformative prior $P(\mu) = 1/\sigma_\mu$ is assumed. The Gaussian integral, $P(\{x_n\}_{n=1}^N|\sigma) = \int P(\{x_n\}_{n=1}^N|\mu, \sigma)P(\mu)\,d\mu$, yields:

$$\log P(\{x_n\}_{n=1}^N|\sigma) = -N \log(\sqrt{2\pi}\sigma) - \frac{S}{2\sigma^2} + \log\frac{\sqrt{2\pi}\sigma/\sqrt{N}}{\sigma_\mu}. \tag{14}$$

The first two terms are the best fit log likelihood (*i.e.*, the log likelihood with $\mu = \bar{x}$). The last term is the log of the 'Occam factor' which penalizes smaller values of σ. When we differentiate the log evidence with respect to $\log \sigma$, to find the most probable σ, the additional volume factor $(\sigma/\sqrt{N})$ shifts the maximum from σ_N to

$$\sigma_{N-1} = \sqrt{S/(N-1)} \tag{15}$$

Intuitively, the denominator $(N-1)$ counts the number of noise measurements contained in the quantity $S = \sum_n (x_n - \bar{x})^2$. The sum contains N residuals-squared, but there are only $(N-1)$ effective noise measurements because the determination of one parameter μ from the data causes one dimension of noise to be gobbled up in unavoidable over-fitting. Figure 1b shows the marginalized likelihood as a function of σ along with the likelihood as a function of σ with μ fixed to its most probable value, $\bar{x}$.

The final inference we might wish to make is 'given the data, what is μ?' To answer this, we marginalize over σ and obtain the posterior marginal distribution of μ, which is a Student t-distribution:

$$P(\mu|D) \propto 1/\left(N(\mu - \bar{x})^2 + S\right)^{(N-1)/2}. \tag{16}$$

3 An Approximating Ensemble

I now illustrate the concept of ensemble learning by free energy minimization by fitting an approximating ensemble $Q(\mu, \sigma)$ to the posterior distribution (8-9). Let us make the single assumption that the approximating ensemble is separable in the form $Q(\mu, \sigma) = Q_\mu(\mu)Q_\sigma(\sigma)$. No restrictions on the functional form of $Q_\mu(\mu)$ and $Q_\sigma(\sigma)$ are made.

We write down a variational free energy,

$$F(Q) = -\int d\mu \, d\sigma \, Q_\mu(\mu) Q_\sigma(\sigma) \log \frac{P(D|\mu,\sigma)P(\mu,\sigma)}{Q_\mu(\mu)Q_\sigma(\sigma)}. \tag{17}$$

We can find the optimal separable distribution Q by considering separately the optimization of F over $Q_\mu(\mu)$ for fixed $Q_\sigma(\sigma)$, and then the optimization of $Q_\sigma(\sigma)$ for fixed $Q_\mu(\mu)$.

3.1   Optimization of $Q_\mu(\mu)$

As a functional of $Q_\mu(\mu)$, F is:

$$
\begin{aligned}
F &= -\int d\mu \, Q_\mu(\mu) \left[\left[\int d\sigma \, Q_\sigma(\sigma) \log P(D|\mu,\sigma) + \log[P(\mu)/Q(\mu)] \right] + \text{const.} \right. \\
&= \int d\mu \, Q_\mu(\mu) \left[\left[\int d\sigma \, Q_\sigma(\sigma) N\beta \frac{1}{2}(\mu - \bar{x})^2 + \log Q(\mu) \right] + \text{const.}' \right.
\end{aligned}
$$

The dependence on Q_σ thus collapses down to a dependence simply on the mean $\bar{\beta} \equiv \int d\sigma \, Q_\sigma(\sigma) 1/\sigma^2$.

Now we can recognize the function $-N\bar{\beta}\frac{1}{2}(\mu - \bar{x})^2$ as the log of a Gaussian identical to the posterior distribution for a particular value of $\beta = \bar{\beta}$. Since a divergence $\int Q \log(Q/P)$ is minimized by setting $Q = P$, we can immediately write down the distribution $Q_\mu^{\text{opt}}(\mu)$ that minimizes F for fixed Q_σ:

$$Q_\mu^{\text{opt}}(\mu) = P(\mu|D, \bar{\beta}, \mathcal{H}) = \text{Normal}(\mu; \bar{x}, \sigma^2_{\mu|D}). \tag{18}$$

where $\sigma^2_{\mu|D} = 1/(N\bar{\beta})$.

3.2   Optimization of $Q_\sigma(\sigma)$

As a functional of $Q_\sigma(\sigma)$, F is (neglecting additive constants):

$$
\begin{aligned}
F(Q) &= -\int d\sigma \, Q_\sigma(\sigma) \left[\int d\mu \, Q_\mu(\mu) \log P(D|\mu,\sigma) + \log[P(\sigma)/Q_\sigma(\sigma)] \right] \\
&= \int d\sigma \, Q_\sigma(\sigma) \left[(N\sigma^2_{\mu|D} + S)\beta/2 - (\tfrac{N}{2} - 1) \log \beta + \log Q_\sigma(\sigma) \right]
\end{aligned}
$$

where the integral over μ is performed assuming $Q_\mu(\mu) = Q_\mu^{\text{opt}}(\mu)$. Here, the β-dependent expression in the brackets can be recognized as the log of a gamma distribution over β (see equation (3)), giving as the distribution that minimizes F for fixed Q_μ: $Q_\sigma^{\text{opt}}(\beta) = \Gamma(\beta; b', c')$, with $1/b' = \frac{1}{2}(N\sigma^2_{\mu|D} + S)$ and $c' = N/2$.

3.3   Joint optimum $Q_\mu(\mu)Q_\sigma(\sigma)$

We now have an implicit equation for the optimal approximating ensemble, with $\sigma^2_{\mu|D} = 1/(N\bar{\beta})$, and $\bar{\beta} = b'c'$. The solution is:

$$1/\bar{\beta} = S/(N - 1) \tag{19}$$

Thus we obtain, by ensemble learning, an approximation to the posterior that agrees nicely with the conventional estimators. The approximate posterior distribution over β is a gamma distribution with mean $\bar{\beta}$ corresponding to a variance of $\sigma^2 = S/(N-1) = \sigma_{N-1}^2$. And the approximate posterior distribution over μ is a Gaussian with mean $\bar{x}$ and standard deviation $\sigma_{N-1}/\sqrt{N}$.

Acknowledgements

I thank Radford Neal and Geoff Hinton for helpful discussions.

References

Box, G. E. P., and Tiao, G. C. (1973) *Bayesian inference in statistical analysis.* Addison–Wesley.

Dayan, P., Hinton, G. E., Neal, R. M., and Zemel, R. S. (1995) The Helmholtz machine. *Neural Computation.* to appear.

Feynman, R. P. (1972) *Statistical Mechanics.* W. A. Benjamin, Inc.

Hinton, G. E., and van Camp, D., (1993) Keeping neural networks simple by minimizing the description length of the weights. In: *Proceedings of COLT-93.*

Hinton, G. E., and Zemel, R. S. (1994) Autoencoders, minimum description length and Helmholtz free energy. In *Advances in Neural Information Processing Systems 6*, ed. by J. D. Cowan, G. Tesauro, and J. Alspector, San Mateo, California. Morgan Kaufmann.

MacKay, D. J. C. (1992) Bayesian interpolation. *Neural Computation* **4** (3): 415–447.

MacKay, D. J. C., (1995a) Ensemble learning and evidence maximization. submitted to NIPS*95.

MacKay, D. J. C. (1995b) Free energy minimization algorithm for decoding and cryptanalysis. *Electronics Letters* **31** (6): 446–447.

Neal, R. M., and Hinton, G. E. (1993) A new view of the EM algorithm that justifies incremental and other variants. *Biometrika.* submitted.

A Constructive Algorithm for Building a Feed-Forward Neural Network

C. Campbell, S. Coombes
Dept. of Engineering Mathematics, Bristol University
Bristol BS8 1TR, United Kingdom

A. Surkan
Dept. of Computer Science, University of Nebraska
Lincoln, Nebraska 68588, United States

Abstract

We propose an efficient procedure for constructing and training a feed-forward neural network with a single hidden layer. The procedure can also be extended to construct neural networks with binary-valued weights.

1 The Algorithm

In this paper we will outline a new constructive procedure for generating feed-forward neural networks with a single hidden layer. The networks generated have an economical structure and generalise well. The network performs binary classification for input vectors with binary-valued components.

Let us consider a neural network with N input nodes labeled by index j and one output node. Suppose we wish to map inputs ξ_j^μ onto a set of targets η^μ, where μ is the pattern index and η^μ has components ± 1. Weights leading from input j to a hidden node i will be denoted W_{ij}. If we use a ± 1 updating function for the hidden nodes then an input vector S_j will give an internal representation $S_i = sign(\sum_j W_{ij} S_j - T_i)$ where T_i is the threshold at hidden node i. We will define the $sign$-function as having an output of $+1$ if its argument is greater than or equal to zero and -1 otherwise.

For binary classification tasks the patterns belong to two sets: patterns with target $\eta^\mu = 1$ (the set P^+) and those with target $\eta^\mu = -1$ (the set P^-). For binary inputs (quantised ± 1) it is always possible to find a set of weights and thresholds which will correctly store *all* the patterns belonging to one of these sets and at least one member belonging to the other set [1]. For example, suppose pattern $\mu = 1$ has target $+1$. If we use weights $W_{ij} = \xi_j^1$ and a threshold $T_i = N$ then $S_i = sign(\sum_j W_{ij} S_j - T_i)$ gives an output $S_i = +1$ if S_j is equal to ξ_j^1 and -1 otherwise. Usually it is possible to do better than this and store a number of members of P^+ in addition to all the P^-. A set of weights and thresholds which correctly store all the P^+ patterns and some of the P^- will be said to induce a $\oplus$-dichotomy while a $\ominus$-dichotomy will correspond to correct storage of all the P^- patterns and some of the P^+. This idea of a $\oplus$ or $\ominus$-dichotomy of the training set has been used by several authors [1, 4] to develop constructive learning techniques. In this paper we propose a new constructive algorithm for binary classification which is a variant on the constructive procedures proposed in [4]. For the latter algorithms all the hidden nodes performed dichotomies of the training set whereas in this algorithm the procedure starts (step **1**) without a dichotomy and then grows a set of further hidden nodes performing $\oplus$ and $\ominus$- dichotomies. This leads to a more economical architecture compared to the algorithms outlined in [4].

To generate a feed-forward neural network with a single hidden layer we proceed as follows. Let P_i^+ and P_i^- be the pattern sets at hidden node i then:

1. We attempt to split the training data with a separating hyperplane (weights and a threshold) using any algorithm for single-layered learning, for example the Minover algorithm (see below). If the dataset is linearly separable (all patterns stored) we terminate, otherwise the weights and threshold for the best solution (storing most patterns) are retained as the weights and threshold for the first hidden node ($i = 1$).

2. We construct a set of hidden nodes inducing $\oplus$-dichotomies. For hidden node i the training set P_i^+ is equal to the original P^+ whereas P_i^- only consists of members of P^- previously unstored either by the hidden node constructed in **1** or at earlier hidden nodes inducing a $\oplus$-dichotomy. We iterate this step until all patterns belonging to P^- are stored.

3. Similarly we construct a set of hidden nodes inducing $\ominus$-dichotomies. For hidden node i the training set P_i^- is equal to the original P^- whereas P_i^+ only consists of members of P^+ previously unstored either at the first hidden node constructed in **1** or at earlier hidden nodes inducing a $\ominus$-dichotomy. We iterate this step until all patterns belonging to P^+ are stored.

4. We use a threshold at the output node equal to the difference between the number of hidden nodes inducing $\oplus$-dichotomies and the number inducing $\ominus$-dichotomies.

To implement this algorithm we need an efficient procedure for obtaining the dichotomies at each hidden node. To obtain a $\oplus$-dichotomy we use the following procedure:

(1) We use an algorithm (e.g. the Minover algorithm outlined below) to find a set weights W_{ij} between the input nodes j and hidden node i. If the pattern set is not linearly separable the algorithm attempts to find the best linearly separable subset storing most patterns.

(2) Using these weights we now calculate $m_i^\mu = \sum_j W_{ij} \xi_j^\mu$ for all the patterns belonging to P_i^+ and P_i^-. Among those patterns belonging to the set P_i^- we find the pattern with the largest value of m_i^μ. Suppose this is pattern $\mu = \lambda$ then we set the threshold at i equal to m_i^λ i.e. $T_i = m_i^\lambda$.

(3) For the set of patterns belonging to P_i^+ we then find if there are any patterns such that m^μ is less than or equal to m^λ i.e. $m_i^\mu \leq m_i^\lambda$. If there are no patterns in P_i^+ with m_i^μ less than or equal to m_i^λ then we have finished training the weights and thresholds leading into hidden node i and we proceed to step (5) below. However, if there are patterns in P_i^+ satisfying this inequality then we find that pattern in P_i^+ which has the smallest value of m^μ. Let us suppose this is pattern $\mu = \nu$.

(4) Among those vectors belonging to P_i^- with $m_i^\mu \geq m_i^\nu$ we find the pattern with the largest value of m^μ and assign it the value $+1$ (i.e. this pattern moves from P_i^- to P_i^+). With the new sets P_i^+ and P_i^- we return to step (1) to find a new set of weights and thresholds.

(5) We have now obtained a $\oplus$-dichotomy. For the remaining members of

P_i^- the sums $\sum_j W_{ij}\xi_j^\mu$ are less than the threshold T_i whereas for patterns belonging to P_i^+ these sums are greater than T_i. However, this dichotomy may not be the best solution and consequently we can proceed with further training to maximise the number of patterns in P_i^- which are stored correctly. To do this we record the number of P_i^- patterns which were correctly stored (and associated weights and thresholds). We then discard these correctly stored P_i^- patterns and use the unstored P_i^- and the original P_i^+ as our training set, repeating steps (1)-(4). Eventually we will exhaust the entire P_i^- set and we choose the solution which stored the largest number of P_i^- patterns as the set of weights and threshold for this hidden node. Step (5) was found to improve generalisation for some problems but for others (e.g. the Shift Detection problem below) improvement was minimal.

To obtain a $\ominus$-dichotomy we follow a very similar procedure. In step (2) we find the pattern with the smallest value of m_i^μ (for $\mu \in P_i^+$) and set the threshold T_i equal to this value of m^μ. If the pattern sets are not linearly separable we search through the P_i^+ to find the pattern which is least well stored, switch its target-value $+1 \rightarrow -1$ and iterate the sequence until a separation of the two sets is achieved.

Some patterns can lie in the hyperplanes found in the $\oplus$ and $\ominus$-dichotomies (for example in step (2) the pattern $\mu = \lambda$ lies in the hyperplane). Since we use the convention $sign(0) = +1$ it is necessary to offset the thresholds in step (2) of the $\oplus$-dichotomy by a positive quantity $T_i \rightarrow T_i + \delta$. For binary inputs quantised ± 1 and binary weights we can set $\delta = 1$, while for real weights δ should be a very small quantity.

Any algorithm for single-layered learning could be in step (1) (e.g. the Perceptron rule). In the simulations below we have used the Minover algorithm [2] i.e. to implement the mapping $\xi_j^\mu \rightarrow \eta_i^\mu$ we start with random weights and iteratively adjust the weights according to the following:

1. At iteration t find the pattern $\mu = \lambda$ such that $\eta_i^\mu \sum_j W_{ij}^t \xi_j^\mu$ is minimal.

2. If $\eta_i^\lambda \sum_j W_{ij}^t \xi_j^\lambda \leq 0$ then update the weights according to $W_{ij}^{t+1} = W_{ij}^t + \eta_i^\lambda \xi_j^\lambda$ and return to step 1, else stop.

Since the dataset may not be linearly separable we use a "pocket" version of Minover recording and subsequently using the best solution storing most patterns. This rule can also be used to find solutions with binary-valued weights (quantised ± 1) which can exhibit superior generalisation for Boolean problems [4] (we illustrate this improvement with the Mirror Symmetry and Shift Detection problems outlined below). To obtain binary weight solutions we clipped the weights to ± 1.

2 Simulations

We will illustrate the performance of the algorithm for two Boolean problems, Mirror Symmetry and Shift Detection. As an example of a more real-life application we also report on performance for digital channel equalisation.

2.1 Mirror Symmetry Problem. For the Mirror Symmetry problem the output of the network is 1 if the input bit string is exactly symmetrical about its centre, otherwise the output is -1. To investigate generalisation ability we generated 100 training patterns such that the first half of the input bit string was randomly constructed from ±1 with both components selected with a 50% probability (the network had 30 input nodes). We define the generalisation rate as performance on a test set (of 1000 patterns) drawn randomly from the same pattern distibution as the training set (but excluding training patterns). For randomly constructed inputs the output will be −1 with a high probability. Consequently we randomly chose the two target values with 50% probability and determined the symmetry of the input strings accordingly. On the test set generalisation 67.4 ± 6.3% with an average 2.4 hidden nodes generated (for a sample of 20 networks). Using binary weights generalisation improved with a performance of 87.4 ± 4.5% (and and average 7.6 hidden nodes generated) for a sample of 20 networks.

2.2 Shift Detection Problem. In the Shift Detection problem we consider a network with 20 input nodes and one output node. The first 10 input nodes are a randomly constructed pattern with components ±1 and the second set of 10 input nodes is given the same pattern set circularly shifted by one bit to the left (target=+1) or right (target=−1). In our simulations we trained the network with 100 patterns and tested generalisation performance on 1000 examples drawn from the remaining patterns (samples of 200 networks were used). On the test set we achieved generalisation of 85.6±3.5% (and an average 3.1 hidden nodes generated) for a sample of 50 networks and using the Minover algorithm. Generalisation improved to 91.8 ± 3.8% using binary-valued weights (with an average 8.8 hidden nodes generated). This performance can be compared to 67.3 ± 5.7% for Backpropagation using the same dataset.

2.3 Channel Equalisation. Digital communications channels are affected by intersymbol interference and additive white noise and neural network equalisers have been suggested as a means of removing these effects by approximating the inverse of the transfer function of the channel . Since our algorithm has binary outputs it could be used in this application. We constructed a model channel with a transfer function $H(z) = 0.26 + 0.93z^{-1} + 0.26z^{-2}$ and signal to noise ratio per bit equal to 10dB. The equaliser had 11 taps and an AD converter (thermometer coding) prefacing the neural network (300 training patterns were used and a test set of 1000 patterns). The proposed algorithm had an error of 2.0% on the test set. After experimenting with Back-Propagation to find the optimal number of hidden nodes we obtained a best performance of 3.2%. For Cascade Correlation we obtained 2.8%.

References

[1] M. Marchand, M. Golea and P. Rujan, *Europhysics Letters* **11**(1990)487-492; M. frean, *Neural Computation* **2**(1990)198-209.

[2] W. Krauth and M. Mezard, *J. Phys.* **A20**(1987)L745-L752.

[3] S. J. Nowlan and G.E. Hinton, *Neural Computation* **4**(1992)473-493.

[4] C. Campbell and C. Perez Vicente, *Neural Computation* **7** (to appear); C.Campbell and C. Perez Vicente, "Constructing Feed-Forward Neural Networks for Binary Classification Tasks" Proceedings ESANN'95, D. Facto Publications, Brussels, 1995, p. 241-246.

DENSITY ESTIMATION USING SOFM
AND ADAPTIVE KERNELS

S.H. Lokerse, L.P.J. Veelenturf, J.G. Beltman
University of Twente, department of electrical engineering
Drienerlolaan 5, P.O. Box 217, 7500 AE Enschede, The Netherlands
e-mail: lokerse@nt.el.utwente.nl; l.p.j.veelenturf@el.utwente.nl
tel: 053-892842   fax: 053-334701

Abstract. Classification based on the Bayes criterion minimizes the probability of classification error. In order to apply this criterion, one has to know the probability densities of each class of data. Based on the Parzen density estimation, a new method is derived. The number of operations involved in the Parzen estimation is reduced by using the vector quantization property of Self Organizing Feature Mapping (SOFM). The width of the kernels is made dependent on the variance of the samples in the clusters.

1. Introduction

Using the Parzen estimation and the Gaussian kernel the probability density in the $\Re^d$ of one class is estimated by

$$p(x) = \frac{1}{m(2\pi h^2)^{\frac{d}{2}}} \sum_{i=1}^{m} \exp\left(-\frac{(x-x_i)^t(x-x_i)}{2h^2}\right) \tag{1}$$

where $\{ x_i, 0<i<m \}$ denote the available samples in the class and h the kernel-width. For a Gaussian density function and a Gaussian kernel the optimal kernel-width h_{opt} is [1]:

$$h_{opt} = \sigma m^{-1/(d+4)} \tag{2}$$

where σ^2 is the average marginal variance of the trainset:

$$\sigma^2 = \frac{\sum_{i=1}^{d} \sigma_i^2}{d} \tag{3}$$

2. Principle of the method

The principle of the proposed method consists of two steps:

1)   The number of kernels is reduced by using the vector quantization property of SOFM. Each class is quantisized using a class-specific codebook.

2)   The kernel width is made adaptive to the variance of the trainsamples in the clusters. A cluster is a subspace that is represented by a weightvector (codebook-vector) of the neural network.

2.1. Vector Quantization

The Kohonen Learning Algorithm [2] is used to perform the vector quantization. After vector quantization the Parzen estimate can be approximated by:

$$p(x) = \frac{1}{(2\pi)^{\frac{d}{2}}} \sum_{i=1}^{n} \frac{m_i}{mh^d} \exp\left(-\frac{(x-w_i)'(x-w_i)}{2h^2}\right) \qquad (4)$$

where n is the number of codebook-vectors and m_i is the number of trainsamples in cluster i. Cluster i is represented by codebook-vector (the weight-vector in the neural network) w_i. The m kernels in the Parzen estimation are replaced by n kernels, centered on the codebook-vectors w_i.

2.2. Adaptive Kernel width

After vector quantization the average marginal variance in each cluster can be calculated:

$$\sigma_i^2 = \frac{\sum_{x \in T_i} (x-w_i)'(x-w_i)}{m_i d} \qquad (5)$$

where T_i is the set of trainsamples in cluster i. If the trainsamples are uniform distributed, then the variance in a cluster as function of n is

$$\sigma_i(n) = \sigma n^{-1/d} \qquad (6)$$

where σ^2 is the variance of all the trainsamples. The kernel width in cluster i, h_i, can now be expressed as

$$h_i = \alpha \sigma_i \qquad (7)$$

where α is a correction factor to let have h_i the same dependence on the number of kernels n as h_{opt} (2):

$$\alpha \sigma n^{-1/d} = n^{-1/d+4} \Rightarrow \alpha = n^{4/d(d+4)} \qquad (8)$$

If the number of codebook-vectors is increased, then the kernel-width calculated as in equation (7) will loose accuracy in two ways. First, it is not possible to calculate the variance in a cluster accurately using a few trainsamples. Second, in some clusters there will only be one trainsample. The codebook-vector will be the same as this trainsample, resulting in zero marginal variances in this cluster. There is in fact no vector quantization in this cluster, so the kernel width in this cluster should be h_{opt}: The optimal kernel-width of a Parzen estimation. Considering this two points leads to the following equation for h_i:

$$h_i^2 = \frac{\alpha^2 \sigma_i^2 (n_i - 1) + h_{opt}^2}{n_i} \tag{9}$$

A small addition is to use the covariance-matrix of samples in a cluster for selecting the shape of the kernels. Simulations showed that the use of marginal variances (not averaged as before) gives the best results.

Vector Quantization of the trainsamples, followed by using kernels with kernel-widths calculated by equation (9) leads to the QAK (Quantisized Adaptive Kernel) estimation.

3. Results

Simulations have been carried out on the 'satimage' database [3]. The database consists of the multi-spectral values of pixels in 3x3 neighbourhoods in a satellite image. Each sample in the database is labeled as one of the six possible vegetations (cotton crop, red soil, grey soil, ...). The database contains 6435 samples with 36 attributes. A 36-dimensional density estimate for six classes are calculated using the following methods:

1. Quantisized Kernel (QK) Estimation, according to equation (4)
2. Inertia Rated Vector Quantization (IRVQ) [4]. The differences with the QAK-estimate are, that the kernel-width is calculated by

$$h_i^2 = \frac{3\sigma_i^2}{2 \ln 2} \tag{10}$$

   and that the number of samples in a subspace, m_i, not is taken into account.
3. Kohonen classification [2]. Vector quantization is performed the same as in the QAK-estimate, but the clusters are labeled as one of the classes by majority voting using the trainsamples of all the classes.
4. Maximum Likelihood Estimate [5]. This method uses an iterative procedure to get a decomposition of the distribution into a number of normal distributions.
5. QAK-estimation, using the marginal variances in a cluster. The proposed method.

The simulations have been carried out using a varying number of kernels (codebook-vectors or normal distributions) and are repeated 5 times using different train- and testsamples. For the trainset always 90% of the 6435 available samples are randomly selected, the other 10% form the testset. Table 1 shows the classification results and the standard deviation of the 5 methods.

number of Kernels	QK		IRVQ		Kohonen		MLE		QAK	
	μ	σ	μ	σ	μ	σ	μ	σ	μ	σ
1	79.0	7.7	77.5	5.7	76.9	5.5	78.8	4.4	79.0	7.7
4	84.3	3.9	83.6	5.2	82.6	4.1	83.9	0.9	83.9	3.1
9	87.0	1.4	85.9	3.1	85.7	2.4	86.3	3.5	86.2	2.9
16	88.4	1.5	88.1	0.7	85.8	1.3	86.9	2.5	88.6	0.3
25	89.2	2.1	87.9	0.3	87.3	2.7	88.7	3.3	89.3	1.1
36	89.2	1.7	88.9	0.6	88.3	0.3	89.3	2.9	89.4	0.8
49	89.4	2.3	89.1	0.2	89.1	1.4	90.2	2.8	90.4	1.8

Table 1 Percentage correct classification

The differences are small, but the QAK method combines a good classification with a small standard deviation. In this simulation the training time of the 'quantisized' methods were approximately two times shorter than the MLE-method. The number of operations during classification for the Kohonen method is less than for the other methods (no kernels have to be calculated).

4. Conclusions

The QAK-estimate of the probablity density function shows good results on the 'satimage'-database. Currently experiments are carried out on other databases.

References

[1] B.W. Silverman, *Density Estimation for Statistics and Data Analysis*, Chapman and Hall, 1986.

[2] L.P.J. Veelenturf, *Analysis and Applications of Artificial Neural Networks,* Prentice Hall, 1995.

[3] P.M. Murphy and D.W. Aha. Uci repository of machine learning databases. Irvine, University of California, Department of Information and Computer Science (anonymous ftp to ics.uci.edu in pub/machine-learning database).

[4] J.L. Voz, M. Verleysen, P. Thissen, and J.D. Legat. Suboptimal Bayesian classification by vector quantization with small clusters. In *Proceedings ESANN* , pages 153-160, Brussel 1995.

[5] K. Fukunaga, *Introduction to Statistical Pattern Recognition*, 2nd edition, Academic Press, 1990.

Maximum likelihood estimates for Markov networks using inhomogeneous Markov chains

Hermann von Hasseln

Department of Neural Information Processing, University of Ulm

Ulm, Germany

May 18, 1995

Abstract

Given an undirected graph with n nodes, and any probability distribution over the n two–valued nodes. Which is the closest (in the sense of the information–divergence) probability distribution, defining a Markov network on the given graph?

Solving this task is equivalent to finding the maximum likelihood estimate in the set of probability disributions which define a Markov network on the given graph. Therefore it is given by the "M–step" of the Expectation–Minimization–(EM) algorithm. Termed in Amari's information–geometric framework, the M–step is the m–projection ("m–step") of the given or observed distribution on the set M of statistical models.

In the field of probabilistic expert systems, the natural approach to knowledge representation in Markov or Bayes networks is based on conditional probabilities. However, the conditional probability approach to Markov networks suffers from serious consistency problems. We present an algorithm, which uses inconsistent conditional probabilities in an iterative way as transistion probabilities in an inhomogeneous Markov chain. It is shown that this algorithm converges to the m–projection on the set of Gibbs–distributions of a given graph.

1 Introduction

Alternating or dual minimization is a general technique for maximum likelihood (ML) estimation, described in the information–geometric framework ([6], [1]). This technique has recently gained a lot of attraction in the learning theory of probabilistic neural networks and related fields. Dual minimization, or the em-algorithm, as termed by Amari ([1]) emphasizing the two orthogonal projections (e- and m-projection), minimizes iteratively the information–divergence (I-Divergence) between two convex manifolds of probability distributions. These orthogonal projections suggest a much higher convergence speed. Amari developed in [1] (see also the references therein) a unified information–geometric framework of em with applications e.g. to Boltzmann–machines, stochastic feedforward nets, hidden Markov models and mixture of experts. Amari showed that EM and em are equivalent in most cases, or, to be more specific, are equivalent in the m-aximization–step, and equivalent in the e-xpectation-step under specific circumstances. The application of dual minimization to

Boltzmann learning is given in [4] and in the general information–geometric framework of *em* in [2]. Since Boltzmann–machines[1] are special cases of Markov random fields (or Markov networks) confined to second–order potentials, it is clear that *em* can also be used for Markov networks on any graph or hypergraph with potentials of any order.

In this paper we give an algorithm, which uses inconsistent conditional probabilities as transistion probabilities in an inhomogeneous Markov chain (MC). Clearly, the stationary distribution of this MC is **not** specified by the local characteristics as is the case, e.g. for the Gibbs–sampler. We show that the inhomogeneous MC, when started with a PD π which does not define a Markov network on a given graph $\mathcal{G}$, converges to the m–projection of π on the manifold of Markov networks on $\mathcal{G}$. The m–projection minimizes the I–Divergence $\mathcal{I}(\pi; p) = \sum_{\mathbf{x}} \pi(\mathbf{x}) \ln \frac{\pi(\mathbf{x})}{p(\mathbf{x})}$ for all PD's p, which define a Markov network on $\mathcal{G}$.

2 Notation

Let μ be a given discrete probability distribution (PD) over n two–valued random variables $\mathbf{X} = (X_1, \ldots, X_n)$. We then have a set Ω (the configurations) of 2^n values for the PD, which we denote by $\mu(\mathbf{X} = \mathbf{x}) = \mathcal{P}r(\mathbf{X} = \mathbf{x})$, or, for short: $\mu(\mathbf{x})$. Let $A \subseteq \Lambda = \{1, \ldots, n\}$. For $A \neq \emptyset$ we define the marginals of $\mu(\mathbf{x})$ as: $\mu_A(\mathbf{x}) = \mathcal{P}r(X_i = x_i)_{i \in A}$, and $\mu_\emptyset = 1$. With S we denote the (open) simplex of these PD's over n variables. For the discussion of Markov networks[2] we define **local and global characteristics**: Let $\mathcal{G}$ be any graph over the n nodes, and let $\partial = \{\partial a : a \in \Lambda\}$ be the system of neighborhoods on $\mathcal{G}$. Note, that $a \notin \partial a$, and that neighborhoods are symmetric: $a \in \partial b \Leftrightarrow b \in \partial a$; $a, b \in \Lambda$. Also, define $\bar{a} = \{a\} \cup \partial a$. A subset $C \subset \Lambda$ is called a **clique**, if for every pair $\{a, b\} \subset C$, $a \neq b$) we have $a \in \partial b$. The set of all cliques of $\mathcal{G}$ will be denoted by C. Let

$$\mu_{\{a\}}^\Lambda(\mathbf{x}) = \mu_a^\Lambda(\mathbf{x}) = \mathcal{P}r(X_a = x_a \mid X_b = x_b; b \in \Lambda \backslash a) \tag{1}$$

be the **local characteristics** of node $a \in \Lambda$ of $\mathcal{G}$. The numbers

$$\mu_{\{a\}}^{\bar{a}}(\mathbf{x}) = \mu_a^{\bar{a}}(\mathbf{x}) = \mathcal{P}r(X_a = x_a \mid X_b = x_b; b \in \partial a) \tag{2}$$

are called the **global characteristics** of node $a \in \Lambda$ of $\mathcal{G}$. If μ is a strictly positive PD over the Variables $\mathbf{X}$ and ∂ is the neighborhoodsystem of $\mathcal{G}$, we say that, if the equations $\mu_a^\Lambda(\mathbf{x}) = \mu_a^{\bar{a}}(\mathbf{x})$ ($\forall a \in \Lambda, \mathbf{x} \in \Omega$) hold, that μ defines a **Markov network** on $\mathcal{G}$.

3 The m–projection as an inhomogeneous Markov chain

We define transitions matrices for an updating process on a given graph $\mathcal{G}$ over n nodes:

---

[1] When we speak of a Boltzmann–machine, we speak of the stationary Gibbs–Boltzmann distribution on the stochastic units of this network.

[2] We speak of Markov networks as graphical representations of Markov random fields.

Definition 3.1 *Let $\mu \in S$ be a strictly positive, discrete PD, and $\mu_a^{\bar{a}}(x)$ the global characteristics. Let $p(a) \geq 0$, $\sum_{a=1}^n p(a) = 1$, be a PD of visting each node $a \in \Lambda$. The $2^n \times 2^n$-dimensional matrix $\mathcal{U}^{(\mu)}(x', x) = Pr(x' \to x)$, given by*

$$\mathcal{U}^{(\mu)}(x', x) = \begin{cases} 0, & \text{if } x' \text{ and } x \text{ differ in more} \\ & \text{than one component,} \\ p(a)\,\mu_a^{\bar{a}}(x), & \text{if } x' \text{ and } x \text{ differ only} \\ & \text{in the } a^{th} \text{ component,} \\ 1 - \sum_{a=1}^n p(a)\,\mu_a^{\bar{a}}(x), & \text{if } x' \text{ und } x \text{ differ} \\ & \text{in no component,} \end{cases}$$

is called a **Boltzmann–matrix**.

It is easy to see that these matrices are **stochastic, irreducible** and **primitive**, which, in turn, means that there exists an unique left eigenvector (the Perron–Frobenius–eigenvector) of the matrix $\mathcal{U}^{(\mu)}$ with eigenvalue 1. This eigenvector is, of course, the stationary distribution of the MC, **but not a Gibss–Boltzmann distribution**, unless we have the Markov properties $\mu_a^{\Lambda}(x) = \mu_a^{\bar{a}}(x)$. For the following we need the definition of two important sets ([1],[7]): Let $\pi \in S$ be a PD. The set

$$\mathcal{Q}_{\pi}^{(A)} = \{p : \ p_a = \pi_a, \ \forall a \in A\}$$

is the set of all PD's p, whose marginals up to order A coincide with those of of the PD π. The set

$$\mathcal{P}_{\pi}^{(A)} = \{p : \ \ln p = \sum_{b \subseteq \Lambda \setminus \emptyset} G_p^b \mathcal{X}_b - \Psi, \ \ G_p^b = G_{\pi}^b, \ b \not\subseteq A\}$$

is the set of PD's p whose expansion coefficients[3] of order $b \not\subseteq A$ coincide with those of the PD π. The following theorem lies at the heart of the dual minimization technique, and is essentially due to [1] and [6][4] (the proof can be found in [8]):

Theorem 3.1 *Let $A = \{a : \ a \subseteq \Lambda\}$ be a set of subsets of Λ and $\{\pi_a(x)\}$, $a \in A$ the set of marginals of a given PD $\pi \in S$, and $\mathcal{G}$ a graph with the set $C = A$ of cliques. The PD $p^{\star}$, given by the two equivalent projections*

$$\mathcal{I}(p^{\star}; u) = \min_{q \in \mathcal{Q}_{\pi}^{(A)}} \mathcal{I}(q; u) \ \ \Longleftrightarrow \ \ \mathcal{I}(\pi; p^{\star}) = \min_{p \in \mathcal{P}_{\pi}^{(A)}} \mathcal{I}(\pi; p)$$

(u is the uniform PD), defines a Markov network on $\mathcal{G}$.

Now we can formulate our main theorem.

Theorem 3.2 *Given a graph $\mathcal{G}$ over n nodes and any PD $\pi \in S$ over $\mathcal{G}$, not necessarily defining a Markov network. Let $\mathcal{U}^{(r)}$ ($r = 0, 1, 2, \ldots$) be a*

---

[3]The coefficient Ψ is normalization of p, which in statistical physiscs is the negative of the *free energy.*

[4]See also [2], [4]. The formulation given here stems from [7],[8] and Laura Martignon and Alfredo Iusem (personal communication).

210

Boltzmann–matrix given by the PD $\mu^{(r)}$ as in Defintion 3.1. With $\mu^{(0)} = \pi$, the iterations

$$\mu^{(r+1)} = \mu^{(r)} \, \mathcal{U}^{(r)} \tag{3}$$

converges to the m–projection $p^\star$ of π on the manifold of Markov networks on $\mathcal{G}$, given by $\mathcal{I}(\pi; p^\star) = \min_{p \in \mathcal{P}_u^{(\mathcal{C})}} \mathcal{I}(\pi; p)$.

PROOF. We give a short sketch of the proof ([8]). The matrix multiplication in eq.(3) can be written as: $\mu^{(r+1)}(\mathbf{x}) = \mu^{(r)}(\mathbf{x}) \sum_{a=1}^n p(a) \frac{(\mu^{(r)})_a^{\bar{a}}(\mathbf{x})}{(\mu^{(r)})_a^{\Lambda}(\mathbf{x})}$. If we choose $p(a) = 1$ cyclically for all nodes $a \in \Lambda$ (which makes no difference in the inhomogeneous MC), we can construct a two–step–IPFP[5] ([3]), which projects the PD $\mu^{(s)}$ on the set $\mathcal{P}_u^{(\{\bar{a}, \Lambda \backslash a\})}$ by Theorem 3.1 (u is the uniform PD). These cyclically projections then converges to $p^\star$ in the non–empty set $\bigcap_{a=1}^n \mathcal{P}^{(\{\bar{a}, \Lambda \backslash a\})} = \mathcal{P}^{(\mathcal{C})}$, by Csiszár's Theorem 3.2 in [5]. ∎

4 Aknowledgements

I am grateful to L. Martignon, A. Iusem and G. Palm for valuable discussions on this work, which is part of my dissertation. This work was partially supported by the *Forschungsverbund Neuroinformatik des Landes Baden–Württemberg*.

References

[1] S Amari. Information Geometry of the EM and *em* Algorithms for Neural Networks. Technical report, Department of Mathematical Engineering and Information Physics, Faculty of Engineering,University of Tokyo, April 1994. to appear in Neural Networks.

[2] S Amari, K Kurata, and H Nagaoka. Information Geometry of Boltzmann machines. *IEEE Transactions on Neural Networks*, 3(2):260–271, March 1992.

[3] Y Bishop, S Fienberg, and P Holland. *Discrete multivariate analysis*. The MIT Press, Cambridge, MA, Cambridge, Massachussets,London, 10th edition, 1989.

[4] W Byrne. Alternating Minimization and Boltzmann Machine Learning. *IEEE Transactions on Neural Networks*, 3:612–620, 1992.

[5] I Csiszár. I-Divergence Geometry of Probability Distributions and Minimization Problems. *The Annals of Probability*, 3(1):146–158, 1975.

[6] I Csiszár and G Tusnády. Information Geometry and Alternating Minimization Problems. In *Statistics & Decision, Supplement Issue No.1*, pages 205–237. R. Oldenburg Verlag, München, 1984.

[7] L Martignon, H von Hasseln, S Grün, A Aertsen, and G Palm. Detecting higher–order interactions among the spiking events in a group of Neurons. *Biological Cybernetics*, 1995. to appear.

[8] H von Hasseln. *A consisteny algorithm for Markov networks*. Dep. of Neural Information Processing, Univ. of Ulm, Germany, 1995. Dissertation in preparation.

---

[5]IPFP = iterative proportional fitting procedure.

Statistical Pattern Recognition - Posters

Density Estimation as a Preprocessing Step for Constructive Algorithms

J.C. Lemm,[*] V. Beiu[†], and J.G. Taylor

Department of Mathematics, King's College London
Strand, London WC2R 2LS, UK

Abstract

The aim of the paper is to combine the generalization power of classical density estimation techniques with the efficiency of VLSI–friendly, constructive algorithms. This is important for VLSI–implementations of classification networks in the case of noisy data to avoid the well–known "overfitting" effects. The method consists of three steps 1) estimation of the probability densities for each class, 2) discretization of the input space and 3) describing the resulting classification regions using only easy implementable boolean AND and OR gates and comparisons. The "noisy spirals" classification problem, a noisy variant of the "two spirals" benchmark, is used for demonstration.

Introduction: Classical neural network architectures with McCulloch–Pitts–like neurons are not well suited for VLSI–implementations. They require multiplications which are quite complex operations in terms of boolean gates, and they usually have a very high connectivity, compared to a limited fan–in VLSI–implementation.

To enable effective hardware realizations of classifiers or function approximators there exist especially designed algorithms which build networks with simple nodes, limited fan–in, and a local connectivity structure [3, 2]. These algorithms are constructive, i.e. they explicitly build up a network architecture adapted to the problem [4, 8]. In contrast, classical neural net algorithms like backpropagation fix the architecture and only train the weights.

Many of the VLSI–friendly algorithms have been developed with regard to their efficiency, but not with regard to their generalization performance. Especially in the case of noisy data the generalization behaviour of a network should be controlled to avoid "overfitting" of the data (or "training of the noise") by too complex networks.

To increase the generalization behaviour of VLSI–friendly algorithms we show how to combine them with a density estimating preprocessing step. We assume a situation where building the net ("learning") is allowed to take some time, but execution of the net should be very fast, so it has to be realized as VLSI–circuit.

Density estimation: We formulate the strategy for classification problems. In this case the optimal strategy is to choose the class C_i with maximal probability

---

[*]on leave from Institut für Theoretische Physik I, Universität Münster. J.L. was supported by the British–German ARC–programme.

[†]on leave of absence from the "Politehnica" University of Bucharest. V.B. is a HC&M Research Fellow of the EC and is financed by the Commission of the European Communities under contract ERBCHBICT941741.

$P(C_i|x) \propto P(x|C_i)P(C_i)$, where x stands for the data point which has to be classified. To apply this criteria, besides the apriori probabilities $P(C_i)$, the likelihoods $P(x|C_i)$ must be known. In the case of a finite number of possible values for x, the likelihoods can be estimated by sampling during the training phase. In the case of a continuous variable x one can never sample all the values of x. Estimating $P(x|C_i)$ from a set of examples $\{x_i, C(x_i)\}$, where $C(x_i)$ denotes the class of x_i, can be done with a variety of methods known under names like "density estimation", "nonparametric regression", or "smoothing techniques". They include kernel methods, k–nearest neighbors, splines and orthogonal series estimators [7]. We use here a gaussian kernel or radial basis function [6] to approximate the likelihood on the basis of n_i d–dimensional data x_j for each class C_i by

$$\widehat{P}(x|C_i) = \frac{1}{n_i \widehat{\sigma}^d (2\pi)^{d/2}} \sum_{j=1}^{n_i} \exp\left(-\frac{1}{2\widehat{\sigma}^2}(x - x_j)^T(x - x_j)\right). \tag{1}$$

Having the estimation $\widehat{P}(x|C_i)$ for the "true" density $P(x|C_i)$ we can map to every point x of the input space the corresponding class $\widehat{C}(x)$ with maximal probability $P(C_i|x)$. The resulting classification regions can now be described by a VLSI–friendly algorithm.

A Constructive Algorithm: The principle of the constructive algorithm we use here is to divide the classification regions in as few as possible d–dimensional boxes with borders parallel to the coordinate lines [2]. Each such box is characterized by $2d$ numbers, i.e. its lower and upper bounds in each of the d dimensions. For the sake of simplicity we explain the algorithm for finding the boxes in the two–dimensional case. A number of starting boxes (=rectangles) for the algorithm are generated by discretising $\widehat{P}(x|C_i)$ on a rectangular mesh. This defines at the same time a minimal length scale and limits the maximal possible number of rectangles and therefore the complexity of the resulting net. The algorithm tries to find a small number of rectangles in the following way: It starts by enlarging a small rectangle given by the discretization step in all directions simultaneously, without leaving the region belonging to the same class as the starting point. If a rectangle cannot be enlarged further in one dimension, the border in this direction is frozen and it is enlarged only in the remaining dimension as far as possible. After the borders of a rectangle are fixed the process continues with the next small rectangle not yet processed.

This allows the construction of a VLSI–friendly net, describing the classification regions using only comparisons, which can be efficiently implemented [1], and boolean AND and OR gates. Every rectangle is represented by a group of four comparisons and three AND gates

$$\left((x^{(1)} > A) \text{ AND } (x^{(1)} < B) \text{ AND } (x^{(2)} > C) \text{ AND } (x^{(2)} < D)\right), \tag{2}$$

where $x^{(i)}, i = 1, 2$ are the components of the vector x. The total region belonging to a specific class is obtained by combining all the rectangles by OR gates. To obtain a limited fan–in the OR gates can be cascaded.

An Example: To demonstrate the method we use a noisy variant of the well-known "two spirals" problem [5]. Two probability densities were generated by superimposing Gaussians with variance $\sigma^2 = 0.1$ centered at 485 points on each

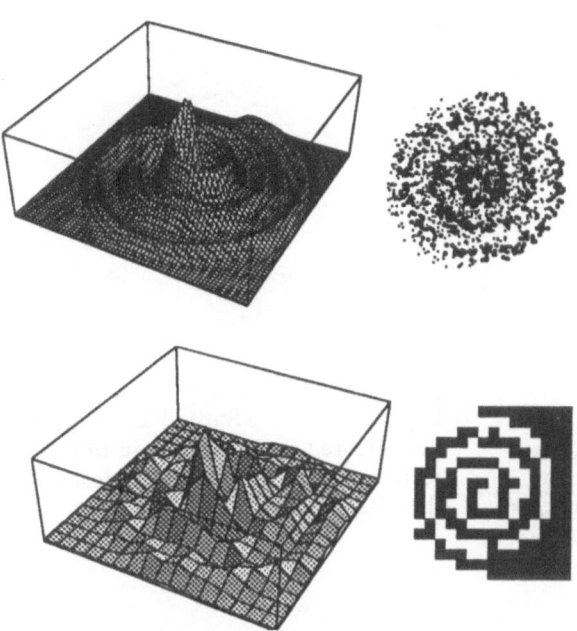

Figure 1: (left) The noisy spiral: $P(x|C_1)$ generated by a superposition of Gaussians with variance $\sigma^2 = 0.1$.

Figure 2: (right) 1000 examples drawn from $P(x|C_1)$ of Fig.1 (circles) and 1000 examples drawn from $P(x|C_2)$ (points).

Figure 3a: (left) Reconstruction: $\widehat{P}(x|C_1)$ of the initial probability distribution $P(x|C_1)$ of Fig.1 with $\widehat{\sigma}^2 = 0.05$ after the 1000 examples of Fig.2 on a 16×16 mesh.

Figure 3b: (right) Classification regions obtained from $\widehat{P}(x|C_1)$ of Fig.3a and the corresponding $\widehat{P}(x|C_2)$.

of two interlocking spirals. (See Fig.1.) Fig.3a shows the discretized probability estimation on a 16×16 and 64×64 mesh constructed from 1000 randomly drawn examples, and Fig.3b the corresponding classification areas. The algorithm has been systematically tested for 1.) varying values of the reconstruction variance $\widehat{\sigma}^2$, 2.) varying number of training examples, and 3.) for mesh sizes of 16×16, 32×32, and 64×64. Two variables have been measured for every network: 1.) The generalization error E as the mean number of wrong classifications for 1000 newly generated test examples. 2.) The complexity of the resulting net as the number N of rectangles needed to describe the classification regions. In Figs.4a and 4b we present the results for the 64×64 mesh because there the effects of "overfitting" can be seen most clearly. The Fig.4a shows that about 100–200 examples are enough to obtain reasonable low error measures. In the region above 200 examples three regions can easily be classified. 1.) A large reconstruction width (standard deviation) $\widehat{\sigma}$ naturally leads to small networks (small N) but to large errors. In this "underfitting" area an *increase* in complexity N is coupled to a *decrease* of the error E. 2.) There is an optimal range for the reconstruction width of about $\widehat{\sigma}=0.5$ to 0.3 where the error is *minimal* and both, the error and the complexity are nearly *independent* from the number of examples and the reconstruction width. 3.) The "overfitting" area is at small reconstruction widths $\widehat{\sigma} < 0.2$ where the large 64×64 mesh already allows for "fitting of the noise". Here an *increase* of the complexity N is linked to an *increase* of the generalization error. The complexity N is proportional to the number of examples, because the network tends to memorize every example individually.

We should mention that for practical purposes a 16×16 mesh is more

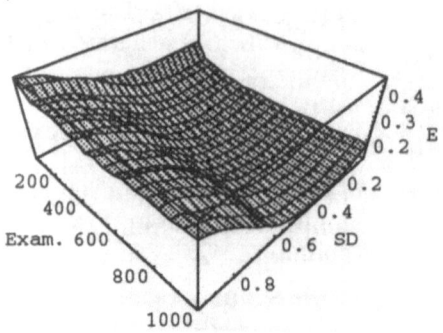

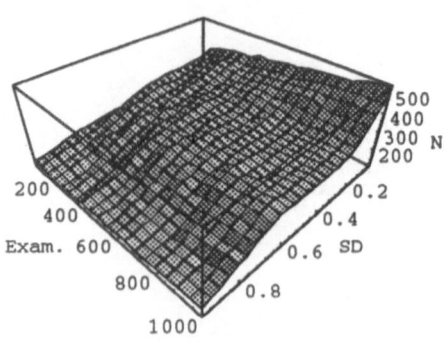

Figure 4a: Generalization error measured with 1000 new input data after n(=Exam.) training data using $\hat{\sigma}(=SD)$ as standard deviation for reconstruction on a 64×64 mesh.

Figure 4b: Same as Fig.4a, but plotted against the number of rectangles(= N) needed to describe the classification regions.

appropriate for the "noisy spirals" if one allows slightly larger errors, because it produces much smaller nets ($N < 80$) with still reasonable minimal error ($E_{Min}(16 \times 16) = 0.201$ vs. $E_{Min}(64 \times 64) = 0.1295$) and shows nearly no "overfitting" phenomena.

References

[1] Beiu, V., Peperstraete, J.A., Vandewalle, J. and Lauwereins, R. (1993). Efficient Decomposition of Comparison and Its Applications. In Verleysen, M.(ed.): ESANN'93, D facto, Brussels, Belgium, 1993, 45-50.

[2] Beiu, V., Peperstraete, J.A., Vandewalle J., and Lauwereins, R. (1994) Learning from Examples and VLSI Implementation of Neural Networks. In R. Trappl (ed.): Cybernetics and System Research '94, World Scientific Publishing, Singapore, 1994, 1767-1774.

[3] Beiu, V. and Taylor, J.G. (1995). Optimal VLSI Implementations of Neural Networks: VLSI-Friendly Learning Algorithms. *The Applied Decision Technologies ADT'95*, London, UK, 3-5 April, 1995.

[4] Fiesler, E. (1994) Comparative Bibliography of Ontogenic Neural Networks. ICANN94, Proceedings, Springer, London, 793–796.

[5] Lang K.J., and Witbrock M.J. (1988). Learning to tell two spirals apart. *Proc. Connectionists Models Summer School*, Morgan Kauffmann, San Mateo, 52–59.

[6] Poggio, T., and Girosi, F.A. (1990). Networks for Approximation and Learning. *Proceedings of the IEEE* **78**, September 1990, 1481–1497.

[7] Silverman, B.W. (1986) *Density Estimation*. Chapman&Hall, London.

[8] Śmieja, F.J. (1993). Neural Network Constructive Algorithm: Trading Generalisation for Learning Efficiency? *Circuits, Systems, and Signal Processing*, **12**(2), 331–374.

MUTUAL INFORMATION NEURAL NETWORKS: A NEW CONNECTIONIST PARADIGM FOR DYNAMIC PATTERN RECOGNITION TASKS

G.Rigoll , J. Rottland
Gerhard-Mercator-University Duisburg
Faculty of Electrical Engineering
Department of Computer Science
Bismarckstr. 90
Duisburg, Germany

e-mail: {rigoll,rottland}@fb9-ti.uni-duisburg.de
http://www.fb9-ti.uni-duisburg.de

This paper presents a new probabilistic neural network paradigm for dynamic pattern recognition problems. The new approach includes the following innovations: 1) It is based on a self-organizing learning approach using information theory principles 2) The neuron activations are interpreted as probabilities and represent a probabilistic decision boundary in the feature space 3) A combination of unsupervised and supervised learning algorithms can be used to train the network weights 4) The neuron probabilities can be further refined by corrective training methods leading to a joint optimization of both, the weights and the neuron probabilities 5) The neural network can process dynamic, time varying patterns of arbitrary length 6) The output activations of the neural network can be evaluated directly or optionally treated as the input to other probabilistic pattern recognition algorithms, e.g. Hidden Markov Models. This combination leads to a powerful hybrid pattern recognition approach.

A one-layer version of the network has been developed and tested using various speech and character recognition tasks. Multilayer versions of the network are possible, too. In the one-layer version, the network parameters are represented by two matrices. Matrix G consists of the weights g_{ij} between the input layer of dimension I and the output layer of dimension J. The network is trained on the recognition of N units (e.g. words or phonemes, or characters) $w_1 ... w_N$. The second matrix P consists of the probabilities, that a neuron $y_j, j = 1 ... J$ fires, if a feature vector $\underline{x}$, representing a symbol w_n has been presented to the input layer of the net. Thus, the matrix P contains the conditional probabilities $p(y_j \mid w_n)$, for all J neurons and N recognition symbols. Fig. 1 shows the one-layer version of the neural network.

Training of the network weights in matrix G is performed by using a self-organizing information theory-based training approach proposed by the first author in [1]. This training approach automatically maximizes the mutual information

$$I = - \sum_{j=1}^{J} P(y_j) \cdot \log P(y_j) + \sum_{n=1}^{N} P(w_n) \cdot \sum_{j=1}^{J} P(y_j \mid w_n) \cdot \log P(y_j \mid w_n) \qquad (1)$$

between the input symbols w_n and the output symbols y_j. It can be shown, that such a training algorithm replaces the standard criterion in the backpropagation algorithm of minimum error between output labels y_j and input labels w_n by a maximum joint information criterion between those symbols. It can be also shown that this novel training algorithm can be interpreted as a probabilistic version of the LVQ training paradigm. After training of the weights in G is finished, estimates of the conditional probabilities in P can be obtained by evaluating the training data labeled by the trained neural network.

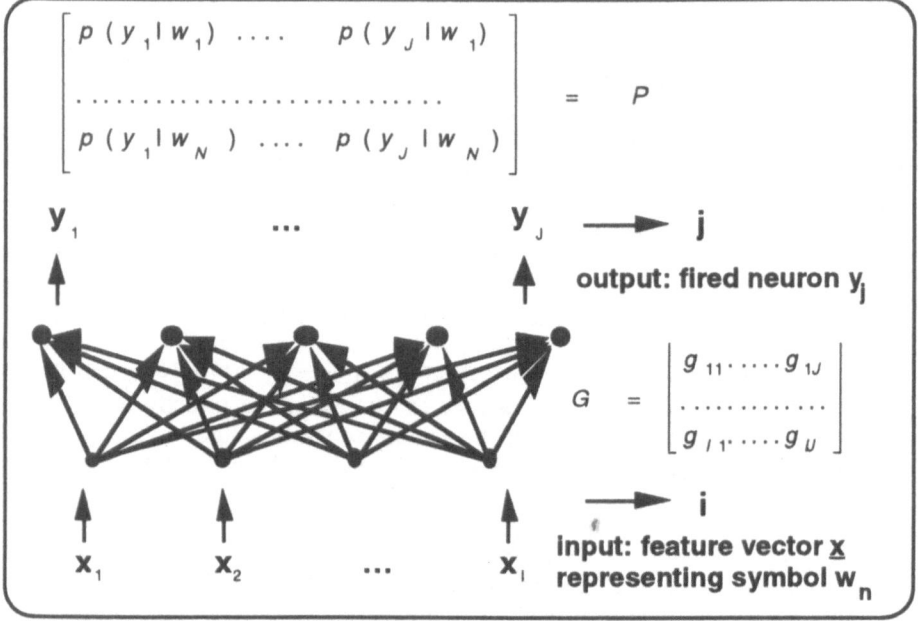

Fig. 1: One-layer version of the mutual information neural network

The result of the training procedure is the formation of complex decision boundaries in the feature space, mapping the firing behavior of the neurons to this feature space. These firing areas are then again described by probabilities, mapping the firing behavior of the neurons to the occurrence of the symbols w_n. In this way, complex probabilistic classification boundaries for the pattern recognition problem are created, as illustrated in Fig. 2. The creation of these probabilistic boundaries is guided by the maximum mutual information principle in a self-organizing way. The complexity can be increased and refined in a self-organizing way, if necessary, by enlarging the number of neurons.

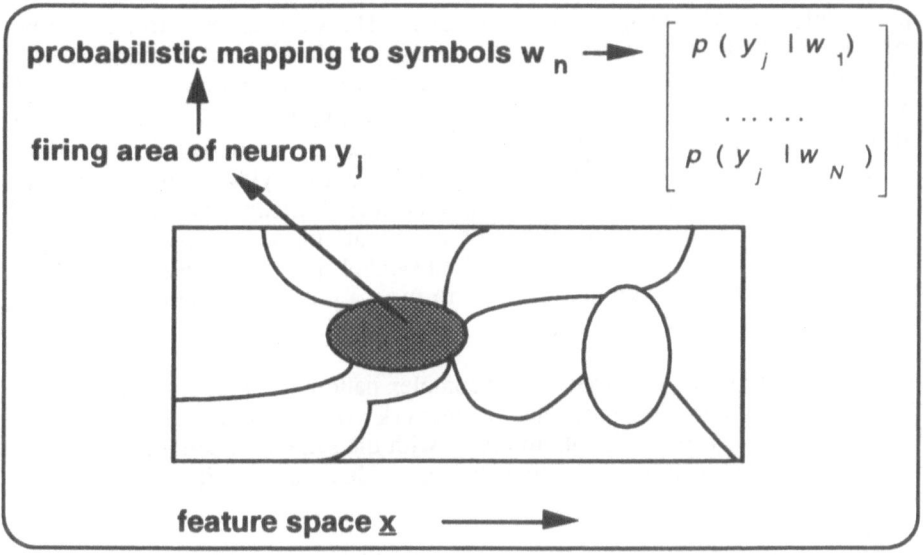

Fig. 2: Probabilistic classification boundaries in the feature space

In recognition mode, the neural network operates as follows: Each presentation of a feature vector $\underline{x}$ to the input layer of the network leads to the firing of one specific neuron y_j in the output layer. The presentation of a dynamic pattern, representing an unknown symbol w_n, and consisting of the string of feature vectors $\underline{x}_1, \underline{x}_2, ..., \underline{x}_T$, leads to the generation of a corresponding sequence of firing neurons $y_1, y_2, ..., y_T$. The goal is to find the symbol w_n, that maximizes the probability $p(w_n | y_1, y_2, ..., y_T)$, i.e. one looks for

$$\max_n \; p(w_n | y_1, y_2, ..., y_T) \equiv \max_n \; p(y_1, y_2, ..., y_T | w_n) \cdot p(w_n) \tag{2}$$

$$\equiv \max_n \; p(y_1 | w_n) \cdot p(y_2 | w_n) \cdot \; .... \; \cdot p(y_T | w_n) \cdot p(w_n) \tag{3}$$

In Eq. (3), the values $p(y_t | w_n)$ can be obtained from matrix P, if the output string $y_1...y_t...y_T$ is known. The probability $p(w_n)$ is the the apriori probability of symbol w_n. Thus, Eq. (3) has to be evaluated for the given string $y_1...y_T$ and all possible symbols w_n in order to find w_{n*} which maximizes this equation. In this way, any dynamic pattern, consisting of a string of feature vectors $\underline{x}_1...\underline{x}_T$ with arbitrary length T can be evaluated by this neural network paradigm. It can be also shown, that this approach can be easily extended to the recognition of more complex dynamic patterns, e.g. word models consisting of phonemes in the case of speech recognition. It can be demonstrated, that in this case even continuous word recognition including grammatical constraints between words is possible with this neural network paradigm, using the Viterbi algorithm, making this approach very flexible and promising. It can be further shown, that various supervised learning cycles can be added to the previous learning phase of the neural net in order to significantly increase recognition performance: The weights in matrix G can be further refined by an iterative supervised learning procedure, where the firing of each neuron y_j is investigated by looking at the known symbol w_n in the training data. If $p(y_j | w_n)$ is not a maximum value for this neuron, then the closest neuron y_i, which fulfills this condition is forced to fire instead, by modifying the weights in matrix G. The probabilities in matrix P can be refined by trying to maximize the recognition rate of the training data, using a corrective training method: For each misclassification of a symbol, the probability of the wrong symbol in P is decreased, while the probability of the desired symbol is increased. Both supervised learning cycles can be combined in order to form a joint optimization of the NN parameters in matrices G and P.

One of the most efficient approaches for complex pattern recognition problems can be obtained by combining these neural networks with Hidden Markov Models (HMM). In this way, it is possible to create with the same connectionist approach either a pure neural pattern recognition system, or alternatively a hybrid NN/HMM pattern recognition system.

The new paradigm has been extensively tested with three standardized databases. These were the "NIST Special Database 3" for handwritten segmented characters, the "Resource Management" (RM) Database, a database for continuous speech recognition and parts of the "Wall Street Journal" (WSJ) database, a large vocabulary continuous speech recognition database. The new method has been compared to powerful competing methods, namely LVQ and a standard discrete HMM-based system for the writer-independent recognition of handwritten characters. The LVQ approach led to a recognition rate of 75%, using gray-scaled feature vectors. Wavelets as feature vectors resulted in a recognition rate of 77%. The standard discrete HMM-based system had a recognition rate of 74%, which could be improved by using the hybrid NN/HMM system to 80%.
To evaluate the efficiency of the new paradigm on speech, a standard discrete HMM-based system has been compared with a NN/HMM hybrid system. The mutual information neural network combined with a HMM reduced the error-rate on the RM-evaluation tests by 12.3% - 25.4%, depending on the test set used. In comparison to other systems developed for this internationally well-known and highly competitive recognition task, which are mainly based on the use of continuous parameter HMM's [2][3], this system is already less than 1% below the recognition rate of the best continuous parameter systems, and there are still various improvements in progress. It may be by now the best discrete parameter system for this task. Preliminary research using a single test of the WSJ-database showed similar results.

[1] G. Rigoll. *Maximum Mutual Information Neural Networks for Hybrid Connectionist-HMM Speech Recognition.* IEEE-Trans. Speech Audio Processing, Vol. 2, No. 1, Jan. 1994, pp. 175-184.

[2] P.C. Woodland, S.J. Young. *The HTK tied-state continuous speech recognizer.* Proc. Eurospeech, 1993, pp. 2207- 2210.

[3] H. Bourlard, N. Morgan. *Neural Networks for statistical inference inference generalizations with applications to speech recognition.* Proc. IEEE-IJCNN, 1991, pp. 242-247.

Output Coding and Modularity for Multi-Class Problems

André Pastoors
Tom Heskes

Department of Medical Physics and Biophysics, University of Nijmegen
RWCP[*] Novel Functions SNN[†] Laboratory
Nijmegen, The Netherlands

Abstract

We study backpropagation networks learning classification problems with multiple classes $k > 3$. The common way to code the output of a network is the one-per-class (OPC) method, where one bit is assigned to each class. A technique called error-correcting output coding (ECOC) converts the k-class learning problem into a large number of two-class learning problems. We propose to use modular architectures as a way to decorrelate the (redundant) network outputs. Various modular architectures are tested on an artificial problem. We conclude that ECOC only improves upon OPC when combined with a sufficiently modular approach.

1   Introduction

In information theory it is customary to add some redundancy to the code vectors to be transmitted in order to be able to reconstruct the original vector, even if some bits are unknown or distorted. This can be achieved by adding some bits to the original code to enlarge the Hamming distance d between the resulting code vectors. It can be shown that the maximal number of errors that can be corrected is $\lfloor \frac{d-1}{2} \rfloor$. The one-per class method (OPC), with as many outputs as classes, yields a highly redundant code: it uses k instead of the minimal $^k \log 2$ outputs, with k the number of classes. However, the Hamming distance between the code vectors is 2 and thus it is impossible to correct an incorrect bit using the correct bits. An alternative is a technique called error-correcting output coding (ECOC). Code vectors for the ECOC method are even more redundant, but are separated by a larger Hamming distance. For each number of classes $k > 3$ there exists an optimal set of code vectors of length $2^{k-1} - 1$. ECOC can also be viewed as some kind of voting scheme where each bit represents a different vote and yields a different hypothesis. The ECOC strategy works best if the different hypotheses are as independent as possible.

Both the OPC and the ECOC method can be applied to neural classifiers such as multi-layered perceptrons trained with backpropagation. Simulation studies have shown that on some problems the ECOC strategy yields significantly better results [2]. However, in these cases the networks with ECOC

---

[*]Real World Computing Partnership
[†]Foundation for Neural Networks

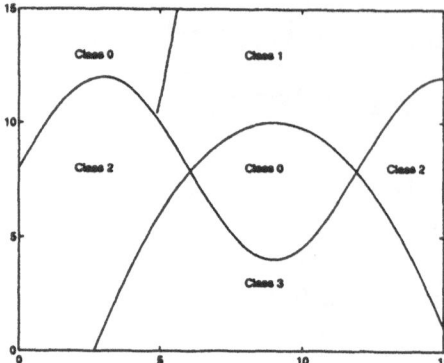

Class	OPC bits				extra bits		
0	1	0	0	0	1	1	1
1	0	1	0	0	1	0	0
2	0	0	1	0	0	1	0
3	0	0	0	1	0	0	1

Figure 1: Classification problem.   Table 1: Error-correcting output code.

required a large amount of hidden units and computation time, probably to decorrelate the votes of the different output bits. An easier way to avoid correlations is to use a modular approach, e.g. by using separate networks for each output bit. For quite different reasons such a modular approach has been suggested in combination with OPC [1]. In this paper we will compare different codings and architectures on an artificial problem with $k = 4$ classes.

2   Simulation paradigm

Our artificial learning problem, taken from [4], is sketched in figure 1. There are two continuously valued inputs and 4 classes. Tabel 1 shows the optimal error-correcting output code for a 4-class problem: the first 4 bits identical to the one-per-class code with 3 additional bits representing class combinations. Network outputs are assigned to the class with the closest code vector according the Hamming distance $d(\vec{o}, \vec{c}_i) \equiv \sum_{j=1}^{N} |o_j - c_{ij}|$, where $\vec{o}$ denotes the network output, $\vec{c}_i$ the code vector for class i, and N the number of outputs. We compare 7 different architectures: not only ordinary OPC, ordinary ECOC, modular OPC, and modular ECOC (top row in figure 2), but also architectures with an intermediate level of modularity: four modules, five modules, and two blocks (bottom row). These architectures differ in the extent and the way they try to decorrelate the various bits.

In our simulations a multi-layered perceptron is trained using a conjugate-gradient implementation of backpropagation. Each architecture is trained on 50 training sets of 60 patterns, randomly drawn from the four-class distribution sketched in figure 1. To prevent overfitting, training is stopped when the performance on a large validation set of 540 randomly drawn patterns seriously starts to deteriorate. All modules are trained and stopped separately. The optimal number of hidden units is determined through cross-validation and appears to be around 5 hidden units per output for blocks of hidden units connected to several outputs and 8 hidden units for modules connected to just one output. The generalization performance of each run is measured on account of the entire grid $[0, 15] \times [0, 15]$ with a resolution of 0.2.

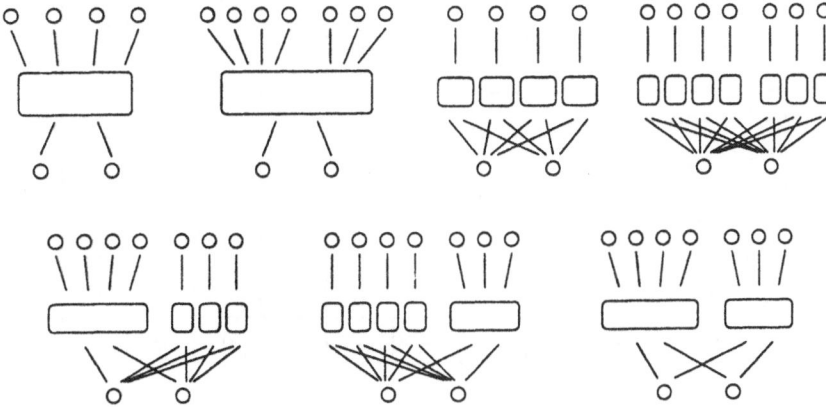

Figure 2: The various architectures compared in our simulations. Each oval represents an independent block of hidden units.

3 Discussion of results

The results of our simulations are summarized in table 2 where the various architectures are ranked according to their percentage of correct classifications on the independent test set. Two methods are called significantly different if in at least 35 of the 50 runs one method outperforms the other (binomial test with $p = 0.01$). Our results imply that ECOC yields a significant improvement over standard OPC if combined with a modular architecture. This modular architecture should be chosen such that the extra output bits are as independent as possible, both from the bits representing OPC and from each other. With less modularity ECOC no longer outperforms OPC, more modularity does not help but does not harm either.

The hidden units of a multi-layered perceptron build hyperplanes to separate the different classes. With all outputs connected to one block of hidden units, the extra output units will have a strong tendency to use the same hyperplanes as the first three units. The high correlation between the extra bits and the OPC bits makes it difficult to correct errors as well as to spoil correct classifications. This is easily confirmed by our simulations where the extra bits in the ordinary ECOC architecture correct and spoil only 0.08% and 0.06%, respectively, of the classifications made solely on account of the first four bits.

It is much less clear why it helps to give each extra output bit its own module. Extra bits that share their hidden units ("two blocks" and "five modules") correct about 2.5% and spoil about 1.5% of the classifications suggested by the OPC bits. ECOC architectures with separate modules for the extra bits ("four modules" and "modular ECOC") still spoil about 1.5%, yet correct about 4.0% of the OPC suggestions. No such an improvement in performance is seen when the first four bits have their own separate module. This apparent difference between the first four bits and the last three bits may be explained by the fact that for a four-class problem the code representing OPC is even more redundant than the code assigned to the last three bits.

	(1)	(2)	(3)	(4)	(5)	(6)	(7)
(1) four modules	0	0	+	+	+	+	+
(2) modular ECOC	0	0	+	+	+	+	+
(3) two blocks	-	-	0	0	0	0	+
(4) five modules	-	-	0	0	0	0	+
(5) ordinary ECOC	-	-	0	0	0	0	0
(6) ordinary OPC	-	-	0	0	0	0	0
(7) modular OPC	-	-	-	-	0	0	0

Table 2: Ranking and significance of differences between the architectures.

4 Conclusion and further directions

The higher the number of classes, the more one can win by using a highly redundant error-correcting output code instead of the standard one-per class code. We have shown that even on a problem with only 4 classes ECOC outperforms OPC, provided that it is combined with a modular architecture to decorrelate the extra bits from the OPC bits and from each other. We expect to find that classification problems with many classes will also benefit from a modular approach.

The effect of ECOC may be described in terms of bias and variance [3, 4]. For neural networks, bias and variance are more or less interchangeable, since by increasing the number of hidden units, we can always build a less biased model at the price of a higher variance. ECOC, especially when combined with a modular approach, requires a larger number of network weights than OPC. This may lead to a lower bias, while the increase in variance due to the larger number of weights is compensated by a decrease due to the combination of multiple votes. Further studies should substantiate these arguments.

References

[1] R. Anand, K. Mehrotra, C. Mohan, and S. Ranka. Efficient classification for multiclass problems using modular neural networks. *IEEE Transactions on Neural Networks*, 6:117–124, 1995.

[2] T. Dietterich and G. Bakiri. Solving multiclass learning problems via error-correcting output codes. *Journal of Artificial Intelligence Research*, 2:263–286, 1995.

[3] S. Geman, E. Bienenstock, and R. Doursat. Neural networks and the bias/variance dilemma. *Neural Computation*, 4:1–58, 1992.

[4] E. Kong and T. Dietterich. Error-correcting output coding corrects bias and variance. Technical report, Oregon State University, Corvallis, OR, 1995.

Applications of Neural Networks - Orals

Optimal Training for Neural Network Applied to Nuclear Technology

Ö. Ciftcioglu

Istanbul Technical University, Electrical Engineering Faculty,
ITU, Istanbul, Turkey.

E. Türkcan

Netherlands Energy Research Foundation ECN,
P.O. Box 1 NL-1755 ZG, Petten, The Netherlands

Abstract

Adaptive training of feedforward neural networks by Kalman filtering is described. Adaptive training is particularly important in estimation by neural network in real-time environment where the trained network is used for system estimation while the network is further trained by means of the information provided by the experienced/exercised ongoing operation. As result of this, neural network adapts itself to a changing environment to perform its mission without recourse to re-training. The performance of the training method is demonstrated by means of actual process signals from a nuclear power plant.

1 Introduction

The neural network training can broadly be divided into two categories as optimization and adaptive filtering. Classical BP is a gradient descent method of optimization executed iteratively with implicit bounds on the distance moved in the search direction in the weight space fixed via the learning rate, i.e., gain and the momentum term. As one can replace a gradient-descent algorithm with advanced optimization methods, stochastic approximation in on-line and/or off-line adaptive filtering can as well be replaced with an advanced stochastic algorithm, namely Kalman algorithm. Hence, adaptive form of Kalman algorithm can be used not only for effective and efficient training of feedforward neural networks but also it can be used for training in a continuously changing environment.

2 Neural Network Training by Kalman Filtering

Kalman filtering can be used for adaptive parameter estimation as well as optimal state estimation for a linear dynamic system. In the application of Kalman filtering to neural network training the parameters are the network weights which are considered as states and the estimation of the states is performed using the information applied at the network's input. However this information cannot be used directly as the size

This work is partly funded by the EU for the CEC — Accident Management Support project.

of the covariance matrix is equal to the square of the number of weights involved. To circumvent the problem we partition the global problem into a number of subproblems each of which is at the level of a single neuron. With the exclusion of the sigmoidal non-linearity part, the model is linear with respect to synaptic weights. In the Kalman filtering model the measurement information is not explicitly known. However, the residual noise is given by

$$\underline{e} = \underline{Y}(k) - \underline{C}(k) \cdot \underline{\hat{X}}(k|k-1) \tag{1}$$

and the influence of the single linear subsystem on the global network error can be estimated as $g = -\frac{\partial E}{\partial y}$. The energy (or cost) function E is defined by

$$E = \frac{1}{2} \sum_{n=1}^{N} \sum_{k=1}^{M} [d_n(k) - Y_n(k)]^2 \tag{2}$$

where M is the number of pair of input/output patterns; N, the dimension of the network's output; $d(k)$ and $\underline{Y}(k)$ the network's desired and actual outputs, respectively.

In a feedforward neural network with Kalman filtering each perceptron output is estimated by the Kalman estimator. Because of the nonlinearity introduced by the logistic function $f(x)$, extended Kalman filtering algorithm is required. However, since the analytic form of the nonlinearity is known, in place of the linearized filter algorithm, one can linearize the perceptron model so that a considerable saving in computation is gained. The two approaches are virtually equivalent. However, by means of the latter approach the stability characteristics of the estimator is enhanced which yields faster convergence in training [1].

3   Application to Nuclear Power Plant

Experimental studies are performed using actual measurement data from the Borssele nuclear power plant (NPP) in the Netherlands. Two data sets [2] designed to be A and B, are used in the present investigation. One of the data sets is obtained during normal operation with temporary power reduction for a certain duration and the other one is obtained during a $9\frac{1}{2}$ hours shutdown. The data are sampled with 8 s/s for AC and 1 s/s for DC. Afterwards, the DC data are formed with a sampling rate of 1 s/min. The first data set represents a power dip where the corresponding electric power varied from 475 MWe to 146 MWe and after a certain lapse of time from 146 MWe to 475 MWe back again. For the second data set, the power decreased from 375 MWe to 91 MWe where the data set is obtained during the stretch-out (extended burn-up of the fuel) followed by normal shutdown.

From 32 available signals, the most relevant ones are selected for the test. From the first data set A, 150 patterns are used for the training of the network. From the second data set B, 569 patterns are used for adaptive learning and estimation. The number of input nodes is twelve, the number of hidden layers is two and each hidden layer contains four nodes. The output has only one node which is used for the estimation of the generated electric power. Although commonly the number of hidden layers is taken to be one especially for practical reasons, here it is taken two.

The data set A where there is a power reduction of 70% is illustrated in Fig. 1 and the data set B where stretch-out and shutdown take place is illustrated in Fig. 2. It is important to note that these two operations are rather different mode operations which take place in different times, namely: A (October 10 ,1990), B (February 18, 1991). In Fig. 1, the measured and estimated data after the training are indicated together with the relative estimation errors in per cent the relevant scale being shown at the right side.

The training is performed by means of Kalman parameter estimation procedure which is relatively very fast compared to the weight determination procedure by standard BP algorithm. The determined weights being kept constant, the generated electric power estimation using the data set B in off-line mode is shown in Fig. 2 together with the relative estimation errors in per cent shown in the same way as it is in Fig. 1. From this figure it is seen that estimation errors are increasing as the power is going down. This is what one should expect because of the changes in the system dynamics in the course of time between two experiments. The error is getting even more pronounced when the power goes lower than 150 MWe which is the lower limit used during training the network.

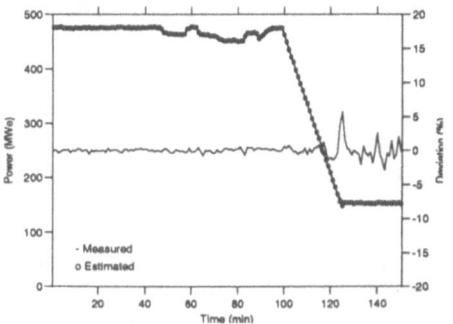

Figure 1: *Data-A obtained during Power dip and used for NN training*

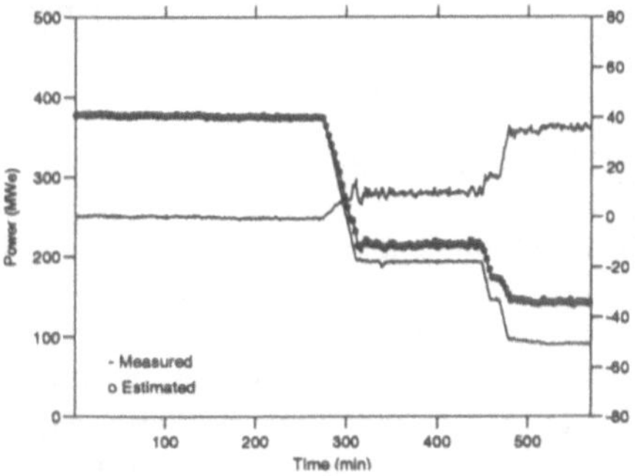

Figure 2: *Data-B obtained during shutdown and used for estimation*

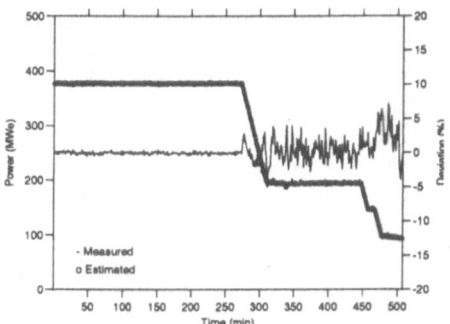
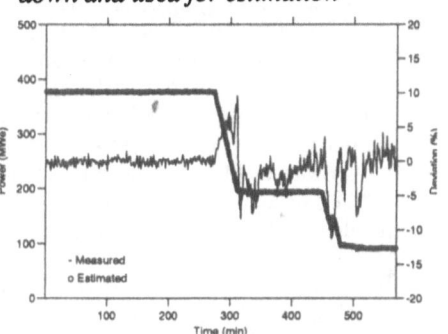

Figure 3: *Estimation of shutdown while performing adaptive training by Kalman*

Figure 4: *Estimation of shutdown by recall (no training)*

Following the studies described above, the data set B is processed by the adaptive realtime estimation and training algorithm which is operated by Kalman filtering methodology. In the case the difference between the measured and estimated value is within prescribed limits the real-time adaptive process remains in progress. Otherwise error message is issued and adaptive process is stopped by equating the Kalman gain to zero while the real-time process continues without interruption.

As result of the processing of the data set B by the adaptive training algorithm, the estimated generated electric power (GEP) and the deviations are shown in Fig. 3. Comparison of Fig. 2 with Fig. 3 reveals the effective estimation of the GEP using the previously trained network. Besides its adaptivity, to test the training performance of the adaptive algorithm, the adaptively trained network is used once more for GEP estimation without adaptive training, that is, keeping the network's weights constant and performing simple estimation commonly termed as 'recall'. The result thus obtained is illustrated in Fig. 4. Comparison of Fig. 2 and Fig. 4 indicates that as result of the adaptive process, the network's training is updated in the meanwhile. The deviations in Figs. 3 and 4 are comparable although in Fig. 4 it is somewhat inferior. On the other hand, the deviations in Fig. 2 and Fig. 4 are substantially different and much improved in the latter. In Fig. 4, it is also interesting to note that as result of adaptive process not only the large deviations seen in Fig. 2 are eliminated but also within the range of operation where deviations are relatively small (i.e., between 0 and 270 min approximately), the deviations still remain comparable while they are within acceptable limits. In another words, the newly gained information is absorbed by the network without significant change on the information already gained before, all information eventually being essentially persistent.

4   Conclusions

It is rather conspicuous that a feedforward neural network trained by using previously obtained measurement data from a dynamic complex system, the status of which is subject to change in the course of the operation, can be used for plant's state estimation. It should be pointed out that network is not maintained with the process information during the time interval between the acquisition of two data sets where, with reference to the system dynamics, the time interval is considerably long in all respects. This gap of missing information is recovered promptly by means of Kalman algorithm as the algorithm retains the retrospective information for a prescribed period of time (like exponential averaging). As the system dynamics to some extend remains the same, the algorithm quickly picks up the essential changes in the dynamics and let the network learn the new status without forgetting the previously gained information. However, since the neural network has a limited capacity for information storage, the previously gained information can be partly released as this is seen in Figs. 2 and 4 by comparing the portions between 0 and 270 min (approx.). As in the adaptive process by Kalman filtering there is no process noise, a process noise is introduced into the Kalman algorithm for the reason of stability. Because of this, the deviations are somewhat higher in Fig. 3 relative to those in Fig. 2 within the portions between 0 and 270 min.

Kalman algorithm in adaptive neural network training provides a fast, robust and reliable estimation. Since the system dynamics information is transformed into error covariance matrix at each perceptron, any significant change in the system dynamics results in accumulated effect of this change at the network's outcomes for immediate recognition. Since each perceptron is individually trained by Kalman algorithm, it is possible to govern the learning of each perceptron under the program control, yielding in control over the whole network's performance (accuracy, sensitivity, robustness, precision etc.), which is desirable and especially required in critical applications.

References

[1] Ciftcioglu Ö. and Türkcan E., *Adaptive Training of Feedforward Neural Networks by Kalman Filtering*, ECN-R--95-001, 1995, ECN, Petten, The Netherlands.

[2] Ciftcioglu Ö. and Türkcan E., *Selection of Hidden Layer Nodes in Neural Networks by Statistical Tests*, EUSIPCO-92, Signal Processing VI: Theories and Applications. J. Van de Walle et al. (Edts.), Elsevier, Amsterdam, 1992.

NONLINEAR PREDICTIVE CONTROL WITH NEURAL MODELS*

H.A.B. te Braake, H.B. Verbruggen

Delft University of Technology, Department of Electrical Engineering

P.O.Box 5031, 2600 GA Delft, The Netherlands

Tel: +31-15-78 33 71, Fax +31-15-62 67 38, Email: h.a.b.tebraake@et.tudelft.nl

H.J.L. van Can

Delft University of Technology, Department of Chemical Engineering

Julianalaan 67, 2628 BC Delft, The Netherlands

Tel: +31-15-78 50 25

1 Introduction

Nonlinear black-box modeling techniques are opening new horizons for modeling and control of nonlinear processes. These kind of models can be used in Model Based Predictive Control (MBPC). These techniques include Wiener Models, Fuzzy Modeling, Recurrent and Feedforward Neural networks and combinations of these. In MBPC, a process model is used to predict process response to alternative controller outputs. There are practically no restrictions with respect to the model structure, so that MBPC can very well deal with process nonlinearities. Model-based predictive control has become an important research area of automatic control theory and, moreover, it has been accepted also in industry [3]. A number of successful applications to industrial processes based on linear techniques has been reported, see [3] for a survey. The ability to handle input and output constraints straightforwardly is one of the reasons for this success.

Contrary to the linear case, application of nonlinear models in a nonlinear MBPC structure is not straightforward. To calculate the controller output a nonlinear, non-convex, optimization problem must be solved at every time instant. The goal of this contribution is to show that neural network modeling can be used as a black-box modeling method for (nonlinear) process modeling and, in combination with MBPC, nonlinear process control. To demonstrate the practical applicability, laboratory experiments are described where MBPC was applied to a nonlinear pressure control problem in a fermentor.

2 Nonlinear Model Based Predictive Control

Predictive controllers are based on the prediction of the future behavior of the process to be controlled (see figure 1). This prediction is obtained from a (non)linear model of the process that is assumed to be available.

---

*Submitted to 3rd SNN Neural Network Symposium, September 14-15, 1995 Nijmegen, The Netherlands

A predictive controller calculates a sequence of future controller outputs $u(k+1), u(k+2), .., u(k+H_p)$ over the prediction horizon H_p, such that the predicted output of the process is as close as possible to a reference trajectory defined by the operator. This control sequence is obtained by the optimization of a certain criterion function which describes the control goals. An example of a criterion function to be minimized is:

$$J = \sum_{i=1}^{H_p} \left(\hat{y}(k+i) + w(k+i) \right)^2 + \beta \left(u(k+i-1) \right)^2 \tag{1}$$

with $\hat{y}$ being the predicted process output, w the desired process behavior (reference trajectory) and u the future control signal, weighted by a scalar β. Such a cost function aims simultaneously at the minimization of the output error and the energy spent on manipulating the process input. Parameter β determines the weighting of these two criteria. Usually, the prediction horizon is

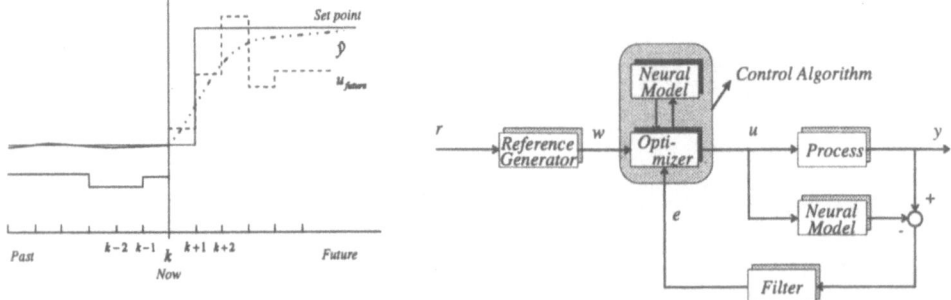

Figure 1: Nonlinear Predictive Controller. Left: Principle of Nonlinear MBPC. Right: Control scheme. r denotes the set point, w the reference signal (reshaped set point), u the calculated control action and e the error between the process and the neural process model.

as large as possible but at least it is larger than the process delay time. To avoid too many calculations and to stabilize non-minimum-phase plants, commonly the control horizon, i.e. the number of future control actions, is defined smaller than the prediction horizon. To construct an 'open loop feedback controller', only the first of the calculated control signals is applied. At every sample time, the optimization is repeated with new measurements, the previously calculated future control actions may be used as initial values for the next optimization.

It is known that a predictive controller based on a linear process model, in the absence of constraints, results in a stable control law which can be calculated analytically, minimizing a predetermined criterion [4], [5]. In the case of MBPC using a neural model, an optimal control action only can be calculated by means of numerical optimization of the criterion function. The resulting controller (optimizer) is a part of the MBPC structure, as depicted in figure 1 (right figure). Based on a neural process model the 'Optimizer' finds an optimal solution of the criterion function and the first element of the resulting control sequence can be applied to the process. Both process inputs and outputs can be subject to operation constraints which should be incorporated in the optimization routine as hard or soft constraints.

Due to the recalculation of the controller outputs at every sampling instant, a combination of 'feedforward' (prediction part) and 'feedback'(recalculation at every time instant) action is present, which is able to react on differences in process and reference requirements and to guarantee a off-set free response. If the predicted and process outputs are not equal (due to noise, disturbances or model-process mismatch) a feedback loop corrects this error. A 'Filter' (see figure 1) is designed to filter out the process noise and to make the loop more stable. A reference generator can be used to reshape the set point to a smoother reference signal, before it is used in the calculation of the control action.

3 Control Experiments

The Neural MBPC principle has been applied to the control of the pressure in a fermentor. For details see [2]. To obtain a neural process model, training data must be available. In the case of the fermentor experiments, the pressure inside the fermentor and the valve position are used as output y and input signal u, resp. The resulting model can be depicted by: $\hat{y}_{k+1} = f_{nn}(y_k, u_k)$, with $f_{nn}(.)$ denoting a standard feedforward neural network with one hidden layer, two inputs (u_k and y_k) and one output (y_{k+1}). Once a 'good' training set is obtained, training of a neural network is nothing else than estimation of the network weights. There are various training algorithms among which backpropagation is the most well known.

However, because it is a gradient descent type of optimization algorithm, it is rather slow and sensitive to local optima. Te Braake and Van Straten [1] for example, proposed a method based on the fixation of the weights between the input and the hidden layer, in such a way a linear-in-the-parameters estimation problem remains, which can be solved with standard least squares techniques. This results in sufficient network structures which are obtained in a fraction of required training time compared to backpropagation.

To find the parameters of a linear black box model, random input signals are often used, because these kind of signals contain all frequencies. The superposition principle and the theory of persistent excitation can be applied. If a nonlinear model must be found, the linear systems theory can no longer be applied. The choise of the training data is then a 'process-dependent' task and depends on the operating conditions of the process.

Once a suitable neural model has been obtained, the model can be used in a Nonlinear MBPC control structure as is shown in figure 1. Two types of constraints are used: input-rate- constraints and input-level-constraints (i.e. the valve position can not be opened or closed more then 90% or 10% resp. and it may not be changed more then 10% each sampling time). In figure 2 the results of both a Neural MBPC controller and a conventional PI controller are shown.

As is shown in this figure, due to it's predictive nature, the controller is capable of tracking the sinusoidal setpoint, even in the presence of constraints. Other experiments, which are not shown here, show that the controller is also capable of handling disturbances and severe process noise.

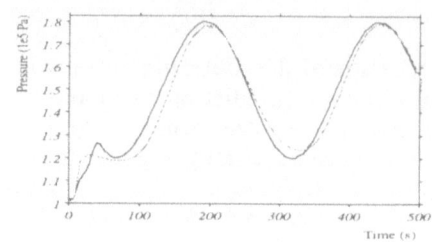

Figure 2: Performance of the neural predicive controller on a sinusoidal setpoint change. .. = set point, .-. = PI and - = Nonlinear MBPC

4 Conclusions

Neural process models provide an effective and convenient way to obtain non-linear models valid over a broad operating range of the process. This makes implementation and tuning of the predictive controller easier than with linear models, which are only valid around a certain operating point. Compared to nonlinear models based on physical knowledge they are easier to identify. Since the modeling part in the MBPC design is the most difficult and time consuming one, significant savings of time and money may be achieved by applying neural networks to model the process.

Neural models can be applied directly in an MBPC structure as numerical predictors forecasting the future behavior of the process. Due to their nonlinear nature, analytical formulation of the control law is not straightforward. Therefore, the MBPC must rely on a good optimization method. In case of complex, multivariable systems, computational limitations may become an issue.

References

[1] Braake, H.A.B. te, G. van Straten (1995). Random Activation weights neural network for fast noniterative training. *Engng. Applic. Artif. Intell*, vol. 8, no. 1, pp. 71-80.

[2] Can, H.J.L. van, H.A.B. te Braake, C. Hellinga, A.J. Krijgsman, H.B. Verbruggen and K.Ch.A.M. Luyben (1994). Design and real-time testing of a neural model predictive controller for a nonlinear system. Accepted for publication in *Chemical Engineering Science*

[3] Richalet, J. (1993). Industrial Applications of Model Based Predictive Control. *Automatica*, vol. 29, pp. 1251-1274.

[4] Soeterboek, R. (1992). Predictive Control; An Unified Approach. First Edition, Prentice Hall International, U.K.

[5] Vries, R.A.J. de and T.J.J. van den Boom (1994), Constrained Predictive Control with guaranteed stability and convex optimization. In: *Proc. Am. Contr. Conf., Baltimore, U.S.A.*, pp. 2842-2846.

SYSTEM IDENTIFICATION WITH ORTHOGONAL BASIS FUNCTIONS AND NEURAL NETWORKS*

Gerard Schram, Michel H.G. Verhaegen, Ardjan Krijgsman

Delft University of Technology, Department of Electrical Engineering

P.O.Box 5031, 2600 GA Delft, The Netherlands

Tel: +31-15785114, Fax +31-15626738, Email: g.schram@et.tudelft.nl

Pejman Djavdan

DSM Services

Mauritslaan 49, 6129 EL Urmond, The Netherlands

Tel: +31-46763942

1 Introduction

For the control of a process, usually the relation between past input-output data of the process and future outputs must be identified. For the identification of nonlinear systems, neural networks can be used [3]. In this context, neural networks are nonlinear black-box models, to be used with convential parameter estimation methods. Two important models are:

- **NNFIR-models:** Neural Network Finite Impulse Response models, which use only past process inputs $u(k - n)$ as inputs for the network;

- **NNARX-models:** Neural Network Auto Regressive with eXogeneous input models, which use past process inputs $u(k - n)$ and past process outputs $y(k - n)$ as inputs for the network.

When using NNFIR-models, the whole dynamic response time should be covered by past inputs, leading to models of large order. On the other hand, an NNFIR model results in a stable simulation model. Including past outputs generally leads to the smaller NNARX models, and are therefore often used. However, NNARX models may become unstable in simulation since past process outputs are then replaced by past model outputs $\hat{y}(k - n)$.

In this paper, a variant of NNFIR models is proposed that yields *stable* as well as *small* models. The idea is to use modified inputs obtained by orthogonal filtering. For this new model structure, an identification procedure is developed and illustrated by the identification of the product quality of a destillation column.

---

*Submitted to 3rd SNN Neural Network Symposium, September 14-15, 1995 Nijmegen, The Netherlands

2   A new identification method

A linear system is usually identified by estimating its Markov parameters, the coefficients of a Laurent series expansion. Alternative expansions in terms of orthogonal functions, such as Laguerre functions, can reduce the number of expansion coefficients considerably for systems with large response time [5]. In general, any stable dynamical system gives rise to a set of orthogonal functions in a natural way [2]. The small number of model parameters is interesting from both aspects of bias (accurate approximation is possible) and variance (few parameters to be estimated from data).

We expect that the use of othogonal basis functions has similar benefits for NNFIR models. It is known that NNFIR models are suited for the approximation of nonlinear dynamical systems [1]. The use of orthogonal functions could reduce the number of required inputs, leading to small-sized models. For tuning the orthogonal functions, we estimate the linear dominant dynamics of the system, using identification techniques for Wiener systems [4]. A Wiener system is a series connection of a linear dynamical system with a static readout map. Summarizing, the new identification method involves:

1. Estimation of the linear dominant dynamics of the nonlinear system [4];

2. Construction of orthogonal functions [2];

3. Approximation of the nonlinear, static mapping from the transformed input space to the output using a feedforward neural network.

3   Identification of a destillation column

In order to control the product quality of a destillation process, the impurity of the product must be known. Since the impurity is only measured infrequently, the goal is to perform predictions based on frequently measured temperatures inside the destillation column. From industry (DSM), we get a set of 1800 samples of temperature and impurity, which will be used as input and output for the model, respectively.

First, the dynamics of the system are estimated using the method in [4]. Two poles, $a_1 = -0.99$ and $a_2 = -0.94$, indicate a dominant 2nd-order process with a large response time. Note that an accurate estimation of the dynamics is important, but not necessary. The estimate has to be as good as possible, but in principle any set of orthogonal functions satisfies [1].

Secondly, the othogonal filters are constructed following [2]. For this purpose, an inner function $G(z)$ is defined representing the dynamics of the process:

$$G(z) = \frac{a_1 a_2 z^2 + (a_1 + a_2)z + 1}{z^2 + (a_1 + a_2)z + a_1 a_2}$$

Then, an orthogonal series expansion based on this inner function is performed. Define $c = -a_1 a_2$ and $b = \frac{-(a_1 + a_2)}{a_1 a_2 + 1}$. Then, the orthogonal expansion is depicted in Figure 1, with $V_0(z)$ given as:

$$z^{-1} V_0(z) = \frac{\sqrt{(1 - c^2)}}{z^2 + b(c - 1)z - c} \left[\frac{\sqrt{(1 - b^2)}}{z - b} \right]$$

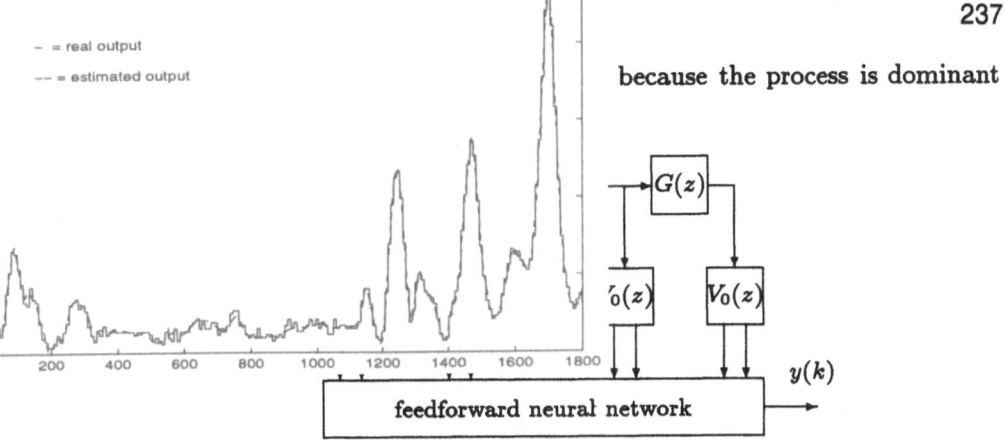

because the process is dominant

Figure 1: Block schematic representation of approximating a 2nd-order nonlinear dynamical system by orthogonal basis functions and a neural network.

For the last step, a simple multilayered network with one layer of sigmoid units is used. The number of inputs equals the number of basis functions. The number of hidden units is chosen by trial and error. The Levenberg-Marquardt optimization method is used for learning. After 50 iterations learning is terminated, since no significant progress is made anymore. For convenience, the Neural Networks using Othogonal Basis Functions are denoted by NNOBF.

The classical approach to this problem is to fit a linear model between inputs and the logaritm of the measured impurity. For comparison, also NNFIR and NNARX models are parametrized. Since the response time is large, the time delays for the NNFIR and NNARX models are set at 8 time steps. The NNARX models use the current input, and one or two sets of past inputs and process outputs, respectively. Table 1 shows the results for the models.

Model	Network size	VAF data set (%)
classical	-	82
NNFIR	6-12-1	95.2
NNARX	3-8-1	98.4
NNARX	5-10-1	98.9
NNOBF	2-8-1	97.2
NNOBF	4-10-1	99.3
NNOBF	6-12-1	99.5

Table 1: Identification results for different models. VAF is the Variance Accounted For, defined by: $VAF = 100\% \cdot \left[1 - (\|y - \hat{y}\|/\|y\|)^2\right]$. A VAF of 100% indicates a perfect model and 0% the mean value of outputs.

The result of the NNFIR model is not satisfactory due to the large response time of the process. As was expected, the results for the NNARX models are better than the NNFIR model. A small NNOBF model, with just 4 basis functions and a neural network with 10 hidden units, already outperforms the NNFIR and NNARX models. Increasing the number of basis functions yields even more accurate NNOBF models. In figure 2, the estimated and real impurity of the destillation column are plotted for the 4-10-1 NNOBF model.

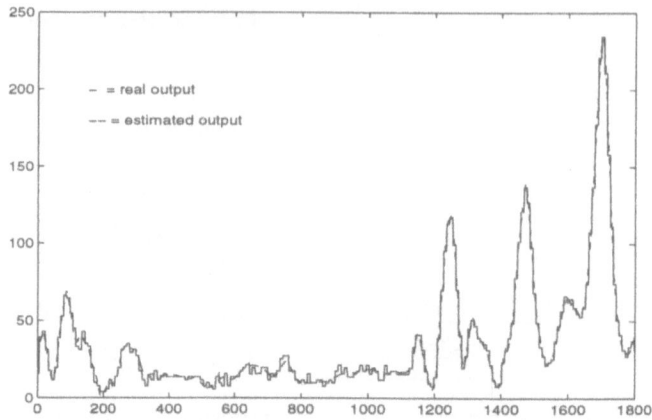

Figure 2: Real and estimated impurity of the distillation column.

The model has to be used for predicting the impurity. However, the data set is too limited for identification *and* validation of the model, because the persistently exciting property does not hold for both sets. In order to circumvent this problem, an on-line version of the new identification method is introduced. The main idea is building up a data set during operation. Simulations indicate that, while the model is updated on-line, the output is predicted accurately (not shown here).

4   Conclusions

The idea of the new method is to approximate a nonlinear dynamical system by a series expansion in terms of orthogonal basis functions followed by a neural network. Identification of a destillation column shows that small and stable models are obtained.

References

[1] Boyd S, Chua LO (1985), Fading memory and the problem of approximating nonlinear operators with Volterra series. *IEEE Transactions on Circuits and Systems*, vol cas-32, no 11, pp 1150-1161.

[2] Heuberger PS, Van den Hof PMJ, Bosgra OH (1995), A generalized orthonormal basis for linear dynamical systems. To appear in: *IEEE Transactions on Automatic Control*.

[3] Sjöberg J, Hjalmarsson H, Ljung L (1994), Neural Networks in System Identification. In: *Preprints 10th IFAC Symposium on System Identification*, Copenhagen, Denmark.

[4] Westwick D, Verhaegen M (1996), Identifying MIMO Wiener systems using Subspace Model Identification Methods. Invited paper for *Signal Processing Special Issue* 1996.

[5] Wahlberg B (1991), System Identification using Laguerre models. *IEEE Transactions on Automatic Control*, vol 36, pp 551-561.

Applications of Neural Networks - Posters

Applications of Neural Networks to pH Control

José M. Aragón, María C. Palancar and José S. Torrecilla.
Dept. of Chemical Engineering. Faculty of Chemistry.
Universidad Complutense de Madrid.
28040-Madrid (Spain).

Abstract

The neutralization of an acidic waste water was controlled by two NNs. One predicts future pH values, the other manipulates an alkaline current.

1 Introduction

The control of pH is a typical nonlinear and asymmetric process [1]. The regulation of pH is required in several processes. For example: the neutralization of acidic industrial streams; the stabilization of waste water; and several microbiological, enzymatic and electrolytic processes.

The control based on neural networks (NN) has proved its great potential to do adaptive and predictive actions. The information reported in the literature in the last years [2-3] shows that the control of pH has been explored by using several control algorithms. Nevertheless, there are not absolute criteria to establish which type of control system is the best one, because it depends on the control specifications required and the nature of the neutralization processes involved in each particular case. We have studied in the last years diverse pH control systems based on PID, adaptive PI and adaptive-predictive models. In this communication we will describe only the most recent results obtained with diverse pH control systems based on neural networks.

2 Methodology

2.1 Neutralization System

The system to be controlled is a neutralization tank for stabilizing an acidic waste water of pH ranging between 5 and 6. The experiments were made at bench scale and they involve the following procedure: a stream containing acetic is fed into a continuous stirred tank reactor; it is fed simultaneously a neutralizing stream that

242

contains an alkali (sodium hydroxide). The pH at the reactor output is periodically measured and transmitted to an on-line computer (sampling time: 2 s). The computer sends periodically the appropriate signal to an automatic valve which regulates the flow rate of the alkaline solution. The response of the system is measured under diverse perturbations of the pH set point and the flow rate of the acidic current.

2.2 Signal Processing and Codes

We have developed several computer codes that perform:

. Input signal processing (management of an A/D board, pH measure).
. Simulation of the neutralization process (based on material balances).
. Input of pH and valve aperture data in the predictive NN.
. Control based on a NN (inverse plant model).
. Output signal processing (management of a D/A board, valve position).
. Learning of both NNs.

2.3 Structure of the Neural Networks

The adaptive-predictive control system here described is based on the combination of two NNs. The first NN is a feedforward network with sigmoid transfer function and backpropagation algorithm. It identifies the neutralization process, and makes a prediction of five future pH values. The control NN is based on the concept of inverse plant model.

The first NN consists of two layers and an input layer (13, 7, and 5 neurons, respectively) with past and present values of pH and valve aperture. The second NN has a similar structure (9 inputs of pH and 4 input of valve apertures, one hidden layer with 7 neurons, and an output layer with 5 future valve apertures).

The number of hidden layers was limited to one to minimize the complexity of the networks. The seven hidden neurons resulted from a compromise between accuracy and generality of the learning procedure to be used in the control system.

Both the predicting and the control NN require a periodic on-line updating of the weights and an off-line learning with experimental data. After learning, the nets give excellent predictions and control.

3 Results

3.1 Optimization of the Neural Networks

For optimizing each one of the NNs and the whole control system, they were studied and tested the following elements:

. The type and number of input and output variables.
. The role of the NNs (identification, prediction, feedforward action).
. The number of neurons in each layer.
. The strategy of learning and the value of the learning rate coefficient.
. The requirements of periodic on-line updating weights.
. The type of control loop (feedback, feedforward, inverse plant model)

3.2 Performance of the Controller

The control loop has been tested with several perturbations of the pH set point and the acid flow rate. The results show that the controller is able to achieve and maintain a pH set point very efficiently in the majority of circumstances. Having a pH set point equal to the neutrality point of the acid/base system is a typical unfavourable condition. Besides, both the neutrality point and the slope of the neutralization curve depend on the nature and concentration of the pair acid/alkali.

An example of the start up of the neutralization operation is shown in Figure 1. The response of pH corresponds to a situation in which the reactor was started-up with an acid mixture of pH = 6.2. The pH set point was fixed in 7. This desired value was reached in less than 3 min. The valve aperture (the dotted curve XV) is also shown.

An example of the roughness of the control system against perturbations of the pH set point is shown in Fig. 2. The incoming acidic current had in this case a constant pH = 5. The pH set point (dotted line) was increased from 5.5 to 10 in successive steps of 0.5 pH units.

References

1. Nahas EP, Hensoon MA, Serborg DE. Nonlinear internal model control strategy for neural network models. Computers Chem. Eng. 1992; 16(12):1039-1057.

2. Bath M, McAvoy TJ. Use of neural nets for dynamic modelling and control of chemical process systems. Computers Chem. Eng. 1990; 14(4/5):573-583.

3. Aldrich C, Sanie J. Comparison of different artificial neural nets. Ind. Eng. Chem. Res. 1995;24:216-224.

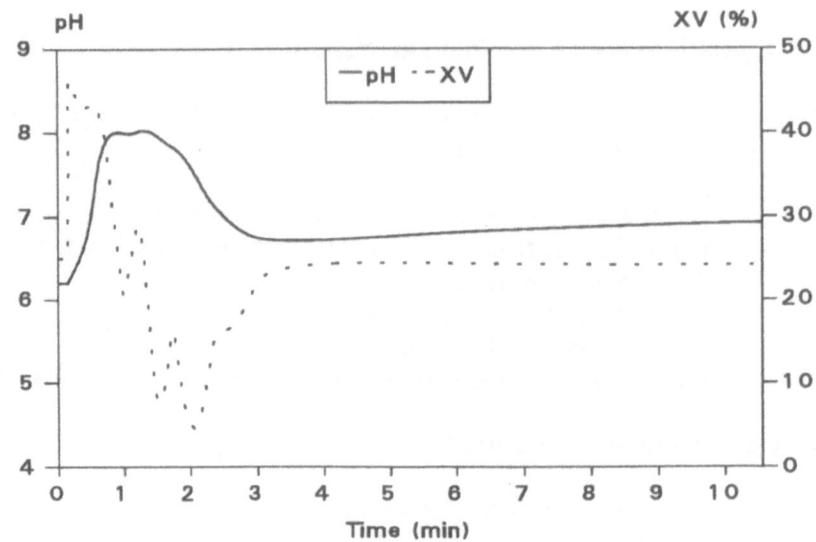

Figure 1. Start-Up

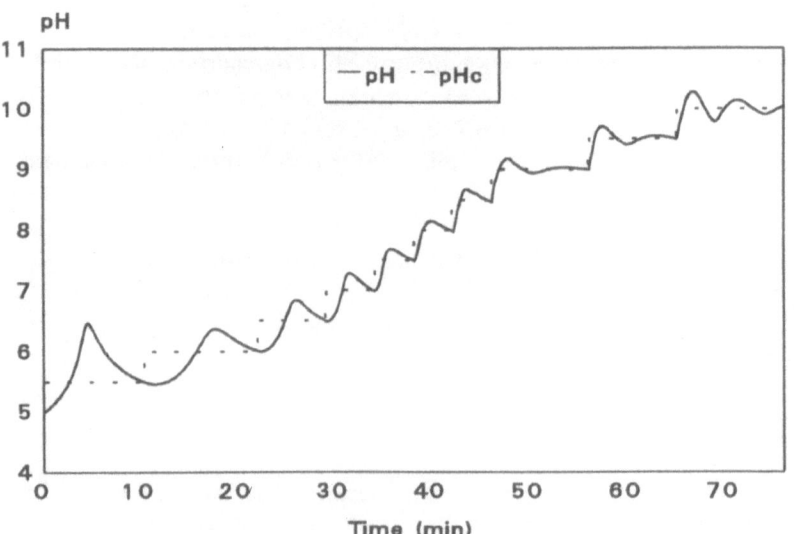

Figure 2. Set Point Perturbations

Appropriate Context Association and Learning Parameters for Word Spotting with Partially Recurrent Neural Networks

D. v. Leeuwen and P. Wittenburg
Max Planck Institute for
Psycholinguistics
P.O. Box 310
6500 AH Nijmegen

M. Poel
University of Twente
Department of Computer Science
P.O. Box 217
7500 AE Enschede

1 Introduction

This paper covers part of a study which concerns the feasibility of real-time word spotting with partially recurrent neural networks (PRNN's). PRNN's have already proven appropriate for other examples of pure sequence recognition [1, 2]. However choices concerning architectural and learning aspects are still hard to make. One of the questions still to be answered, is how these aspects influence the term of memory of a PRNN. This paper tries to obtain some directives regarding architectures and learning algorithms.

2 Context Neuron vs. Context Connection

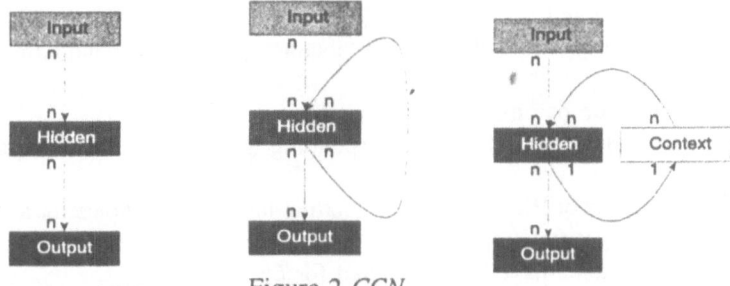

Figure 1 *MLP.*   Figure 2 *CCN.*

The PRNN's treated in this paper may best be described as standard multi layer perceptrons (MLP's) with one hidden layer, which are extended by fully interconnecting the hidden layer. In general, there are two ways to represent recurrent connections and the time delay that is associated with them:

- Either directly, by associating time delay with a special recurrent connection. Such a PRNN will be referred to as context connection network (CCN).
- Or by associating time delay with a special linear neuron [3] which is inserted at the end of a recurrent connection with unity weigth, together with a second connection. Such a PRNN will be referred to as context neuron network (CNN).

Figures 1, 2 and 3 show a MLP, a CCN and a CNN. Although both context associations resolve themselves into mathematically equivalent functionality, the corresponding backpropagation based learning algorithms offer different possibilities. CNN's are usually trained with backpropagation by ignoring the connections with unity weights during backpropagation of errors [3]. CCN's may be trained with truncated backpropagation through time [4, 5, 6]. Doing so, the truncation parameter T determines the depth of the unfolded network used for training, and hence the accuracy of the gradient approximation made. If T equals the length L of the subsequence to be recognized (SS), exact gradient computation is achieved.

Using $T = 2$, both approaches only use the current network activations and the most recent network activations back in time. In that case the only mathematical difference between the two algorithms is that CNN's only use the hidden activations back in time, whereas CCN's use the activations of the entire network back in time.

3 Experiments

In order to get a grip on the relation between the term of memory of networks and various architectural options as well as the relation between the term of memory of networks and various learning options, some investigations have been executed, using artificially generated data. The following architectural parameters determined the complexity and most importantly, the term of memory of the PRNN involved:

- Type of context association (either context neuron or context connection).
- Size of the fully connected hidden layer.

In randomly generated binary input pattern sequences (IPS's), the occurrences of a certain SS with length L are located. The target function indicates these locations. Two learning parameters influenced the obtained performance and term of memory:

- The truncation parameter T (for CCN's).
- The target delay Δt between corresponding input patterns and output patterns.

Future applications of PRNN's to word spotting justifies the choice for a simple pulse function as a target function.

The experiments have been executed for L equal to 2, 3 and 4. Both train and test sets consisted of a randomly generated IPS with approximately 128 occurrences of the SS. Both CNN's and CCN's have been used. For each of the three SS's the number of hidden neurons was chosen in such a way that the network was capable of learning to recognize the SS in question sufficiently well, even for the largest Δt involved. The largest Δt involved equaled L - 1. For CCN's, T varied from 2 to L + 1. In figure 4 an occurrence of SS '1111' has been indicated by 4 target functions with Δt's varying from 0 to 3.

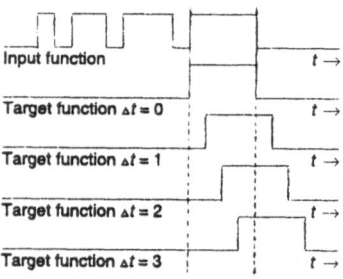

Figure 4 *Target functions for various Δt's.*

As IPS's are generated randomly and an interpolation is expected to be made during learning, the eventual output function is expected to yield a probability. This probability tells us to what extent the SS has been recognized at that specific moment in time. In case Δt equals 0, this probability equals $2^{(\text{Length(Subsequence already recognized)} - L)}$. This expected behavior will be referred to as "Cohort"-like behavior [7]. It is obvious that the expected "Cohort"-like behavior allows the sum of the squared errors (SSE) of a data set to be estimated. In figure 5 the expected output functions corresponding to the target functions in figure 4 have been depicted. Comparison of figures 4 and 5 shows that the expected SSE decreases to 0 if Δt increases to L - 1.

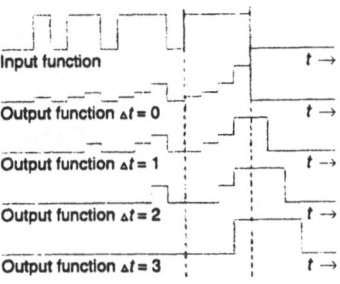

Figure 5 *Expected output functions for various Δt's.*

The SS's equaled '10', '010' and '0010'. The number of hidden units that proved to be sufficient for learning these SS's turned out to be 3, 4 and 8 respectively. The estimated SSE's for Δt equal to 0, equaled 67, 130 and 243 respectively.

4   Results

L	2		3			4			
Δt	0	1	0	1	2	0	1	2	3
CNN	**67**	17	**127**	45	98	**235**	152	96	245
CCN T = 2	**68**	27	**127**	48	134	**235**	181	84	291
CCN T = 3	**68**	1	**125**	46	37	**224**	146	58	37
CCN T = 4	-	-	**124**	45	2	**225**	141	54	62
CCN T = 5	-	-	-	-	-	**224**	142	57	3

Figure 6 *The various minimal SSE's for L equal to 2, 3 and 4.*

Figure 6 presents the SSE's obtained by the various training processes. The bold SSE's obtained with $\Delta t = 0$ meet their expectations, i.e. approximate their estimates.

It seems that the differences in the obtained SSE's between different T's, and hence between CCN's and CNN's, are small for minimal Δt's. This seems promising, as a small T decreases the complexity of learning significantly. An explanation of small differences could be the line of argument which justifies data segmentation and gradient approximation: If the networks' state represents the thusfar processed IPS, a time alignment between the networks' state and the next input pattern resolves itself into a time alignment between the thusfar processed IPS and the next input pattern.

In case Δt equals L - 1, a T that seems sufficient and necessary to obtain a SSE that approximately converges to 0, equals Δt + 2. This too subscribes the expectations, although the necessary magnitude of T needs some explanation as well:

T must be large enough to bridge the gap between input and corresponding output, as the state of the network no longer represents the IPS that corresponds to the networks' output, but represents the IPS that corresponds to some future output. In case T equals $\Delta t + 2$, the networks' state participates in learning until the error of that future output has been propagated back.

5 Conclusions

Summarily, the results give rise to the following conclusions regarding context association and learning parameters:

- The assumption that the networks' state contributes to the obtained term of memory during learning, is subscribed by the results.
- In accordance with their mathematical similarities, CNN's and CCN's with $T = 2$ perform comparably well.
- T is stronger correlated to Δt than to L.
- T should equal at least $\Delta t + 2$.

It is obvious that choices concerning the type of context association and parameters like T and Δt should depend on the desired behavior. For "Cohort"-like behavior a small T (e.g. 2), and hence CNN's, and a small Δt (e.g. 0) seem sufficient, but for a SSE approximately equal to 0, a larger T (e.g. $L + 1$), and hence CCN's, and a larger Δt (e.g. $L - 1$) seem necessary. Applying these attainments to word spotting, "Cohort"-like behavior seems preferable, for then T and Δt seem less dependent on L which tends to vary due to variation in duration of various utterances of one word. Still, a smaller SSE facilitates a more smooth convergence during learning.

Currently, experiments are evaluated which have to point out what demands the behavioral desires mentioned, make with respect to the number of hidden neurons.

References

[1] Hertz J., Krogh A. and Palmer R.G., Introduction to the theory of neural computation, Addison-Wesley Publishing Company, Amsterdam, 1991.

[2] Couwenberg R., Speech Recognition with Recurrent Neural Nets, University of Twente, Department of Electrical Engineering, Enschede, 1990.

[3] Elman J.L., "Finding structure in time", *Cognitive Science*, 14:179-211, 1990.

[4] Minsky M. and Papert S., Perceptrons, Maple Press Company, Cambridge, MA, 1969.

[5] Hinton G.E., Rumelhart D.E., and Williams R.J., "Learning Internal Representations by Error Propagation", *Parallel Distributed Processing*, vol. 1, chap. 8, 1986.

[6] Zipser D., "Subgrouping reduces complexity and speeds up learning in recurrent networks", *Advances in Neural Information Processing Systems II*, pages 638-641, San Mateo, California, 1990.

[7] Marslen-Wilson W.D., "Speech understanding as a psychological process", *J.C. Simon (ed.), Spoken language generation*, Dordrecht: Riedel, 1980.

EEG SIGNAL ANALYSIS USING DYNAMIC TIME WARP TRANSFORMATION AND KOHONEN'S NEURAL NETWORK

F. Cremer and L.P.J. Veelenturf

Research group BSC-NT, Faculty of Electrical Engineering,

University of Twente, PO Box 217, 7500 AE Enschede, The Netherlands

Email: FCremer@nt.el.utwente.nl & L.P.J.Veelenturf@el.utwente.nl

Tel: 053-892842, Fax: 053-334701

Abstract. The detection of one dimensional signals patterns often requires some insensitivity to variations in the length of the pattern, which is to be detected. In these cases the Euclidean distance between the target and an unknown pattern vector is not a usable distance. A new method of dynamic time warping, called DTW Transformation (DTWT) solves this problem. It indicates how the unknown pattern vector and the target differ in length and shape. The resulting time-vector and amplitude-vector of this transformation are used to train two different Kohonen's neural networks to eventually classify the unknown pattern vector. A complete detection system is built on this transformation, which can detect spike-wave complexes (SWC's) in electroencephalograms (EEG's).

1 Dynamic time warping as vector transformation

Dynamic time warping (DTW) is a distance measure, which is a method used to compare two vectors of different size, see [1]. The limitation of this distance measure is that it does not indicate how these vectors differ. The algorithm can be adapted into a transformation. This reversible transformation, called Dynamic Time Warp Transformation (DTWT), tries to map a candidate-vector $\overline{V}_C$, with a size of $L_C = \dim(\overline{V}_C)$, onto a template-vector $\overline{V}_T$, with a size of $L_T = \dim(\overline{V}_T)$. DTWT generates two vectors: a time-step-vector $\overline{V}_S$. and an amplitude vector $\overline{V}_A$. By definition the size of these vector equal to the size of the template: L_T. The time-vector $\overline{T}_T$, the integral of the time-step-vector is defined as:

$$\overline{T}_T(k) = \sum_{i=0}^{k} \overline{T}_S(i), \text{ with } 0 \leq k \leq LT \tag{1}$$

At each point i the value of $\overline{T}s(i)$ (for $0 \leq i < L_T$) is recursively defined by:

$$\overline{V}_S(i) = 0 \quad \text{IF} \quad \min_{S \in \{0,1,2\}} \left\| \overline{V}_T - \overline{V}_C(\overline{T}_T(i-1) + S) \right\| \text{ is at S=2}$$

$$\text{AND } \overline{T}_S(i-1) \neq 0 \tag{2}$$

$$\overline{V}_S(i) = 2 \quad \text{IF} \quad \min_{S \in \{0,1,2\}} \left\| \overline{V}_T - \overline{V}_C(\overline{T}_T(i-1) + S) \right\| \text{ is at S=0}$$

$$\text{AND } \overline{V}_S(i-1) \neq 2 \tag{3}$$

$$\overline{V}_S(i) = 1 \qquad \text{ELSEWHERE} \tag{4}$$

with $\overline{T}_T(-1) = 0$ and $\overline{T}_S(-1) = 1$, by definition

The last condition in equation (2) and (3) ensures that the effective size of the candidate lies between $\dfrac{2}{3} * L_T$ and $2 * L_T$. The amplitude vector is then defined by:

$$\overline{V}_A(i) = \overline{V}_C(\overline{T}_T(i)), \text{ with } 0 \leq i < L_T \tag{5}$$

This transformation can be used e.g. to distinguish between a ramp function and a sinusoid function, see figure 1. A normal sinus is used as template. In normal DTW the distance of both the sinusoid and ramp function are more or less the same, since the amplitude vector are more or less equal. But with the DTWT the time-step-vector is rather different.

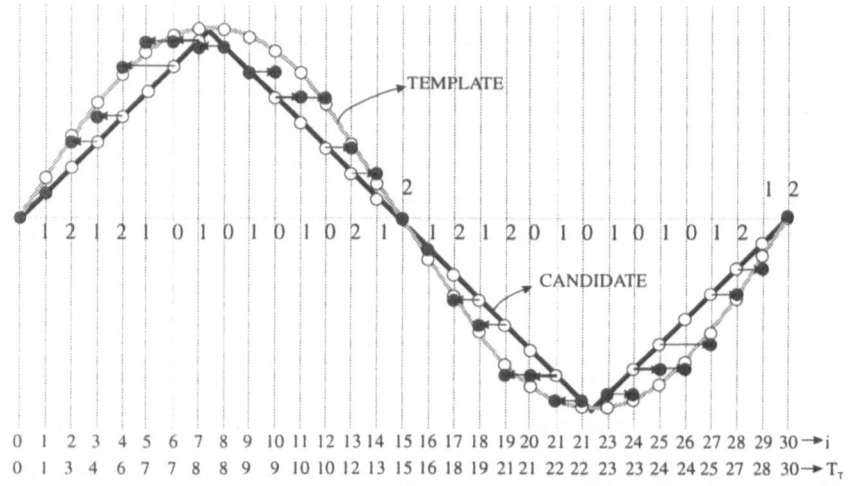

Figure 1: DTW Transformation on a ramp function

The original candidate vector can be reconstructed, with the number of samples missing equal to the number of time-step-vector components with the value 2, with the following equation:

$$\overline{V}_C'(\overline{T}_T(i)) = \overline{V}_A(i), \text{ with } 0 \leq i < L_T \tag{6}$$

With interpolation the algorithm will give continues values for $\overline{T}_S$ between 0.5 and 1.5. This results into a better mapping onto the template vector. In cases where the mapping is optimal, i.e. $\overline{V}_A = \overline{V}_T$, the candidate vector should be equal to:

$$\overline{V}_C'(i) = \overline{V}_T(\overline{T}_T(i)), \text{ with } 0 \le i < \overline{L}_T \tag{7}$$

In that case the candidate is a time distorted version of the template, with the same phase as the template. For $\overline{T}_T(i) = k$, the candidate differs only in size from the template and thus has the same shape.

2 A detection system with DTW Transformation

The problem of spike-wave complex detection, see [3], is a problem where DTWT can be applied. The DTWT is used to generate feature vectors for spike-wave complex detection in electroencephalogram's. These spike-wave complexes are symptoms of epilepsy. Since the time variation is separated from the signal the Euclidean distance can be used.

A candidate vector is generated out of a window of 200 samples of the EEG signal, see figure 2. The DTWT is carried out on this vector. One of the known spike-waves is used as template vector. The resulting time and amplitude vectors are resampled with a factor 3 to reduce the dimension of these vectors.

Kohonen's Neural Network, see [2], is used as a classifier for these vectors. For each vector a separate network is used. In the training phase, the time-vector and amplitude-vector are used to train a 10 by 10 network. In the label phase each networks are independently labeled. Two methods of labeling are used master-slave labeling and majority labeling.

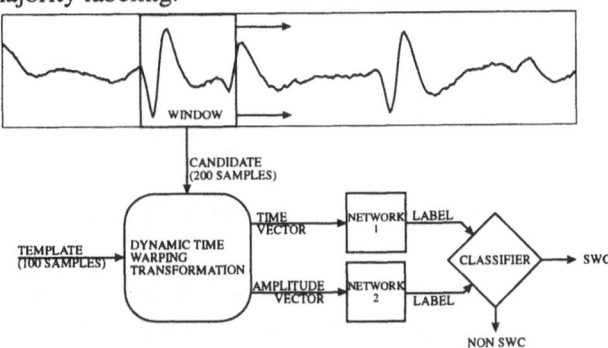

Figure 2: Complete spike-wave complex detection system

If the networks are trained and labeled, detection can begin. The same transformations are carried out as in the training phase. If both the labels of the time and amplitude vector are above a certain threshold the current window position is marked as containing a spike-wave complex.

3 Detection results

The detection system as described in paragraph 2 is used in a simulation to detected spike-wave complexes in one channel of an EEG. It is trained with 50 out 67 marked examples of spike-waves. The detection results for various classification methods, as described in [2], is stated in table 1. Noise is added to input signal to artificially increase the number of examples.

Labeling	Net size	Noise	False alarm	Missed
Master-slave	8x8	1%	6	1
Master-slave	10x10	1%	12	0
Majority labeling	10x10	1%	2	1
Majority labeling	10x10	0%	6	0

Table 1: Detection results

4 Conclusions

DTWT is a useful input transformation for situations where some invariance for time variations is required. The resulting vectors can be successfully classified by Kohonen's neural network. Nearly all information is contained in the time-vector and amplitude-vector, since it can be reconstructed. Classification depends therefore on the complete signal.

5 References

[1]   M. Nadler and E.P. Smith, *Pattern recognition engineering*, Wiley-Interscience Publication, 1993

[2]   L.P.J. Veelenturf, *Analysis and Applications of Artificial Neural Networks*, Prentice Hall, 1995

[3]   J.R. Smith, *Automatic Analysis and Detection of EEG Spikes, IEEE Transactions on biomedical enigineering*, VOL BME-21, 1974

Learning the Equations of Data

By C. M. Roadknight[1], D. Palmer-Brown[1] and G. E. Sanders[2].

1. Parallel Research Group, Department of Computing, The Nottingham Trent
University,
Burton Street, Nottingham NG1 4BU.

2. Department of Life Sciences, The Nottingham Trent University,
Clifton Lane, Nottingham NG11 8NS.

The use of artificial neural networks (ANNs) for data modelling is now established [1, 2]. An extension to this is presented which enables the characteristic equations of complex datasets to be learned.

The effects of environmental factors and man made pollutants on agricultural crops is well known [3], but this is mainly through single factor, dose response experiments. What is largely unknown is the effect of differing levels of multiple factors, such as humidity, temperature and ground level ozone concentrations. Standard statistical techniques for prediction and decision making cope badly with the non-linearity and small sample sizes inherent in this problem. However, prediction can be achieved using an ANN approach. An ANN strategy was chosen because the predictive powers of ANN's are well documented and suitable for this problem [4].

Considerable success has been achieved in predicting damage with this method, demonstrating that the ANN is modelling the environmental and pollution effects. A 3-layer network is trained using a standard back propagation algorithm [5] on a selection of seven-day datasets and then tested on datasets that it has not been trained on. The input layer of the network consists of nodes associated with daily environmental or pollution related factors for each of the seven days during the growth period of a bean plant. Other inputs are encodings of the condition of the plant (number of leaves, level of damage at day 0). This amounts to over a hundred input units. The output for the network is the change in percentage area of leaf damage over the seven days.

The network is trained to a point where it can correctly predict the change in the level of damage over the 7 days in all 54 of the training datasets and it can also predict the level of damage, to a good accuracy (RMS 0.0053), for all 18 of the test datasets (fig 1).

In addition to predicting damage levels, the model is required to provide an interpretation of the interaction of many environmental factors. An algorithm has been devised that simplifies the weights of the ANN into a reasonably concise equation (typically 10-20 terms). In general this equation contains a mixture of both biologically understood rules and less understood indicators of damage. The former give credence to the equation, the

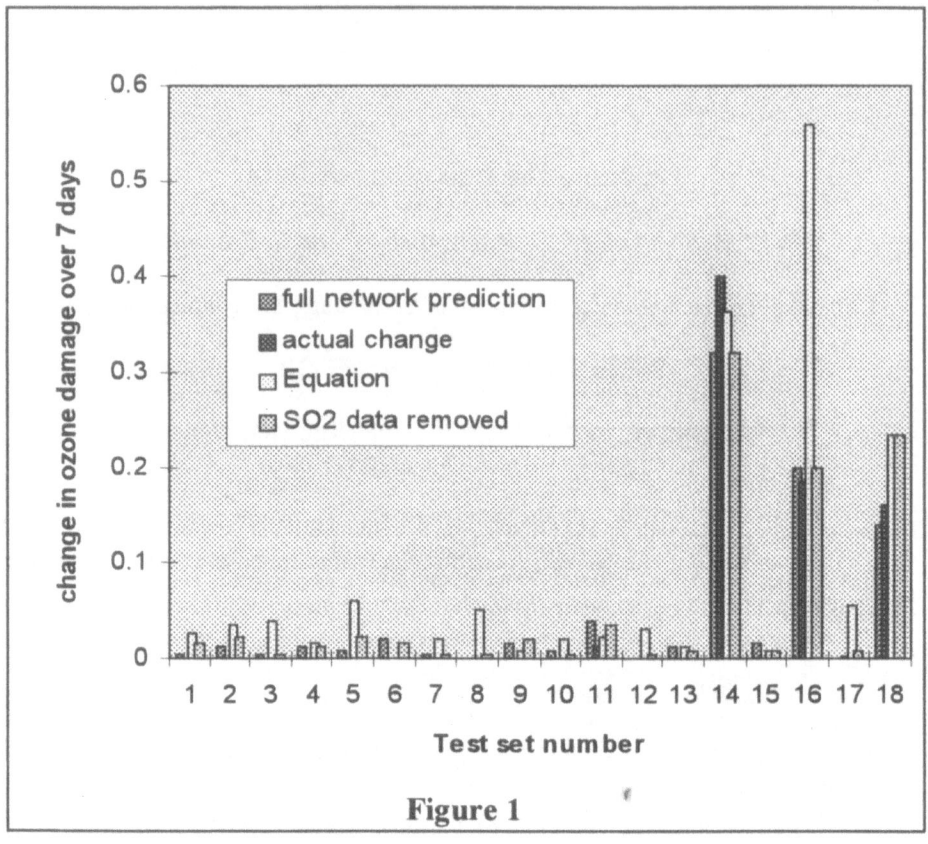

Figure 1

latter can then become the focus of biological experiments. The equation extractor works by building combinations of linear, piecewise models of the predominant neurons in the network. This is performed to any specified level of model complexity.

For the initial data, the following equation was extracted:

F[InitialLeafNumber - MaxSO2@d3 - (2*InitialLeafDamage)] +
F[InitialLeafNumber +InitialLeafDamage - (SumSO2@d2 + MaxSO2@d2 + MaxSO2@d3)]+
F[SumSO2@d1+SumNO2@d2-(MaxSO2@d3+ MaxSO2@d6+ MaxNO2@d6+ SumNO2@d5 + SumRad@d7+MaxNO@d7+InitialLeafDamage)]

The terms in this equation are environmental or pollutant factor levels eg. Number of leaves on plant at day 0 (InitialLeafNumber), highest sulphur dioxide reading on day 3 (MaxSO2@d3), sum of hourly readings of solar radiation during day 7 (SumRad@d7).

This equation contains the primary inputs for 3 hidden units, contained within an activation function (**F**), which is approximated as a straight line between 0 and 1, with a value of 0.5 at F(0).

The algorithm for network processing and equation extraction is as follows:

For the chosen maximum number of hidden nodes
  Set required importance level (H) for weights from hidden units to output unit
  Find all hidden units above H
For the chosen maximum number of input nodes
  Set required importance level (I) for weights from input units to hidden units
  Find all inputs to chosen hidden units above I
Create equation by matching input nodes, approximate weights and sign to source of input node value.

At present the number of terms in the equation (10 to 20) and the number of hidden units contributing to the equation (eg 2 or 3) are decided before extraction. The validity of the equation is tested by running a minimal network consisting only of the weights used to generate the equation. Even though this minimal ANN may contain as few as 1% of the connections of its predecessor, it performs the predictions to a similar standard (fig 1); correctly classifying the 3 large (>0.1) changes, the 2 small changes (>0,<0.01) and the 13 zero changes. The test set 16 prediction is overestimated by this network but it is still classified correctly as large (>0.1). The minimised network can achieve this good level of prediction because it is based on the key fetures of the dataset, as learned by the full network. This suggests that these simple equations can be used as predictive tools and as indicators and interpreters of the source of crop damage.

It is possible for the network to use secondary indicators which obscure the model being sought. One of the best performing equations for predicting ozone related leaf injury contained no ozone level inputs and contained many sulphur dioxide (SO_2) inputs and this seemed biologically improbable. There are three possible explanations:
a. Lower levels of SO_2 are a greater cause of ozone related injury than ozone. This may arise if sulphur dioxide offers a protection from ozone damage. This seemed unlikely
b. Less SO_2 indicates higher levels of ozone. If this is true it is necessary to find out why the ANN used a secondary indicator and not the factor itself.
c. Lower levels of SO_2 indicates higher levels of two or more factors, so sulphur dioxide might fall as a consequence of an increase in ozone levels and an increase in temperature levels, for example. So the network would increase the weight on the one multi factor indicator in preference to the several factors themselves.

To test this, a network was trained without all SO_2 related inputs. There are many possible models that an ANN can learn from one application. Their relevence depends on which aspects of the complete dataset are presented to the network. This ANN trains to a similar standard as it did with sulphur dioxide (fig 1), showing that the role of SO_2 is as a secondary indicator. The equation extracted from this trained ANN is as follows:

3*F[D2(SumOz)+D3(SumOz+MaxOz)-D1(SumRad+MaxRad)-2*(D6(MaxRad))+
(4*InitialLeafDamage)]
+F[3*InitialLeafNumber - D3(MaxNo)]
+ F[3*InitialLeafNumber - 4*InitialLeafDamage]

The profiles of the sulphur dioxide were of small but significant difference whereas the differences between levels of ozone were much greater. The network weights needed to be larger to amplify the differences in the inputs to account for this, so SO_2 appeared more important in the equation. This can be overcome by normalising the input data, to between 0 and 1 but also to a universal mean and universal standard deviation.

In summary, this paper has reported how a method has been devised for extracting working equations from ANN's, so that their 'black box' nature is overcome and greater trust can be placed in their results.

References

[1]   Ripley, B.D. Neural networks and related methods of classification. Journal of the Royal Statistical Society series B **56**, 1994, pp. 409-456.

[2]   Weiss, S.M. and Kulikowski, C.A. Computer Systems that Learn. Morgan Kaufmann. 1991

[3]   Heck, W.W., Taylor, O.C. and Tingley, D.T. (eds). Assessment of Crop Loss from Air Pollutants. Elsevier Applied Science, New York. 1988

[4]   Weigend, A.S., Huberman, B.A. and Rumelhart, D.E. Predicting the future: a connectionist approach. Int. Journal of Neural Systems, vol. 1, 1990, pp.193-209

[5]   Rumelhart, D.E., Hinton, G.E., & Williams, R. J. Learning internal representations by error propagation. In D.E. Rumelhart, & J.L. McClelland (Eds), Parallel distibuted processing: Exploration in the microstructure of cognition (pp. 318-362). 1986. Cambridge, MA: MIT Press

New Digital Hardware Concept for Self-Organizing Feature Maps

Dietmar Ruwisch, Heiko Rahmel and Mathias Bode

Institut für Angewandte Physik, Universität Münster, Germany

Abstract

We present a new, all digital hardware concept for Kohonen self-organizing feature maps. The neurons of this implementation are rather simple computation nodes. Aside from a data bus they are interconnected with next neighbors only. This low connectivity opens up the possibility of integrating a large number of neurons in a digital neuro-chip. Furthermore, a number of those neuro-chips could easily be linked to a neural network of arbitrary size.

1 Introduction

Topographic feature maps are well-known from the neocortex of the brain. However, they are a useful tool in technical applications, e.g. vector quantization or projection tasks[1]. T. Kohonen[2] gave a very efficient algorithm leading to the desired self-organization. However, implementations on conventional computers cannot profit from the parallel structure of this algorithm. Therefore, specialized hardware architectures have been proposed, using digital[3] as well as analog[4] technologies. The architecture presented in this work is characterized by a very low connectivity within the network, aside from a global data bus only next neighboring units are interconnected. Furthermore, these units ("neurons") are rather simple computation nodes, they only need to add, calculate absolute values and devide by a power of two. General multiplications are not required, the devision by a power of two can be easily realized using a shift register.

2 Kohonen's algorithm

When an input feature, $\mathbf{U}$, is presented to the neural network, Kohonen's algorithm demands that a "winner neuron" and neurons in its spatial neighborhood participate in a learning step. The relative amount of learning decreases with increasing spatial distance to the winner neuron. Every neuron, N_i, represents a point, $\mathbf{W}_i$, in feature space. The winner neuron, N_j, is the "best matching unit", i.e. the neuron with

$$\|\mathbf{W}_j - \mathbf{U}\| \leq \|\mathbf{W}_i - \mathbf{U}\| \quad \text{for all } i. \tag{1}$$

The Kohonen learning rule concerns a spatial neighborhood of the winner neuron:

$$\mathbf{W}_i^{new} = \mathbf{W}_i^{old} + \eta(d_{ij})(\mathbf{U} - \mathbf{W}_i^{old}). \tag{2}$$

d_{ij} denotes the spatial distance of a neuron, N_i, to the winner neuron, N_j. The neighborhood function $\eta(d_{ij})$ is restricted to

$$\eta(d_{ij}) \leq 1, \tag{3}$$

which avoids "overshooting" learning. In close neighborhood of the winner neuron learning should be "attractive":

$$0 \leq \eta(d_{ij}) \leq 1 \qquad \text{if} \qquad d_{ij} \leq d_a. \tag{4}$$

In order to speed up self-organization it can be useful to chose "repulsive" interaction for more distant neurons:

$$\eta(d_{ij}) \leq 0 \qquad \text{if } d_a \leq d_{ij} \leq d_r. \tag{5}$$

Learning is spatially restricted to a certain neighborhood of the winner neuron by

$$\eta(d_{ij}) = 0 \qquad \text{if} \qquad d_r < d_{ij}. \tag{6}$$

In the following we will show how such a "mexican-hat"-like learning function $\eta(d_{ij})$ is realized by the presented architecture.

3 Hardware realization

Every neuron stores a vector $\mathbf{W}_i$ in binary manner, e.g. a 16-bit value per component of $\mathbf{W}_i$. The input feature $\mathbf{U}$ is presented to the neurons via data bus. Each neuron calculates the absolute value of the difference $\|\mathbf{W}_i - \mathbf{U}\|$ between its own $\mathbf{W}_i$ an the input feature, $\mathbf{U}$. The minimum of all $\|\mathbf{W}_i - \mathbf{U}\|$ and thus the winner neuron is found by means of binary search[3]. This is a very efficient method working completely in parallel.

A single learning step (eq.2) of the system is devided into a number of substeps. In the first substep, only the winner neuron, N_j, executes the learning rule

$$\mathbf{W}_i^{new} = \mathbf{W}_i^{old} + \frac{\mathbf{U} - \mathbf{W}_i^{old}}{l}. \tag{7}$$

The learning parameter l, $|l| \geq 1$, can be restricted to a power of two, thus division by l can be performed by means of a simple shift register. In the following substeps, additionally those neurons join the set executing the learning rule (eq.7) which are next neighbors of neurons having already participated in learning. This means, a "learning front" propagates through the neural network as shown in fig.1. In the first d_a substeps l is thought to have a positive value, $l = l_a > 0$. This leads to an "attractive" interaction indicated by the "+"-sign in fig.1a–c. After d_a substeps the winner neuron has executed the learning rule (eq.7) d_a times, while neurons in the distance $d_{ij} = d_a - 1$ (manhatten distance) have done this once. In the following substeps, l is thought to have a negative value, $l = l_r < 0$, for all affected neurons. This leads to a "repulsive" interaction sketched by the "−"-sign in fig.1d. When the system has performed d_r substeps altogether, learning is stopped, the system is resetted and a new cycle

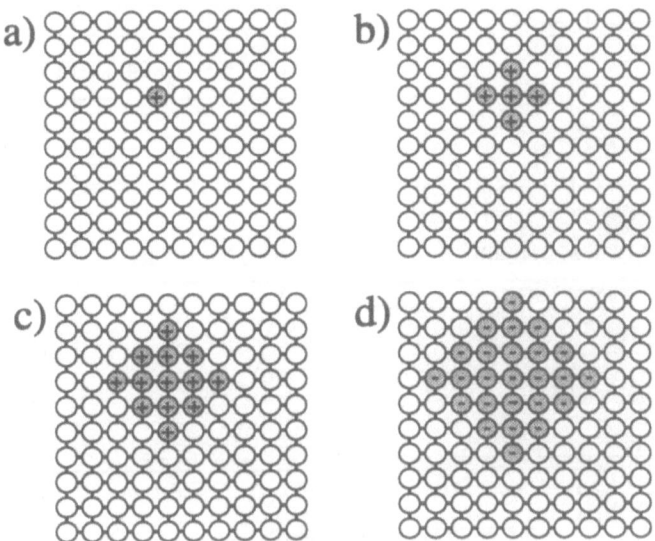

Figure 1: a) In the first substep only the winner neuron, N_j, adapts its W_j. b, c) In the following substeps learning spreads out into the neighborhood of the winner neuron. As in a) "+" indicates "attractive" interaction. d) "Repulsive" interaction in the last substep leads to a mexican-hat-like neighborhood function $\eta(d_{ij})$ (see fig.2). The information "learning front arrived" detected by an edge cell can be used to link a number of single arrays to a larger network.

may begin by presenting a new input feature, $\mathbf{U}$. The neighborhood function $\eta(d_{ij})$ used in eq.2 resulting from this procedure reads

$$\eta(d_{ij}) = 1 - \left(1 - \frac{1}{l_a}\right)^{d_a - d_{ij}} \left(1 - \frac{1}{l_r}\right)^{d_r - d_a} \qquad \text{if} \qquad d_{ij} \le d_a,$$

$$\eta(d_{ij}) = 1 - \left(1 - \frac{1}{l_r}\right)^{d_r - d_{ij}} \qquad \text{if} \quad d_a < d_{ij} \le d_r,$$

$$\eta(d_{ij}) = 0 \qquad \text{if} \qquad d_r < d_{ij}. \qquad (8)$$

The shape of $\eta(d_{ij})$ is shown in fig.2 for different values of l_a, l_r, d_a and d_r. If a monotonic $\eta(d_{ij})$ is required instead of a mexican-hat-like, the repulsive phase can be omitted (i.e., $d_a = d_r$). Note, that this learning procedure operates without the single neurons explicitly knowing their spatial distance, d_{ij}, to the winner neuron.

If an array of neurons is integrated on a semiconductor chip, a number of chips can be linked to a network of arbitrary size in the following way: All chips use the same data bus. In addition they have to be interconnected with their next neighbors. If the learning front reaches the edge of one chip (fig.1d) this edge position is supplied to the respective neighboring chip, in which the propagation of learning will go on. The propagation scenario controlling the learning procedure reveals a striking similarity to the propagation of Calcium

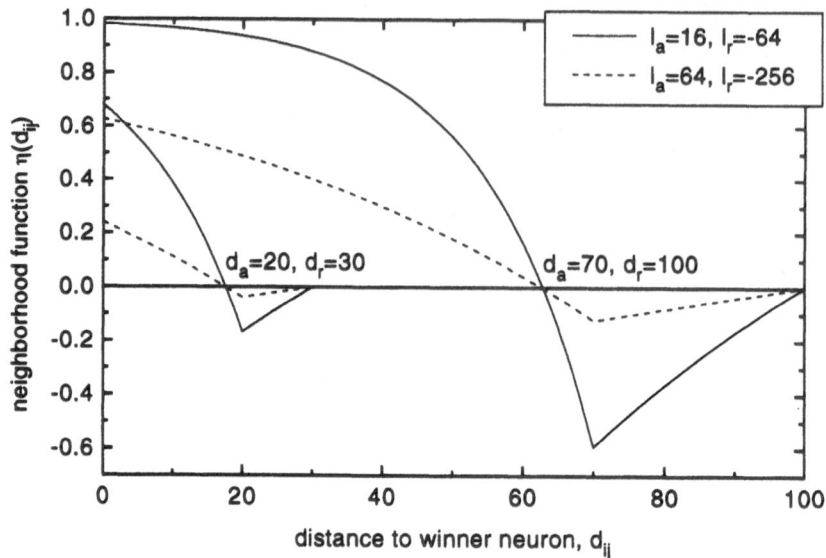

Figure 2: Learning function $\eta(d_{ij})$ for different parameters l_a, l_r, d_a and d_r, that could be used in a network of $100 * 100$ neurons. For the sake of computational simplicity each l-value is restricted to a power of two.

waves within astrocytes of cultured glia[5]. Thus, our hardware architecture may be understood as an "Artificial Neuro-Glial Network".

References

[1] Wienke, D. and Hopke, P.K., Projection of Prim's minimal spanning tree into a Kohonen neural network for identification of airborne particle sources by their multielement trace patterns. In *Analytica Chimica Acta*, **291**, p.1, 1994.

[2] Kohonen, T., *Self-Organization and Associative Memory*, pp. 125–160, Springer Verlag, 1984.

[3] Melton, M.S., Phan, T., Reeves, D.S. and van den Bout, D.E. The TIn-MANN VLSI chip. In *IEEE Transactions on Neural Networks*, **3**, p.375, 1992.

[4] Ruwisch D., Bode M. and Purwins H.-G.,*Parallel Hardware Implementation of Kohonen's Algorithm with an Active Medium*, Neural Networks, **6**, p.1147, 1993.

[5] Nedergaard, M., Direct signaling from astrocytes to neurons in cultures of mammalian brain cells. In *Science*, **263**, p.1768, 1994.

NEURAL NETWORKS IN PRACTICE

Applications of Neural Networks - Orals

Applicability of Artificial Neural Networks in Small and Medium-sized Businesses

E.T. Auée

InnovatieCentrum Midden- en Zuid-Gelderland

1 Introducing InnovatieCentra

The Dutch InnovatieCentra [Innovation Centres] are an initiative of the Dutch government, in response to a report of a committee of Dutch captains of industry. This Commissie Dekker [Dekker committee] concluded that small businesses are very important for the Dutch economy. To survive in the expected global competition, the committee concluded, these small and medium-sized businesses (SMB) have to use modern technology as much as possible. However, knowledge of modern technology proves to be difficult to access by the SMB. They lack the time and capacity to follow modern developments. The InnovatieCentra were founded to bridge this gap. It is their mission to make knowledge available and applicable for small companies, and to stimulate them in using it for innovation. To fulfil this mission, the InnovatieCentra work both reactive, responding to a specific question of a small company, and pro-active, focusing on specific technology.

Knowledge technology is one of several technologies the InnovatieCentra stimulate actively. This technology covers fields such as expert systems, fuzzy logic, case-based reasoning and, of course, artificial neural networks. In this paper, we will focus on the use of neural networks in small businesses. We will look at the fields in which neural networks can be profitable and discuss conditions to be fulfilled in their use. We will also glance at some alternative techniques. We will conclude with some final remarks on applying neural networks in SMB.

2 What are Artificial Neural Networks?

The answer to this question can be partly found by analyzing the name of this technique. Artificial suggests an imitation of an existing example. The term neural network reminds us of biological brains. This is exactly what artificial neural networks are, a mathematical analogy of the human brain. Just like our brains, artificial neural networks consist of neurons and links between these neurons. The analogy continues: artificial neural networks can be used to learn from examples, and to use the acquired knowledge in new situations. It would be premature to fear that we are on the threshold of building humanoids. The artificial neural networks we are talking of nowadays cannot yet model the rich life of a firefly, let alone man.

Though we are not on the level of biological neural networks, artificial neural networks have proved to be worthwhile in many practical business situations. They can extract the underlying experience from collected historical data. Often, these underlying patterns are not easily detected by people. After all, it is known that computers are better in numbers than people. However, patterns hidden in the data can be very useful in steering our businesses. Just look at the paper by Smit in these proceedings. It uses historical data to predict the time and capacity needed to produce a transformer that has never been built before. Or, consider the case of the ink factory. The relationship between formula and physical properties of an ink is learned from a large database of existing inks. They use this relationship to determine the composition of a used ink, thus being able to produce a new ink with the same properties. A final example from the process industry, is the case of the animal feed factory. Animal feed is produced from strongly varying components. This feed must have specific physical properties, no matter what its components are. The neural network had been trained to calculate the required adjustment of various process parameters to reach the desired properties, on the basis of existing results.

These and many other operational examples show us that, in spite of the obvious superiority of its natural examples, artificial neural networks can be used profitably in business applications. Especially when important company knowledge is hidden in historical data, neural networks can help to grasp this hidden experience. However, artificial neural networks are not panaceas, universal remedies for all situations. In the next section we will discuss the requirements to apply neural networks successfully.

3 Requirements for neural networks

Neural networks build a model that represents a relation between input data and output data. Both input and output data can relate to multiple variables. A dataset of good examples, e.g., examples of input data with their related correct or desired output data, is used to train the network. If you train the network with these data, you can use it to predict the correct values of the output data, given certain input data values.

When your company produces plastics, for example, you may have records of process parameters related to the mechanical properties of the plastic. When you use the process parameters as inputs and the properties as outputs, the neural network will learn the influence of the process parameters to the mechanical properties. This representation, then, can be used to predict the mechanical properties, caused by the current process settings. If necessary, these settings can be adjusted following the prediction.

In this paper we will not discuss the way a neural network builds its representation from the data, nor will we elaborate on the model a neural network builds. What is important for our purpose, is to realize that neural networks build a black-box model from the data. This model cannot be translated into something people can understand with their physical intuition. The representation is only suitable for making predictions, not for explaining them.

Because neural networks can learn from examples automatically, they have one property which is especially relevant for SMB s. They are not necessarily very expensive. Standard tools are available, and many companies provide services in the development stage.

3.1 Properties of the problem

Because neural networks build a black-box model, using them only makes sense when the solution to the problem is not yet known. If relevant knowledge is available, it is often a better approach to use this specific knowledge to solve the problem. This approach leads to a glass- box model that can be understood, and thus guarded, by its users. Physical modelling or fuzzy logic may be good alternatives in these situations. Fuzzy logic is especially interesting when only qualitative knowledge of the situation is available.

When the relationships hidden in the available data are not yet known, neural networks may be a means to extract these relations from the data. But didn t we already have a fine technique for doing this, a technique known as statistics? What do neural networks add to this technique? In fact, neural networks are comparable to statistics. Just as in statistics, neural networks express relationships in data. However, there are big differences. Neural networks can cope with very whimsical relationships between multiple variables. Statistics have their limitations in non-linearity and in situations with multiple inputs and outputs. Whenever the data are suitable for statistical techniques, these techniques are preferable. Statistics lead to a better insight into the problem area, neural networks do not.

3.2 Requirements of the data

When the problem meets all requirements for using neural networks, and cannot be solved better with an alternative technique, there is another very important aspect to consider. You will need data to train the neural network. The experience reflected in these data must cover the area in which the neural network must be used. In practice, this means you need data on all relevant inputs and outputs. This is not as simple as it seems as it is often difficult to asess which inputs contribute to the relevant outputs. The neural network will not help you there. Its power to represent the strangest relations causes also its weakness to detect nonsense. A neural network will learn a relation between birthrates and stork population. However, the knowledge it learns from these data is not very useful. The only thing you can do is to examine all the factors that might be relevant and use all available knowledge of the situation that has to be modelled to validate the result. When you do not have historical data on all r elevant factors, a situation we meet frequently, you can still use the data to train a neural network. The predictive performance of the network will, of course, degrade, but may still be worthwhile.

Apart from having to determine which data are relevant for your network and which are not, you must also make sure that you have enough data and that all data are correct. Evidently, you cannot train a neural network with 10 inputs, using five examples. It is very difficult to give a rule of thumb for the number of data required. What the network actually does when building its model, is comparable with interpolation. This means that examples should be available that are characteristic for the problem area. Especially in areas in which the behaviour of the system is known to be irregular, it is important to gather enough data. Naturally it is evenly important to make sure all data are valid. Often when neural networks were developed, this was a problem as data can be easily corrupted by mistakes made by the employees collecting them, or inaccurate sampling devices. A last warning is relevant in this context. It is extremely dangerous to trust predictions based on inputs that are outside the range used to train the network.

When the results of these qualitative analysises are satisfactory, a quantitative experiment should be done. Only through quantitative experiments you will find out about the relations that are really hidden in the data. Also, it will provide information for the selection of specific algorithms. In most cases, during the quantitative analysis and implementation of the neural network, some additional support from a specialist in the field is sensible.

4   Concluding remarks

The operational examples of neural networks prove that this technology can be used profitably. It is a convenient way of extracting information and knowledge from historical data, knowledge that otherwise could not be used. When considering the use of neural networks, the first thing to do is to find out if this technique is the best option. This can be done by analyzing your problem. In a qualitative analysis you focus on the problem to be solved and the relevant knowledge that is already present. In this phase you can also define the exact functionality you expect from the neural network and the organizational and financial consequences of building and using one. The use of neural networks can be very interesting for multinationals as well as SMB's as their application is not necessarily very expensive. The InnovatieCentra offer to support the qualitative analysis, and thus help SMB's to reach a point where well-founded decisions can be made.

Calculation for Client Specific Transformers

H. Brockmeyer

Smit Transformatoren BV

Nijmegen, The Netherlands

Abstract

The industrial transformers manufactured by Smit are large and complex machines. Each delivered transformer is unique, made to meet the specifications of the client. Moreover, production techniques change throughout the years. This makes production planning difficult. Since recently, Smit uses neural networks for the calculation. Neural networks yield better results and they require less effort than the techniques which Smit used previously.

1   Introduction

Since its establishment in 1913, Smit is specialist in the development and the manufacture of large transformers for industry and generation of energy. Transformers of Smit are supplied over the whole world. Their prices vary from a hundred thousand guilders to 15 million guilders. They take 6 months to 2 years manufacturing, including engineering time. Manufacturing time of a transformer tends to decrease rapidly over the years. The transformers are custom designed. Since each client has his own specifications with regard to power, voltage, efficiency etc, each transformer is unique, with its own production time.

2   The problem of production planning

When we start the production of a transformer, only global specifications by the client and the delivery date are available. Since each transformer is different, it is difficult to control the production process. One of the problems is to predict the production time. For this problem, not only the delivery date and budget play a role, but also questions with respect of the planning of the total needed man hours, the progress of the project, the planning between the different departments, the optimal utilization of personnel capacity and the planning of delivery of the raw materials. For these reasons, even with our experience and know-how, production planning remains complicated.

3   Our previous solution

About 15 years ago, we started to systematically sort out the production data of previous projects. The aim was to estimate the production time for a new project as precise as possible on the basis of these data. By matching predicted times and the realized production times, we worked out an *ad hoc* equation, in which all the factors – about 150 parameters – contribute. With this equation we could estimate the production time. However, since production methods and experience of personnel changes continually, and moreover, since each time a different transformer is produced, the equation should kept up-to-date. So the laborious and time-consuming work for the continual adaptation (by hand) of the equation remained.

4 Innovative technology

In 1993, InnovatieCentrum Midden- en Zuid-Gelderland ('IC', an initiative of the Dutch Ministery of Economic Affairs to transfer knowledge into Dutch small and medium sized companies) contacted us to tell about the possibilities of knowledge technology for Smit. In the first meeting, a possible application already has been raised; the possibility to use a neural network for our production planning. We did not had any knowledge with neural networks, but via IC we came in contact with Biologica, a small company specialized in neural network technology.

5 Neural networks

Biologica showed us the possibilities with neural networks. Biologica did some pilot experiments with our data. Their results were better than ours, and thus our interest has been aroused. On Biologica's advice a commercial neural network package has been bought. Using a spread sheet the input of data into the neural network package has been automized. In this way we obtained a user-friendly neural network.

We had several reasons to keep the development and implementation of the neural network within Smit. One is that it was not financially attractive for us to put out this project to others. Another is that for a strategic function as production planning, we do not want to be dependent on others. Finally, a reason is that we see possibilities for other applications of neural networks within Smit.

6 Benefits

Although it is difficult to express the benefits of neural networks in financial terms, the advantages of the neural network solution are clear. On the one hand, with the neural network the quality of the prediction is increased. As a result, we can make a better planning, improve the control of our production processes and realize a higher degree of utilization of our resources. On the other hand the neural network are easier in use. It saves much time in maintenance compared to the equation which we used previously – this has been the initial reason to consider the application of neural networks.

Modelling of Industrial Processes using Natural Computation

A.P. de Weijer

Akzo Nobel Central Research

Arnhem, The Netherlands

June 14, 1995

Abstract

The relatively new field of natural computation has already many useful applications. In this paper the use of multivariate statistics and natural computation to probe process-structure-property relationships is demonstrated. An application of multivariate statistics and natural computation to the relationships between process conditions, physical structure and the (thermo-)mechanical properties of poly(ethylene terephthalate) yarns illustrates their usefulness.

1 Introduction

The core activity of industry is the manufacturing of products from raw materials with specific properties in efficient production processes. The added value of the product is predominantly determined by the technology. Therefore improvements in processability or product performance can be achieved if qualitative relations between the way the product is processed and its properties are available.

An important drawback in the development of a fundamental understanding of industrial processes is that a technology can advance without accurate knowledge of the principles behind the advancement. Since it is important in many modern industries important to keep technologically ahead of competitors, research on fundamental aspects is often given minor importance.

2 The use of experimental data

Recent developments in laboratory information systems and data storage capacities have resulted in the advent of models based on correlation between data-sets. As well as the many new developments in multivariate data analysis (in general), the concept of artificial neural networks has contributed significantly to this rising interest. Despite the large number of published papers on the *modelling* of experimental data, the *interpretation* of experimental data still forms a vital part of every scientific investigation. Large databases are nowadays often routinely generated with data on processes, physical and chemical structure, and properties of materials.

The problem of introducing the structure as an intermediate of process - property relationships is in the selection of structure variables. To make interpretation possible, the structure has to be characterised with a minimum number of quantities. One of the main problems is that the structure variables are often highly inter-correlated. Setting an experimental design on the process variables does not solve this problem since the structure is a *function* of the process conditions. A second problem is that the statistical data analysis is based on correlation, rather than causality. These correlations do not imply causality although they are often interpreted as causalities in the literature. Some authors have addressed this problem in illustrative examples [5] [6]. These situations are in most cases evidently clear and are easily recognised by researchers. However, more subtle examples can be found in complex industrial and environmental data sets in which, in the first instance, the cause of the correlations is unclear.

3 Process - structure - property relationships

The close connection between process conditions, structure and properties can be clearly illustrated by the classical example of the different forms of carbon [7]. In Fig. 1 the phase diagram of carbon is shown. Pure graphite switches to diamond at 100 kbar and 3000° K at useful reaction rates. The application of these process conditions results in crystals with strong covalent bonds. Such carbon crystals are single molecules in which every carbon atom is linked through covalent bonds to four neighbours in a tetrahedral structure. Materials formed this way are often tough and unreactive and are not electrically conductive. Graphite, produced from pure carbon under milder conditions, is another form of solid carbon and is one of the softest materials known. It consists of layers of carbon bound together by covalent bonds into hexagons (α form) or rhombohedral (β form). The hexagonal alpha type can be converted to the beta form by mechanical treatment, and the beta form reverts to the alpha form on heating above 1000°K. One valence electron on each carbon binds the sheets of covalent bonds together. These electrons have such limited mobility between the sheets that the conductivity perpendicular to the layers is about 10^{-4} times that parallel to the layer. By sublimation of pyrolytic graphite at low pressures and temperatures above 2200°K so-called "white" carbon is formed which is a transparent birefringent material.

This classic example illustrates the close inter- relationship between the way a material is produced, its physical structure and the resulting properties. It shows that process, structure and properties are in causal connection to each other.

For efficient improvements in the performance and quality of a product, a profound understanding of these relationships is required. Structure and properties are both functions of process settings whereas the properties are functions of structure (see Fig. 2).

3.1 Statistical vs. physical models

The structure - property relationship is often elaborated in physical models. In addition to such deterministic models, which have traditionally had strong roots in pure and applied research, statistical models play an important role in understanding these relationships. Both model types have their strengths and weaknesses. It is important to realise that these model types are complementary, not mutually exclusive.

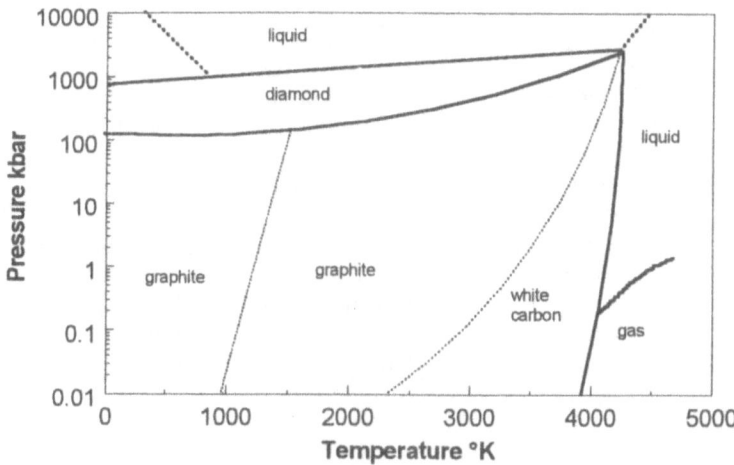

Figure 1: **The phase diagram of carbon. It shows that under different process conditions different forms of carbon are formed.**

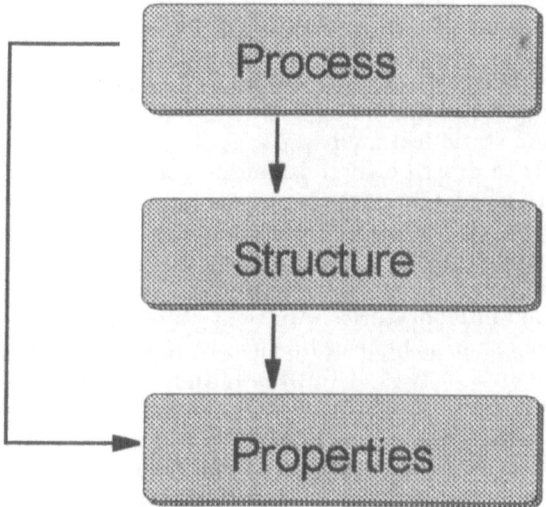

Figure 2: **The forward-relations between process settings, structure and properties of materials.**

Table 1: **Some characteristics of various types of models.**

	Physical model	Statistical model
Principle	physical theory causality	exp. data correlation
Development time	long	short
Experimental effort	low	high
Validation	exp. data	exp. data
Reasoning	in principle transparent	black box
Accuracy	low	high
Predictivity	high	limited

The physical model and the statistical model may be regarded as the two extremes in the modelling of structure-property relationships. The term 'physical' implies the existence of proven basic laws from the natural sciences (physics, chemistry) applied to the material concerned. The aim of physical models is to provide a quantitative explanation for observed phenomena in terms of basic key parameters. Comparison of the predictions from the model with experimental data is often used to validate the model. In general, the development time of a physical model is long. A reason for this is that it is a continuous process of implementation and testing. Physical models often lack accuracy because of simplifications to the underlying equations (to make calculations easier) and because of inconsistency with the non-ideal behaviour of the relation in practice. However once the model has been validated, it can be applied more generally than can statistical models.

When it is not possible to describe the relation in well-defined physical quantities or when accurate predictions are required, a statistical model can be a useful tool for getting insight into the mechanisms that determine properties. The extrapolation capacities of statistical methods are poor since the model is confined to the experimental ranges on which the model is based. Experimentation in pilot plants and in laboratories results in general in a large amount of data production. If the experiments can be effectively and economically done, the use of statistical models instead of (or in addition to) physical models can be considered. A comparison between physical and statistical models is given in Table 1.

Once a statistical model has been developed it can help validate the physical model, since in this case pure relations (relations without variable contributions from other factors) can be derived. Experimentally, such pure relations can not be obtained since all structure parameters are functions of process settings. In a contrary way, validated physical models can help to find the essential factors, i.e. structure parameters, that influence the response that has to be modelled. In general, much experimental data is necessary to calibrate a statistical model and the pre-information obtained from physical models can reduce the number of experiments or measurements that have to be done.

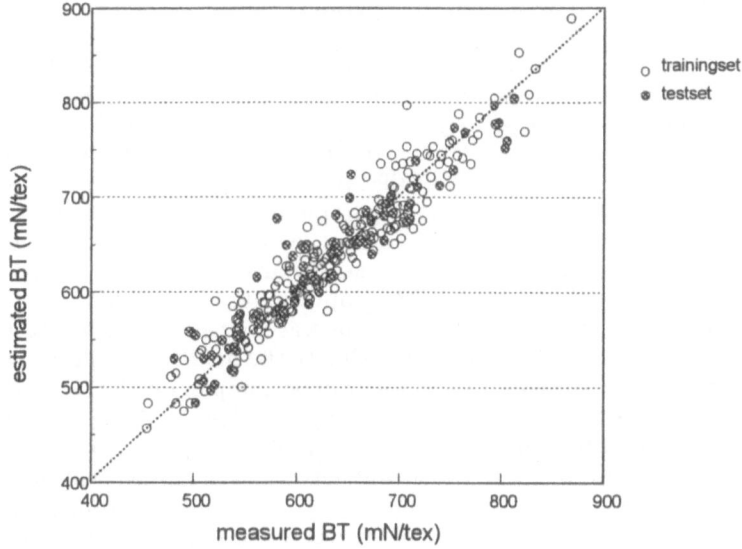

Figure 3: **Estimation of BT by ANN for the test-set and training-set.**

4 Artificial Neural Networks and Genetic Algorithms to obtain quantitative process - structure - property relationships

Artificial Neural Networks (ANNs) are gaining increasing popularity in different applications. Since the breakthrough in 1986, when a learning rule for complex ANNs was discovered [13], applications have frequently been published in the literature [10][3]. An Artificial Neural Network (ANN) was used to model the relation between the physical structure and mechanical properties of poly(ethylene terephthalate) (PET) yarns [8]. The relation was studied on a set of in total 554 drawn yarn samples that came from a variety of applied process conditions. The ANN, consisting of three layers of neurons, was trained using the physical structures and property measurements of these 554 yarns as input and output patterns respectively. All ten properties were fit to the physical structure within an acceptable variance for quantitative use.

4.1 Calibration and validation results

A neural network for the relation between the structure and properties of PET yarns was trained as described above. After 600 iterations a minimum error in an independent testset was found. To illustrate the ability to estimate the measured yarn properties from the physical molecular structure of PET, a plot of the estimated breaking tenacity (BT) versus the measured breaking tenacity for the training-set and test-set is shown in Fig. 3. Comparable results with process parameters as input patterns could be established.

4.2 Modelling of stress-strain curves

The stress-strain curve obtained in uni-axial tensile tests is one of the most important mechanical properties of materials. In general, however, the response is complicated and models closely agreeing with the experimental data up to the point of rupture are not available. In the case described above, some intermediate points and the end-point of the stress-strain curves are used. Since these points are the 'critical' stress-strain parameters, i.e. they determine the shape of the stress-strain curve to a large extent, no arguments can be given as to why it is not possible to predict the stress-strain curve as a whole. In order to test this, an artificial neural network with 5 input units, one hidden layer with 8 neurons and an output layer with 100 neurons was trained. The same test-set and training-set were used to calibrate and validate the model. The effect of each structure parameter, measured on the non-strained sample, on the form of the stress-strain curve can be considered independently. This is done by varying one signal of an input node while the other nodes are set at a constant value. By examination of the output nodes of the neural network, the influence of the pure effect on the stress strain behaviour is examined. The valuable contributions of these exercises to the understanding of structure - property relationships is illustrated in the following example. As a measure of the level of the amorphous orientation, a factor based on sonic modulus measurements is chosen to be one of the structure parameters in the structure representation (Fas). It is generally acknowledged in polymer science and confirmed by the main effects calculated from the ANN that the average polymer chain orientation determines properties to a large extent. The value of Fas can theoretically vary from zero for fully disorientated, to one for perfectly orientated yarns. In practise the orientation parameter varied from 0.69 to 0.83 in this series of drawn yarns. The influence of chain orientation on stress - strain behaviour has been widely discussed in the form of physical and empirical models [11] [12]. The statistical or empirical modelling of stress-strain curves based on a structure representation that is closely related to the well-known two-phase model of semi-crystalline polymers gives the pure contributions that can be used to test the physical models that have been proposed. An example of such a pure contribution, i.e. at the condition of all other structure parameters being constant, is shown in Fig. 4 in which the pure effect of the average polymer chain orientation via sonic modulus measurements (Fas) is depicted.

4.3 Inversion of the forward relations

Another interesting relation for process and product development is the reverse relation to the one discussed, viz. the relation of mechanical properties to physical structure and process parameters. Development of yarns with optimum performance requires detailed insight into the influences of process variations and the structure arrangements of the chain elements on yarn properties. From an industrial point-of-view it is therefore interesting to know which physical structures give the combination of desired mechanical properties. Since this is an ambiguous and NP-complete relation it is not possible to use only a neural network. Hence an optimisation technique (here, a genetic algorithm) is combined with an ANN trained for the forward-relation in order to search for optimal solutions for the reverse relation (Figure 5). In this application each

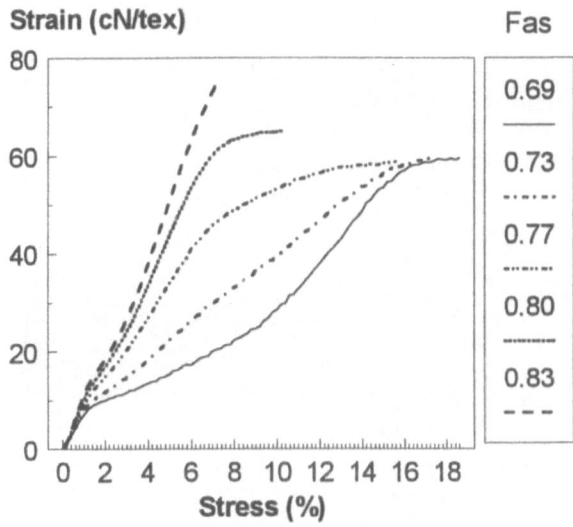

Figure 4: **Variation in stress-strain behaviour as a function of the average polymer chain orientation (Fas) predicted by an artificial neural network. The other variables are set at constant values.**

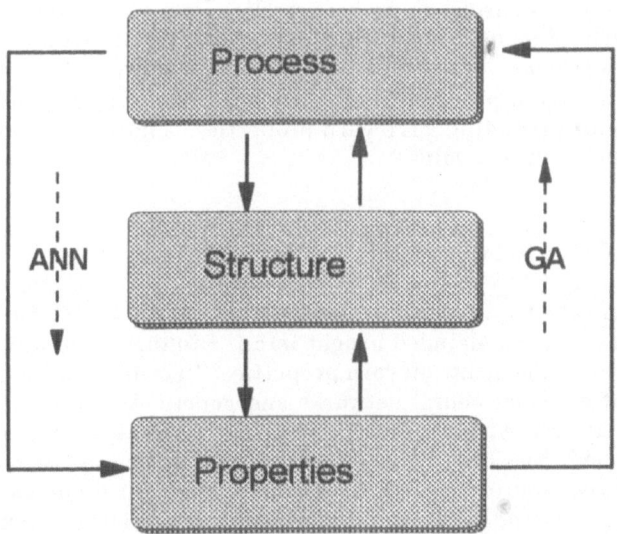

Figure 5: **The forward-relations have been modelled by ANNs. The ambiguous reverse relation is obtained with GAs using the trained and validated ANN for the forward-relation.**

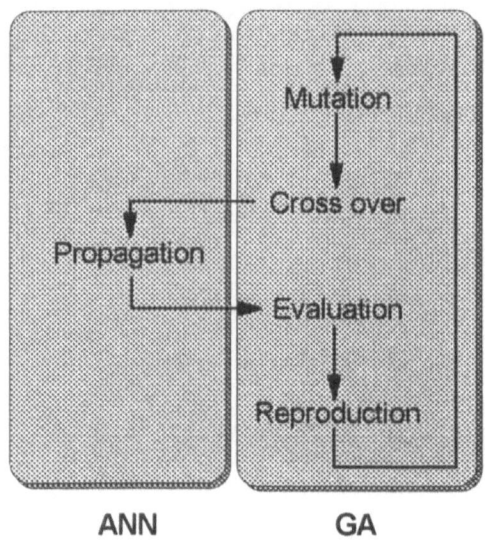

Figure 6: **Modification of the basic execution cycle of GAs. Genes are propagated through the ANN and evaluated.**

gene **s** is a set of five yarn structure characterisation measurements and the evaluation function that partially consits of the trained ANN which predicts the mechanical and shrinkage properties $(\hat{Y}_n)$ from **s**. In contrast to the basic execution cycle of genetic algorithms, our evaluation function is composed of two functions (Figure 6).

The first function propagates all state vectors through an ANN which has been trained for predicting PET yarn properties. The second function g transforms $f(\mathbf{s})$ into a fitness value.

5 Conclusions

The development of yarns with properties that closely match the requirements of customers requires a detailed insight into the influences of process variations and structure arrangements on yarn properties. In order to achieve this, use has been made of artificial neural networks and genetic algorithms. Although the internal structure of ANNs is incomprehensive with respect to the mechanisms of the modelled relations, it is a fast and accurate way to make data and relations between data-sets easily accessible. The trained and validated ANNs, genetic algorithms and background knowledge in the form of texts and figures are implemented in a user-friendly software system. In this way, knowledge concerning the relations between process conditions, physical structure and end-use properties of PET yarns are made accessible to other scientists. Although the discussed relations are mainly confined to synthetic yarns, the concepts are also applicable to other large scale production processes.

References

[1] A. Baines, F.P. Bradbury, and C.W. Suckling. *Research in the chemical industry*. Elsevier, New York, 1969.

[2] C.B. Lucasius. *Towards Genetic Algorithm Methodology in Chemometrics*. PhD thesis, University of Nijmegen, Nijmegen, 1994.

[3] J.R.M. Smits. *Exploring the Possibilities of Applying Artificial Neural Networks on Problems in Analytical Chemistry*. PhD thesis, University of Nijmegen, Nijmegen, 1993.

[4] A.P. de Weijer. *Process - Structure - Property Relationships Obtained with Natural Computation*. PhD thesis, University of Nijmegen, Nijmegen, in press.

[5] E.V. Thomas. A primer on multivariate calibration. *Anal. Chem.*, 66:795A–804A, 1994.

[6] A.K. Dewdney. *200% of Nothing*. Wiley, New York, 1993.

[7] P.W. Atkins. *Physical Chemistry*. Oxford University Press, London, 1984.

[8] A.P. de Weijer, L. Buydens, G. Kateman, and H.M. Heuvel. Neural networks used as a soft-modelling technique for quantitative description of the relation between physical structure and mechanical properties of poly(ethylene terephthalate) yarns. *Chemom. Intell. Lab. Syst.*, 16:77–86, 1992.

[9] A.P. de Weijer, C. Lucasius, L. Buydens, G. Kateman, and H.M. Heuvel. Using genetic algorithms for an artificial neural network inversion *Chemom. Intell. Lab. Syst.*, 20:45–55, 1993.

[10] J. Zupan and J. Gasteiger. Neural networks: a new method for solving chemical problems or just a passing phase. *Anal.Chim.Acta*, 248:1–30, 1991.

[11] M.G. Northolt, A. Roos, and J.H. Kampschreur. Birefringence, compliance, and viscoelasticity of poly(ethylene terephthalate) fibers. *J. Pol. Sci. Pol. Phys.*, 27:1107–1120, 1989.

[12] D.W. van Krevelen. *Properties of polymers*. Elsevier, Amserdam, 1990.

[13] D. Rummelhart, G. Hinton, and R. Williams. Learning representations by back-propagating errors. *Nature*, 323:533–536, 1986.

Neural Network Control for Steel Rolling Mills

Thomas Martinetz
Siemens AG, Corporate R & D
Wittelsbacher Platz 2, 80333 München, Germany

Peter Protzel
FORWISS, Neuro & Fuzzy Group
Am Weichselgarten 7, 91058 Erlangen, Germany

Otto Gramckow, Günter Sörgel
Siemens AG, Processautomation Rolling Mills
Schuhstraße 16, 91050 Erlangen, Germany

Abstract

Worldwide, steel and aluminum production and manufacturing is still one of the major basic industries with a huge amount of material and energy consumption. Hence, optimization of the various process control schemes which are involved can lead to significant savings. Artificial Neural Networks are a new information processing technique which provides a novel approach to process control problems and promises major improvements. Therefore, Siemens together with FORWISS has been studying and developing neural control schemes for a number of different process control problems which occur at hot line rolling mills (Lindhoff et al., 1994). In this paper we give a brief survey of the different control aspects which were tackled with this new approach and comment on their current status.

1 Introduction

Production costs at a hot strip rolling mill could be reduced significantly by reducing the amount of material which is wasted because the produced strip does not meet the quality requirements posed by the customers. The wasted material has to be brought back into the production process by re-melting, which requires a tremendous amount of handling and energy costs.

The most important quality requirements which have to be met concern the shape of the strip. First of all, the customer requires that across the whole strip of about 1km length the desired final thickness is achieved within a tolerance of the order of 0.05 millimeters. In principal, the rolling force control loop is able to achieve this precision. However, after sending the strip into the mill, it takes a certain amount of time until the control loop has changed the preset rolling

---

[1]FORWISS is the German acronym for "Bavarian Research Center for Knowledge-Based Systems".

force to the force which actually yields the right thickness. This amount of time determines the length of the strip head where the thickness is not within the specified tolerance and, hence, determines the amount of material which has to be cut at the head of the strip and is wasted. To reduce this time span, process optimization systems for hot strip mills try to preset the rolling forces such that they lead to the desired thickness from the beginning and the control loop has to make as few adjustments as possible. We will describe how Neural Networks can significantly improve this presetting of the rolling force.

Secondly, the customer requires that across the whole strip the width does not fall below the desired value, which would ruin the strip. To assure this requirement, the strip is currently rolled to a width which is about 12mm above the desired value. After the rolling process, the borders are cut such that the strip exactly obtains the desired width, which is usually between 800mm and 1600mm. An increase in the accuracy of the width control would allow to reduce the margin of 12mm and, hence, would significantly reduce the amount of material which is cut at the borders. To this problem, Neural Networks have also been successfully applied, which is described in Section 3.

In addition to the closest possible thickness and width tolerances, it is becoming increasingly important to be able to exercise an influence on the strip profile as well. Today hot strip mills are required to produce minimum profiles, defined in its simplest form as thickness differences between the middle and the edges of a strip, and to maintain defined profile values within close tolerances during the milling process. This is especially important because the relative strip profile cannot be modified during subsequent cold rolling. Section 4 describes how Neural Networks can be used to improve the existing models and strategies for profile control.

There are a number of other quality requirements which concern the inner structure of the material. In this difficult problem domain, Neural Network approaches have also been applied successfully (Poppe and Martinetz, 1993). In this short paper, however, we concentrate only onto the control schemes which determine the geometry of the strip.

2 Rolling Force Control

Figure 1 shows a sketch of a hot strip rolling mill consisting of four stands. At each stand n a certain relative thickness reduction $\epsilon_n = (d_{n-1} - d_n)/d_{n-1}$ has to be achieved. For this purpose at each stand a rolling force is preset before the slab runs into the mill. To achieve the right thickness reduction within a tolerance of about 0.05 millimeters, the relation between the rolling force F_n and the resulting relative thickness reduction ϵ_n has to be known very accurately to be able to preset the right rolling force; however, this relation depends on many quantities and is difficult to describe. Up to now, physical models of the underlying processes with different parameter settings for different steel qualities have been employed. The achieved accuracy, however, can still be improved significantly. Further, this approach requires tedious bookkeeping of

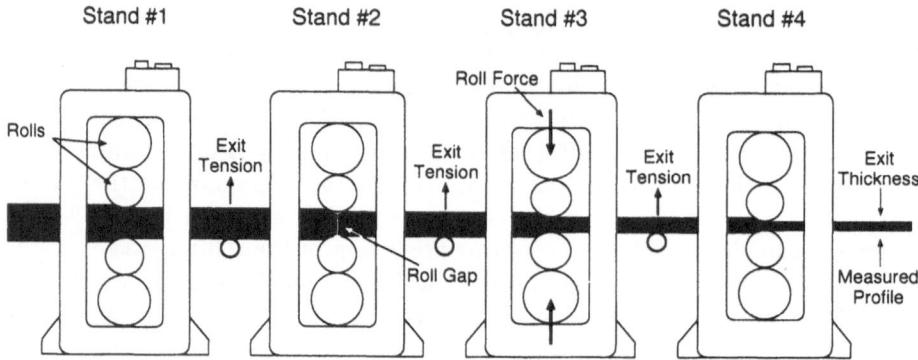

Fig. 1: Sketch of a hot wide strip rolling mill consisting of four stands. At each stand a predetermined thickness reduction of the strip takes place. The final thickness, width, and profile of the strip is measured after the last stand.

the parameter settings for many hundred different steel qualities, and for each new steel quality the model parameters have to be adapted from scratch, which is expensive since it requires to first roll a number of strips with wrong rolling force pre-settings.

The goal of developing a Neural Network approach was not only to improve the estimation accuracy, but also to overcome the weakness of the conventional method of applying different models to different steel qualities, which does not allow to generalize to new materials. With a Neural Network which distinguishes between steel qualities by taking into account the material's chemical composition, it becomes possible to have a single model for different steel qualities. Different materials have different input vectors for the Neural Network, which allows "generalization to new materials and even to get rid completely of the rather artificial category "steel or aluminum quality".

At each stand n, the rolling mill in this case consisted of seven stands with $n = 1, ..., 7$, a Neural Network $N_n(\mathbf{x}_n | \mathbf{w}_n)$ is employed to estimate the rolling force F_n which has to applied at stand n. The input $\mathbf{x}_n$ for the Neural Network, a 25-dimensional vector, contains the concentration of the sixteen most important chemical elements of the material of the strip plus the physical quantities describing the strip when it has reached the stand n, e.g. the strip's width, thickness, temperature, etc. at the respective stand.

The Neural Networks are adaptive through their weights $\mathbf{w}_n$. For achieving good performance it turns out that it is necessary to adapt the networks on-line with each strip which is rolled. The adaptation of the network weights $\mathbf{w}_n$ is performed through gradient descent on the quadratic error $(F_n - N_n(\mathbf{x}_n | \mathbf{w}_n))^2$, which yields

$$\Delta \mathbf{w}_n = \eta (F_n - N_n(\mathbf{x}_n | \mathbf{w}_n)) \frac{\partial N_n(\mathbf{x}_n | \mathbf{w}_n)}{\partial \mathbf{w}_n} \qquad (1)$$

as the adaptation rule.

Stand	1	2	3	4	5	6	7
Conv. method	1005	802	772	763	769	849	859
Neural Net	767	636	557	549	619	736	754
Improvement	+24%	+21%	+28%	+28%	+20%	+13%	+12%

Table 1: The RMS error of the rolling force with the conventional method and the Neural Network approach (in kN). The Neural Network approach is able to reduce the RMS error up to 28%.

The data from 10000 strips, which corresponds roughly to the production of one month, were available for pretraining the seven networks, i.e. each of the seven networks was pretrained with 10000 data pairs $(F_n^{(i)}, \mathbf{x}_n^{(i)})$. After this pretraining, which might be performed in a batch mode, the networks were tested in an on-line mode with 53812 strips. During this simulated on-line test, with each strip the estimation errors of the seven networks were determined and for each network an adaptation step was performed. The on-line test simulates the real application at the mill and yields exactly the rolling force error which would have been achieved if the neural network had been deployed during the five month when the 53812 strips were rolled. At the end of the on-line test the seven RMS errors over the 53812 strips were determined and could then be compared with the errors of the conventional method. The result is shown in Table 1. Averaged over the seven stands the Neural Network approach was able to achieve a reduction of the RMS error of the rolling force of 21%, which counts as a very significant improvement. The neural network approach has now successfully been tested on-line at Krupp Hoesch Stahl AG, Westfalenhütte Dortmund, and has become a major component of a commercially available process optimization system.

3 Width Control

Figure 1 shows the so-called finishing mill. At the finishing mill with its horizontal rolls only the thickness of the strip can be controlled, not its width. The width is determined in a rolling process with vertical rolls, which is performed before the strip runs into the finishing mill. The problem, however, is that during the horizontal rolling the width of the strip does change. To obtain the desired width at the end of the rolling process, it is necessary to estimate the widening of the strip by the horizontal rolling. A good estimate for this widening then enables one to reduce the width of the strip by this amount with the vertical rolling, which then leads to the required final width of the strip after the whole rolling process.

So far, the estimation of this widening has not been satisfying. It is still necessary to add a margin of about 12mm to the desired width to make sure that at the end of the rolling processes the width does not fall below the required value along the whole strip. To increase the accuracy of this widening estimation, a neural network approach has been developed. The input **x** of the neural network, a 24-dimensional vector, contains all the quantities which might influence the widening of the strip during the horizontal rolling. These are, e.g., the temperature of the strip, its thickness, the thickness reduction at each stand, the strip's width, the rolling velocity, the backward and forward tension at each stand etc. The output $N(\mathbf{x}|\mathbf{w})$ of the network is then an estimation for the widening Δb of the strip.

To achieve a good performance for the widening estimation, it is again necessary to adapt the network on-line with each strip which is rolled. The adaptation of the network weights **w** is again performed through gradient descent on the quadratic error $(\Delta b - N(\mathbf{x}|\mathbf{w}))^2$ with an adaptation step according to (1).

For pretraining the network, data pairs $(\Delta b^{(i)}, \mathbf{x}^{(i)})$ from roughly one month production were used. After this pretraining, the network was again tested in an on-line mode, this time with 70306 strips. On these 70306 strips, the Neural Network approach achieved an RMS estimation error of 2.7mm, compared to 3.7mm of the conventional method. This means that compared to the conventional method the Neural Network achieved a reduction of the estimation error of 27%. Reducing the margin of 12mm by just one millimeter translates into savings of about a million dollars per year for a modern hot wide strip rolling mill. The Neural Network approach has now successfully been deployed on-line at Thyssen Stahl AG/Beeckerwehrt for about a year.

4 Profile Control

Figure 2 illustrates the definition of the term "profile" and shows an exaggerated "bending effect" due to the roll separating forces. There are a number of additional effects that influence the strip profile, such as roll bending or roll thermal crown when the roller expands due to the heat transferred from the hot strip. All these effects overlap to produce a very complex overall profile that depends on current settings of the rolling forces and the geometry of the current strip as well as on the recent process history, because the thermal crown is influenced, e.g., by the width of previous strips and the pause time between strips. Thus, it is very difficult to obtain a mathematical model that describes this highly nonlinear and instationary process up to the desired degree of accuracy.

The main difficulty in the context of profile prediction is that an actual measurement of the profile can be conducted only after the last stand. This means, there are no measured values available for the intermediate profiles that could be used to verify or adapt a mathematical profile model for each individual stand. Thus, the current use of mathematical models requires delicate and repeated fine-tuning of various model parameters by hand which is obviously very expensive due to the time and the material wasted.

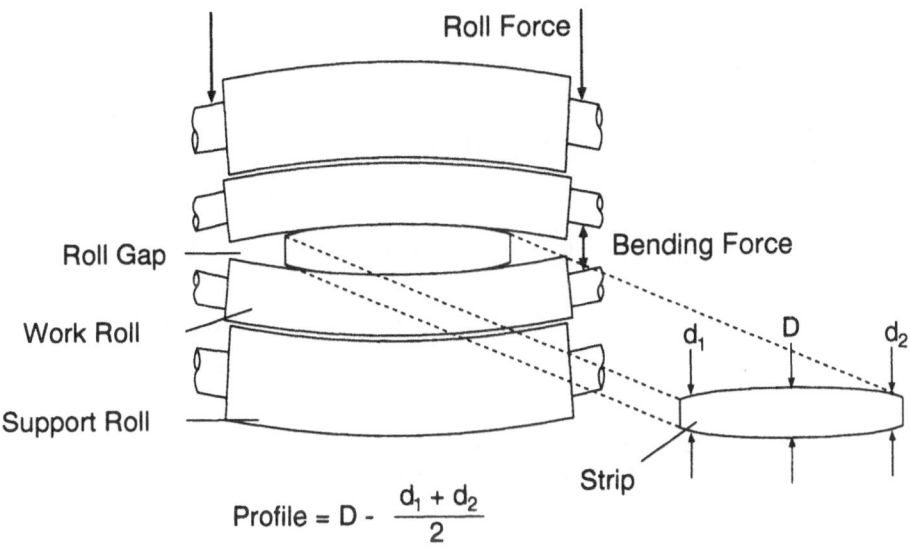

Fig. 2: Definition of the profile as the average thickness difference between the middle and the edges of a strip.

Figure 3 shows one possible approach how Neural Networks can be used in combination with existing mathematical models to improve the prediction capability for the final profile. The roll-gap profile at each stand is predicted by complex mathematical models (MM), which include a model of the roll bending and of the thermal crown of the roll, thus taking the process history into account. The Neural Network is used to combine the information of the MMs with other process parameters and is pretrained and then adapted on-line to provide the functional relationship that predicts the resulting profile.

This combination of Neural Networks and mathematical models allows to preserve the basic knowledge about physical dependencies as expressed by the mathematical models, while gaining an automated fine-tuning capability through the adaptive Neural Network. Preliminary simulation results with actual data obtained during a 3 month period from a 4-stand hot rolling mill aluminum plant indicate that the overall prediction performance can be improved by about 20% compared to the mathematical model alone with hand-tuned parameters. On-line experiments at the mill with the combined MM/NN model are underway to validate our current simulation results.

5   Summary and Outlook

With a number of real-world applications we have shown that neural networks are able to improve the conventional control and optimization schemes for hot line rolling mills. Some of these Neural Network approaches are now major components of commercial process optimization system for rolling mills, others

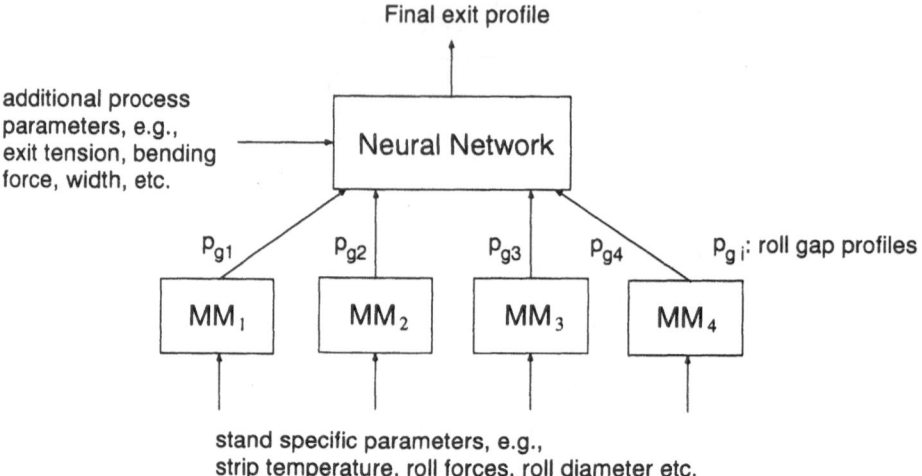

Fig. 3: One possible combination of a Neural Network with different mathematical models (MM) to predict the final profile after the rolling process.

are still tested but already show promising results. Still only the first steps have been made. Many further control, optimization and diagnosis problems in the application domain of steel and aluminum production and manufacturing are still open to be tackled with Neural Network approaches. After the potential of Neural Networks has now been clearly demonstrated, the effort will be further increased with the goal to make Neural Networks a standard technique in process automation for steel and aluminum production and manufacturing.

References

Lindhoff, D., Sörgel, G., Gramckow, O., and Klode, K.-D. (1994). "Erfahrungen beim Einsatz Neuronaler Netze in der Walzwerksautomatisierung". Stahl und Eisen, 114, Heft 4, S. 49-53+208.

Poppe, T. and Martinetz, T. (1993). "Estimating Material Properties for Process Optimization". Proc. of the International Conference on Artificial Neural Networks - ICANN '93, Amsterdam, 13-16 Sep. 1993, Springer Verlag, pp. 795-798.

Condition Monitoring with National Power

John MacIntyre, Professor Peter Smith

School of Computing and Information Systems, University of Sunderland,
Sunderland, England

Abstract

Condition monitoring is a developing discipline in machinery maintenance. Data such as vibration levels (both overall and in terms of frequency spectra), temperature, oil analysis, etc, are acquired from plant, and analyzed to determine the condition of that plant at the time of measurement. Software packages are currently available to allow graphical display of the data, with varying degrees of diagnostic tools available to assist engineers in performing data analysis. Furthermore, some rule-based expert systems are available to perform machinery defect diagnosis; again there are varying degrees of automation and human interaction in these packages. However, these systems only deal successfully with clearly defined problems within a narrow band of parameters; they are notably unsuccessful at coping with contradictory, incomplete, or "noisy" data - just the type of data found in many real-world applications.

This paper describes the implementation of an off-line condition monitoring system at Blyth Power Station, one of the stations owned by National Power in the United Kingdom. It explains the application area and the type of data acquired. The paper then goes on to describe the neural network models which have been developed to analyze condition monitoring data.

Keywords: Condition monitoring, data collection and analysis, expert systems, neural networks.

1 Introduction

Blyth Power Station is part of National Power, the major electricity generating company in the United Kingdom. The Station occupies a 241 acre site on the North East coast of England, approximately 15 miles north of Newcastle-upon-Tyne. The site comprises two stations, Blyth 'A' and Blyth 'B', with a combined generating capacity of 1,180 megawatts. Blyth 'A' was commissioned in June, 1960, and Blyth 'B' in September, 1966. Blyth 'A' consists of four 120MW generating units, and Blyth 'B' two 350MW units. The age of the plant, on-going reductions in maintenance resources, and increased commercial pressures as a result of privatisation, have combined to create a requirement within the Station to move away from the traditional, manpower-intensive strategies of Planned or Breakdown Maintenance, towards a Condition Based Maintenance policy for critical areas of auxiliary plant [1]. The Station management recognised this requirement, and in

September, 1992, entered into a collaborative agreement with the University of Sunderland to establish a SERC (Science and Engineering Research Council) funded project whose aim would be to develop and implement an intelligent condition monitoring system for use within the Station.

2 Condition Monitoring

Blyth Power Station has been involved in the implementation of off-line condition monitoring on various items of critical auxiliary plant since January, 1993. Condition monitoring as implemented at Blyth involves the capture of various types of data from items of plant which can be analyzed to give a diagnosis of the operating condition of the machine. Typically, data captured include vibration levels, temperatures, oil samples, and various process parameters such as pressures and load currents. This data is collected off-line, using a portable data collection instrument - the PL31b-01, manufactured by Diagnostic Instruments Ltd., in Scotland - for the vibration, temperatures etc, and a simple syringe kit for oil samples. Data collected with the portable instrument is then automatically uploaded into a specialised predictive maintenance database package from the American company, Entek Scientific Corporation, which allows graphical plotting and trending of the information, and provides various analysis aids such as spectral alarm levels and frequency identification techniques. Oil samples are sent to a specialist laboratory and specific values from the resulting reports are manually entered into the database. Previous papers [2,3] describe in detail the operation of the condition monitoring system at Blyth.

Once all the information is in the predictive maintenance database, it must be analyzed to produce a diagnosis of the condition of each machine being monitored, with the objective of either performing maintenance before the machine fails, or deferring planned maintenance which is deemed to be unnecessary. Although the software does provide some sophisticated tools to assist in analysis, this task essentially relies on the experience and expertise of a maintenance engineer in interpreting the data. Consequently, many condition monitoring systems suffer, either through human error, or through lack of an engineer with sufficient experience to analyze the data. Nevertheless, the use of condition monitoring has been successful in both identifying defects prior to failure, thus allowing maintenance action to be taken, and also in determining when planned maintenance actions could be deferred. This has generated significant financial benefit for the Station, with a substantial return on investment in hardware and software.

2.1 Typical Condition Monitoring Data

Figure 1 below shows a vibration spectrum captured from the inboard bearing on a Primary Air Fan. The spectrum is obtained by performing a Fast Fourier Transform (FFT) on the time domain vibration signal, after a band pass filter and envelope filter have been applied to it. The spectrum plots vibration frequencies against amplitudes, in this case in acceleration (g). In this spectrum, a large peak is visible at the frequency specific to an outer race defect (ORD). Inspection of the bearing confirmed that an outer race defect existed, and the bearing was replaced.

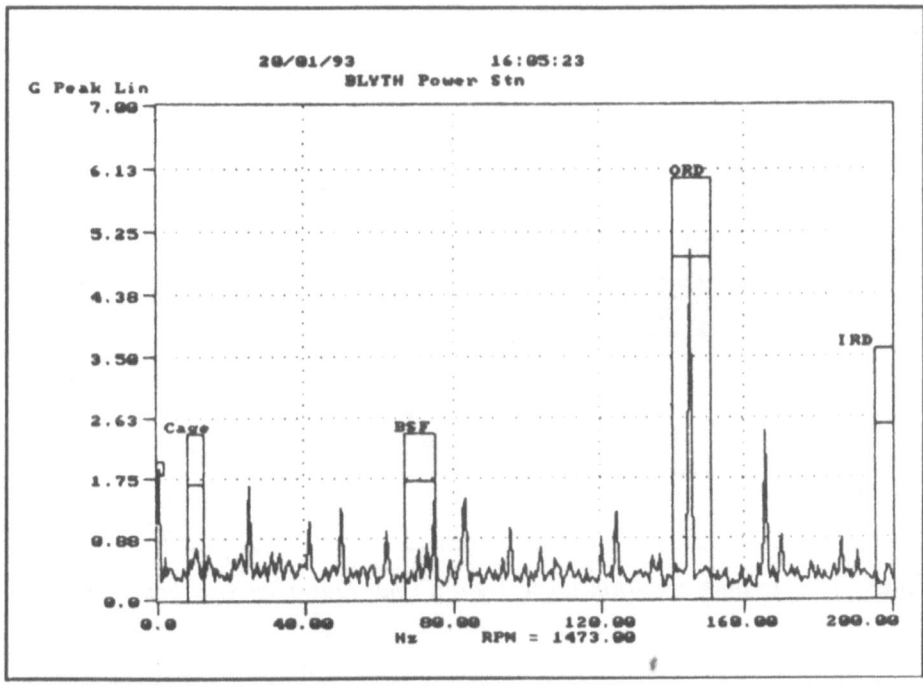

Figure 1 - Vibration spectrum of Primary Air Fan bearing, clearly showing outer race defect

The above example shows a clear and well-defined spectra with distinct characteristics relating to the bearings. However, in many cases the data is far more difficult to analyze, due to background noise, low frequencies and amplitudes, difficulty of access for measurements etc. Figure 2 shows a vibration spectrum acquired from a Coal Pulveriser Gearbox; the general level of noise, plus the fact the defect frequencies are in the low end of the spectrum and therefore can be masked by the background vibration, make this a much more complex problem for analysis [3,4].

In the case of the Coal Pulveriser Gearbox, samples of the lubricating oil are taken and spectrographically and chemically analyzed, and certain values from the resulting data are trended [5]. Figure 3 shows the trend plot of the ferrous content (parts per million) of the oil from one such gearbox. This information is used to help in diagnosing the condition of the gearbox, alongside vibration data such as is shown in Figure 2.

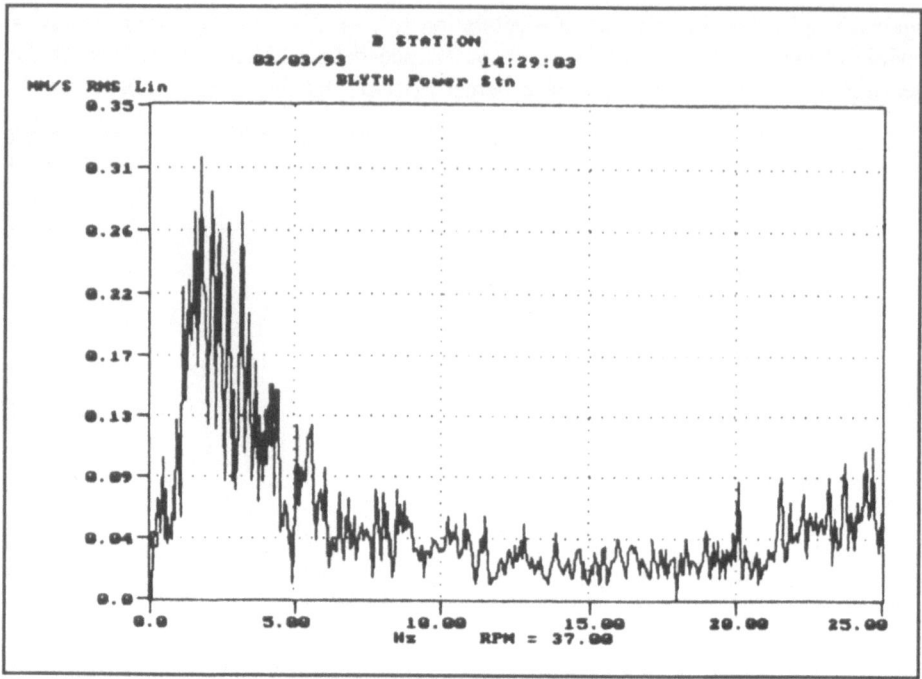

Figure 2 - Vibration spectrum taken from a Coal Pulveriser Gearbox

3 Application of Neural Networks in Data Analysis and Diagnosis

A large part of the data analysis function in condition monitoring involves diagnostic and decision-making tasks, and it is natural that attempts have been made to apply artificial intelligence techniques to automate this process. The main area of interest and applications development in AI has been (until recently) that of expert systems. However, in the last few years, attention has been turning to the potential for neural networks to deal with some of the more complex data analysis tasks. Burrows [6] identifies the need for automated systems to improved the decision-making process in plant maintenance; Broomhead and Jones [7], and Harris [8] also identify that model-based or knowledge-based systems have difficulty in data analysis and diagnosis where complex or chaotic data patterns are experienced.

A number of proprietary software systems have been developed using knowledge-based approaches for condition monitoring diagnostics. Some of these rely on a high degree of inter-activity with the user in terms of question-and-answer input, where the user's answers provide the condition statement of an IF {condition} THEN {statement} clause in the rule base. Milne [9] gives case studies of knowledge-based approaches to process control and machinery diagnostics using condition monitoring techniques, both on- and off-line, and also states that there is a need for automation of the data analysis process.

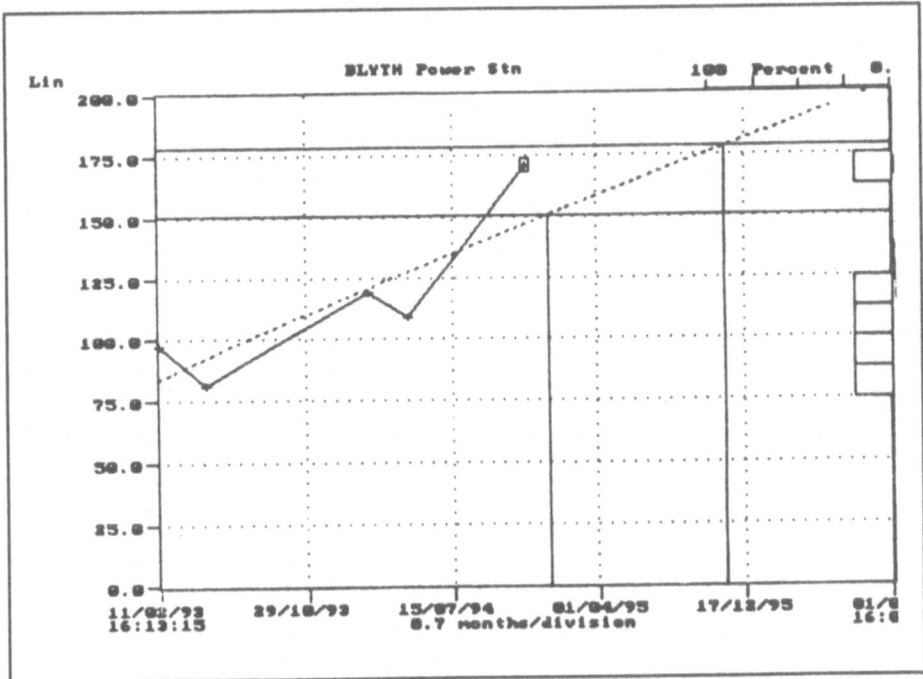

Figure 3 - Trend plot of the ferrous debris (parts per million) in the Coal Pulveriser Gearbox lubricating oil

3.1 Development of the Neural Bearing Analyzer

At Blyth Power Station, a neural network model has been developed for the analysis of vibration spectra from the Primary Air Fan bearings. This network is based on the Multi-Layer Perceptron (MLP) architecture, using the back propagation training algorithm. The use of back propagation allows the adjustment of weights in the neural connections in multiple layers; this is critical if a network is to solve non-linearly separable problems [10,11]. The MLP architecture using back propagation relies on a technique referred to as "supervised training", in which an input vector (a set of numerical values derived from the data) is presented to the neural network along with a target output vector (a numerical representation of the desired output from the network). The actual output is compared to the target output for each input vector, and the root mean squared (RMS) error is calculated. This error is then propagated backwards through the neural connections, and the process is repeated until the RMS error is within an acceptable threshold, typically 0.001 [10,11,12].

The Neural Bearing Analyzer (NBA) model was developed through several versions. Training data sets were constructed from real data collected from Station machinery, where the target output could be confidently generated as a result of known bearing condition due to inspection or replacement of bearings. Diagnostic Consultants Limited, a UK condition monitoring consultancy firm, assisted in production of the target values. Testing of the models was performed both with real data acquired from the Primary Air Fans, some of which had known conditions from inspection, and also with artificially-generated data. The network's classifications were compared with the classifications of the consultant condition monitoring engineer.

The earliest models involved presentation of the full vibration spectrum (400 datum points) to the network as an input vector, and a relatively large output set, with separate classes for each of the bearing components, and different levels of defect severity. This architecture proved to be unsatisfactory, due to very long training times and frequent failure to converge to a satisfactory error value. Later versions used a simplified output set, with condition estimates for the bearing as a whole, in order to reduce the possibility of misclassification across very restrictive decision boundaries. However, the size of the input vector still resulted in long training times and problems in convergence. It was felt, therefore, that the input vector needed to be substantially reduced in dimensionality if a satisfactory model was to be developed. It was also recognised that the values which would be used as inputs to the network would need to be carefully selected so as not to lose any important discriminatory information from the raw data.

Various techniques were examined for use in the 'pre-processing module' of the Neural Bearing Analyzer. A simple expert system approach, where rules based on the bearing type, geometry and running speed of the shaft would identify bearing defect frequencies to be selected as inputs, was investigated. However, testing showed that this approach was, in itself, too simplistic for the data, as features were present which were important in classifying particular types of defect, but which were not selected by the expert system. Rather than develop a set of bearing-specific rules (which would prevent the model being used in a generic application), an additional technique was sought which would be able to pick out significant variables within the data set. The technique eventually chosen for this task was Principal Components Analysis, which is a mathematical process which identifies variables which are responsible for the greatest degree of variation within a given data set [16]. Testing showed that this technique could identify variables which would not be selected by the simple expert system, but which were important discriminatory features for classifying defects.

Consequently, the two techniques, i.e. the simple expert system and principal components analysis, were combined to produce the pre-processing module for the Neural Bearing Analyzer. In addition, a small expert system was produced to post-process the output of the neural network into a diagnostic report, giving advice to the maintenance engineers. Therefore, the Neural Bearing Analyzer is a distinctly hybrid model, combining expert systems, neural networks, and 'traditional' mathematical techniques. Figure 4 shows a schematic representation of the Neural Bearing Analyzer model.

3.2 Application of the Kohonen Network

An alternative approach to the application of neural networks in condition monitoring has utilised another neural network architecture, namely the Kohonen network, which is an example of a self-organizing map using "unsupervised

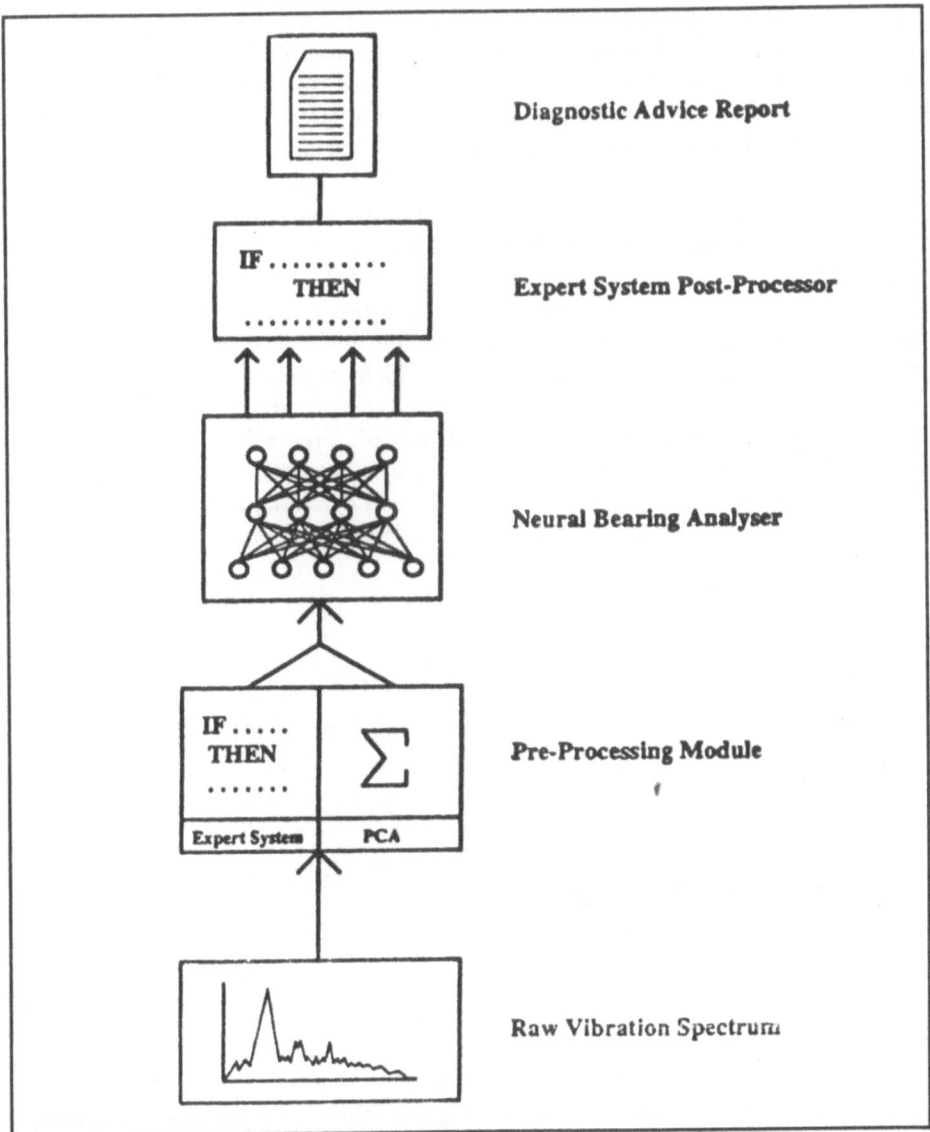

Figure 4 - The hybrid Neural Bearing Analyzer in use at Blyth Power Station

learning". Here the output layer is a grid of interconnected nodes, all of which are fully connected to the input vector. A competitive learning algorithm is used in which the output neurons compete with each other to present the highest value in the output vector. In this case, frequency spectra values are used as vector inputs and the Kohonen network is trained with vibration data from a machine in good condition. In its monitoring mode, vibration frequencies are presented to the network and the winning node identified. The error distance between the input and the winning node is calculated and used as a measure of the machine's "health". As a fault develops the input vector will move away from the nodes in the network, and thus the error distance will increase giving a warning that a fault is developing. This system, however, can only detect a fault; the lack of diagnosis represents a significant restriction in its use [15]. An extension of this system enables limited diagnosis by training the network with examples of vibration data from a machine with a known fault, as well as normal or "healthy" data. The network is then labelled so that when monitoring, the output indicates that a trained fault type has been detected. It is, however, very difficult to collect sufficient representative data from machines in poor condition. Unless the machine can be set up with faults specifically for data collection purposes, it is practically impossible to get a full training set which is representative of all possible fault data. Figure 6(a) below represents a two-dimensional Kohonen network which has identified a fault, although at this stage the fault cannot be diagnosed - the network simply indicates that the input vector has moved outside of the normal "space". Figure 6(b) depicts a similar network with labelled nodes, which can be used to identify the fault by finding the nearest labelled node to the input vector. This type of network has been successfully applied to the detection and diagnosis of a variety of condition monitoring and welding applications [13].

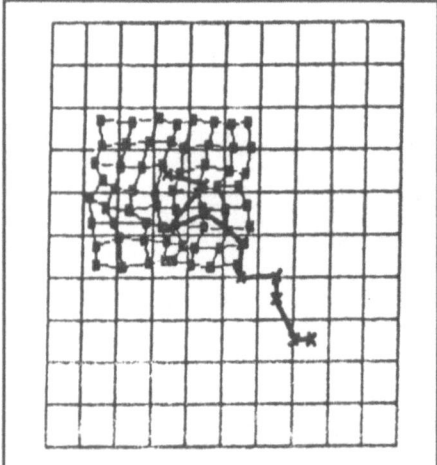 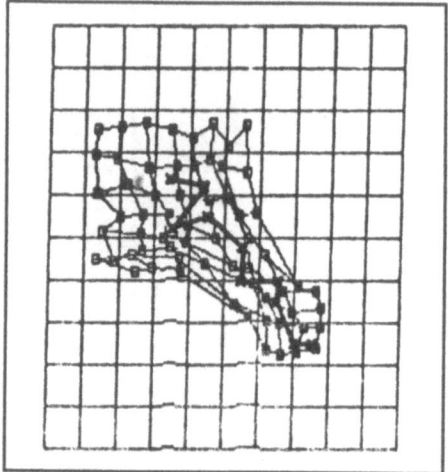

Figure 5(a) - Two-dimensional Kohonen network identifying "abnormal" input vector

Figure 5(b) - Fault identification using a labelled Kohonen network

3.3 Further Developments

The authors are currently examining the use of Radial Basis Function networks as an alternative to the Multi-Layer Perceptron in the Neural Bearing Analyzer, as well as Auto-Associative Networks as an alternative to the Kohonen network model discussed above. The authors are also involved in a EUREKA project, NEURAL-

MAINE (EUREKA Project EU1250) [17] which aims to develop on-line, real-time diagnosis capabilites for large rotating plant (e.g. turbines, milling plant, winding gear) based on neural networks.

4 Conclusions

In practical terms, condition monitoring has been shown in its implementation at Blyth Power Station to be an extremely valuable technique for the area of industrial plant maintenance, and can generate significant financial benefits. The task of analyzing condition monitoring data, which is difficult and subjective, offers an application area for neural networks with potential in engineering and manufacturing industry. Work carried out by the authors to date indicates that this potential can be realised through the correct design and implementation of appropriate neural network and hybrid artificial intelligence architectures.

References

1. A. M. R. Flegg, "Profitable Condition Monitoring within National Power", Machine Monitoring Systems Ltd., Chesham, Buckinghamshire, 1990.

2. J. MacIntyre, "Development of an Off-Line Condition Monitoring System at a Coal-Fired Power Station", *Condition Monitor*, No. 75, March 1993.

3. J. MacIntyre, P. Smith, C. Wiblin, "Development and Implementation of a Condition Monitoring System for Off-Line Monitoring of Auxiliary Plant at a Coal-Fired Power Station", Proceedings of 5th International Congress on Condition Monitoring and Diagnostic Engineering Management (COMADEM), Bristol, England, 1993.

4. Schlumberger Instruments, "Frequency Response Analysis", Technical Report No. 010/83, Instruments Division, Victoria Road, Farnborough, Hampshire GU14 7PW, 1983.

5. G. Duvall, "Lubrication Oil Condition Monitoring", *Condition Monitor*, No. 79, July 1993.

6. J. Burrows, "Strategies, Techniques and Tools for Improving the Decision-Making Process of Plant Maintenance", Proceedings of 4th International Congress on Condition Monitoring and Diagnostic Engineering Management (COMADEM), Nimes, France, 1992.

7. D. Broomhead, R. Jones, "Condition Monitoring and Failure Prediction in Chaos", Institute of Electrical Engineers Colloquium on Advanced Vibration Measurements, Techniques for the Early Prediction of Failure, London, England, 1992.

8. T. Harris, "Neural Networks and their Application to Diagnostics and Control", Proceedings of 4th International Congress on Condition Monitoring and Diagnostic Engineering Management (COMADEM), Nimes, France, 1992.

9. R. Milne, "Amethyst: Rotating Machinery Condition Monitoring", Proceedings of the American Artificial Intelligence (AAI 90) Conference, Washington, USA, 1990.

10. G. E. Hinton, "How Neural Networks Learn from Experience", *Scientific American*, September, 1992.

11. J. Dayhoff, Neural Network Architectures - An Introduction, Van Nostrand Reinhold, 1990.

12. T. Harris, "An Introduction to Neural Networks", Proceedings of the 6th International Conference on Joining of Materials (JOM-6), Helsingor, Denmark, 1993.

13. T. Harris, "Neural Networks in Machine Health Monitoring", *Professional Engineering*, July/August, 1993.

14. T. Kohonen, "An Introduction to Neural Computing", *Neural Networks*, Vol. 1, 1988.

15. T. Harris, J. MacIntyre, P. Smith, "Neural Networks and their Application to Vibration Analysis", Proceedings of the Structural Dynamics and Vibration Symposium, New Orleans, USA, 1994.

16. I. Joliffe, "Discarding Variables in a Principal Components Analysis II: Real Data", Journal of Applied Statistics, Vol. 22, 1972.

17. T. Harris, J. MacIntyre *et al*, "NEURAL-MAINE: Intelligent On-Line Multiple Sensor Diagnostics for Complex Machinery", Proceedings of the 8th International Congress on Condition Monitoring and Diagnostic Engineering Management (COMADEM), Kingston, Canada, 1995.

The Electronic Nose for Process Control

Mark A Collins[1] & Laurent Moy[2]

[1]Neural Computer Sciences Southampton UK, [2]AlphaMOS, Toulouse, France

Considerable commercial interest has recently been given to the electronic nose technology. Integrating it with neural networks offers the potential for use in the process control industries, controlling processes on the basis of odour assessments.

Introduction

Odour sensing is a very important task carried out in industry. The olfactory quality of a final product is one of the most important parameters that can be assessed and is performed daily in many research and quality control labs. Food chemistry, cosmetic (perfumes, fragrances etc.), plastic industries (polymers, packaging, etc.) and environmental control are just a few examples. Knowing that the odour of a product is 'good' or 'bad' is a fairly subjective measure (even though the odour itself maybe composed of known molecular entities) and takes many years to acquire and learn. With the increasing need to have automated processes, these manual, expert steps can cause process delays as well as complete loss of product if the time to perform the assessment is too long.
An automatic system to classify, detect concentrations, etc., of odours would obviously be extremely valuable. Odour quality assessment could then be performed, on line, at all times and in an objective and standardised manner. Despite the commercial usefulness of odour sensing no instrument yet exists that can mimic the human sense of smell in terms of sensitivity, selectivity, response time and sample size.

Electronic Noses

In recent years however many researchers have attempted to create *artificial electronic* noses, with varying degrees of success. An artificial electronic nose is an instrument capable of mimicking at least some of the functionality of the human sense of smell and should be able to detect both simple and complex odours. Electronic noses are typically comprised of an array of gas sensors, signal transduction circuitry and a pattern recognition engine [1].
The system setup shown in Figure 1 is for the FOX2000 electronic nose (AlphaMOS, France)

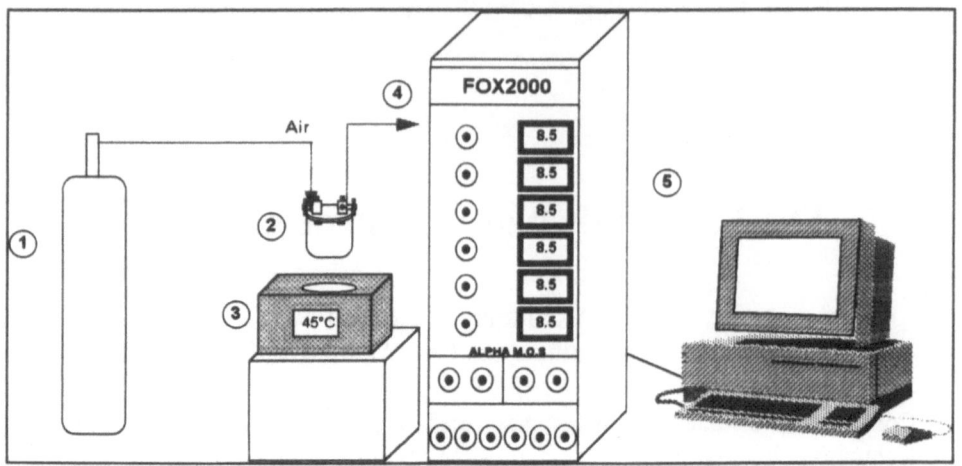

Figure 1 Typical Electronic Nose System

① Synthetic air (optional) and humidity regulator
② Glass container with the odorant sample or autosampler OdorScanner (50 vials)
③ Measurement chamber to control the temperature of the headspace (RT - 205°C)
④ Electronic nose with 6, 12 or 18 sensor channels
⑤ Microcomputer for storing and processing sensor data

Neural Networks and the Electronic Nose

The system described above is capable of detecting odours from industrial processes. In the presence of the sample contained in a glass vial, the sensor resistance either decreases or increases depending on the sensor type. The combination of up to 18 sensor responses is then used to report the fingerprint of the sample. Using the gas sensor array to sample, for example, the headspace of a beer fermentation results in a complex sensor pattern (Figure 2.0) that can be evaluated to provide a number of important product features, e.g. alcohol content, off flavours, and even the identity of the product. This variation and fluctuation of odours from natural products (fish, meat, coffee, cheese) and synthetics (plastics, deodorants, perfumes) results in a characteristic sensor signal pattern that can be associated with a particular odour. ANNs can learn to recognise the sensor pattern, taking into account their complex nature, subtlety and even noise and drift in the signals. In use, the network is trained using a set of known examples with their appropriate odour classes or labels. Following the training, previously unseen samples can be rapidly classified as 'good', 'bad' or 'indifferent', as type A, B, C or even as concentrations e.g. 20%.

Traditional techniques such as Principal components analysis, cluster analysis and multiple linear regression, have been used to analyse the sensor data. However, since the sensor response is often non-linear, and the sensor patterns are noisy, and subject to drift these techniques often perform poorly. Furthermore the results of the analyses themselves need to be interpreted by experts, so the potential to have a decision made in real time is lost.

Using the Neural Network for Discriminating the Off Flavours of Beer

Data was acquired from the FOX2000 instrument for standard beers and `off` flavoured beers (spiked with diacetyl).). A single time slice from the sensor pattern was used in order to provide a dataset of responses for 6 sensors. Sample data was collected in triplicate for 4 different types of beer. In other cases the whole response curve has been be used, or a particular time window. The more information represented to the network the better (usually) it is at discriminating and coping with any noise in the input patterns.

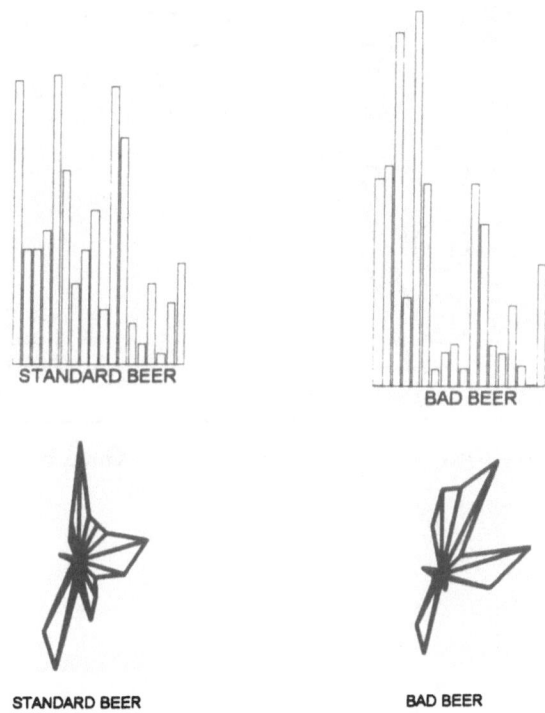

Figure 2 Typical Sensor patterns (Histogram and Radar Plot) from the electronic nose

A simple multi layer perceptron was used to obtain a ANN model for the beers. Each input node represented a single value for a 10 second time slice from each of 6 sensors. Two output nodes were used to represent 'Good' and 'Bad' decisions. (Figure 3.0) Data was normalised by assuming that the maximum sensor response was 1, and the network trained with Quick Propagation algorithm using the Neudesk Neural Network development tool (NCS, Southampton , UK). The training error after 300 epochs was 0.0001, and testing of the trained network with unseen data showed a 99.99% success in recognising the test cases of standard and adulterated beers.

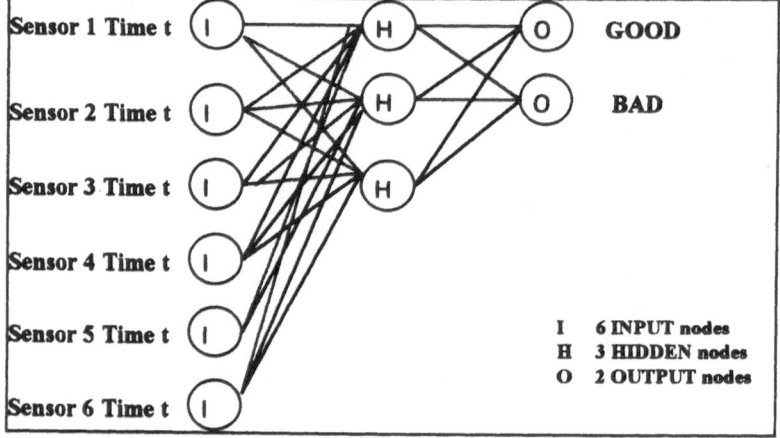

Figure 3 Neural Network Topology for Recognising Good/Bad Flavoured Beers

The use of a neural network run time engine (NeuRun, NCS Southampton, UK) allows the ANN component to be closely integrated into the FOX200 instrument software. Sensor data can be transferred to the neural network software for training and validation. Once the system is trained the user only has to press a button to have the current signal processed by the network and the answer displayed back, in real time if required (Figure 4.0) Technically this is achieved by utilising the Windows communication protocol DDE (Dynamic Data Exchange) between the Fox2000 interface developed in LabView (National Instruments) and NeuRun.

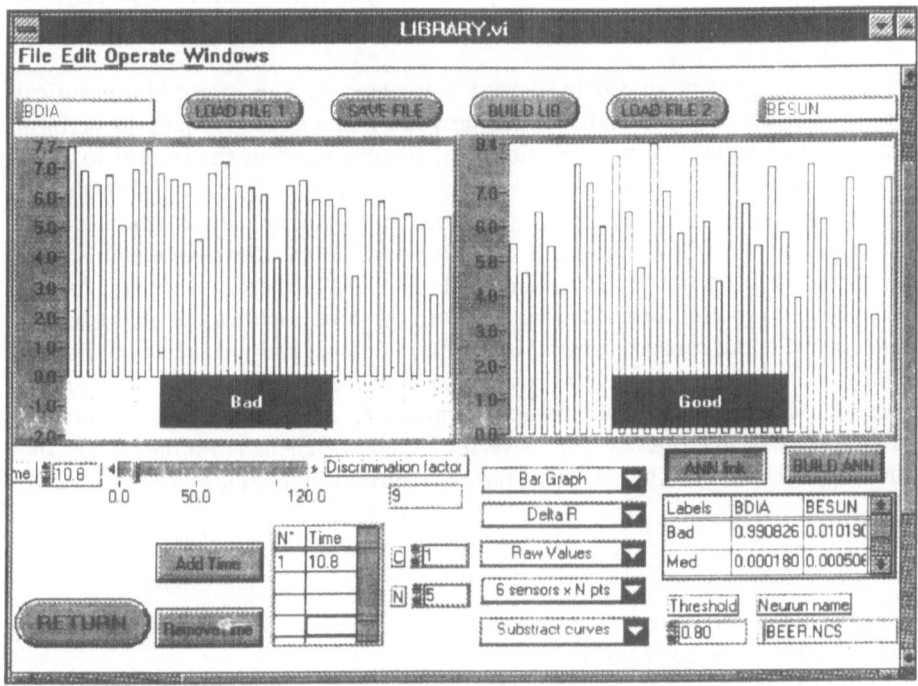

Figure 4 Real Time ANN decsion making with the FOX2000 Instrument

Summary

Integration of ANNs with the electronic nose allows the instrument to be used for on-line real time decision making which can then be used in process control loop. The alternative to the network approach would be some form of statistics e.g. Principal Components Analysis or simply looking at sensor patterns and interpreting them by eye. This of course demands that the operator be very skilled in data interpretation, which may limit the wider use of the instrument. The advantages and benefits of the partnership between the technologies are:

- Sensor data is complex and difficult to model - ANNs are very good at solving non-linear, multidimensional data sets

302

- There is the potential to more closely mimic the human olfactory system through trainability and generalisation
- Once system is trained it can be used by anyone and it is possible to have available the whole of a company's quality control expertise On-Line 24 hours a day.
- The ANN can pick up the subtle differences between 'good' and 'bad' products where other techniques would fail or be confused by for example variations in raw materials
- ANNs may be re-trained and 'topped-up' with data
- ANN systems may be rapidly developed in comparison to traditional statistics or expert systems

As the technology is more widely adopted, automated systems, trained on historical process data and using neural network decision making will be able to perform assessments of product or process odour and involve these in the control loop. In the food industry for example, expert taste and odour panels can be assisted to make consistent products batch after batch, year after year [2]. The marriage between the ANN and the electronic nose will also be valuable the environmental sector, controlling air conditioning for example if poor air quality is detected. As the sensor technology itself develops into smaller and more robust units, the potential to have smart odour sensors in many industrial processes and products will become a reality. Consumer goods such as ovens and microwaves, the automotive sectors, aerospace, defence and environmental control are areas where the integrated, smaller, ANN enabled device will have most impact. There is no doubt that the ANN component allows the electronic nose to realise its potential as a decision tool.

References

1. Gardner JW and Bartlett.PN. A brief history of electronic noses. Sensors and Actuators 1994;B(18-19):22-221

2. Collins MA. Intelligent Electronic Noses- Not to be Sniffed at?. Food Tech Europe 1995;5:150-154

Automatic Quality Control of Roofing Tiles

Dr.-Ing. Hans-Joachim Kolb
Dipl.-Ing. Joachim Wagner,
MEDAV Digitale Signalverarbeitung GmbH
Uttenreuth, Germany

Nature of the Task

It is standard practice, not only in the German roofing tile industry, to test all products at the end of the production process. The traditional test for roofing tiles comprises a sound test to detect cracks and a surface inspection. These tests are currently performed by sorting personnel.

Attempts at automating the test and inspection process in the past often failed due to the inadequacy of the classification and measuring techniques or the typically required test speed of 45 units per minute per line. Automatic testing will firstly ensure constant high quality and secondly make the operator independent of sorting personnel.

Subjective and Objective Testing

Audible and visual inspection of the roofing tiles is normally carried out by semi-skilled employees. However, these testers are therefore largely responsible for quality control. The monotony of the work cycle leads to fatigue and produces changes of mood as well as a not to be neglected scattering of test quality. In addition, the important task of continuously observing the stability of the production process cannot be performed. No precise feedback is received for the statistical process check.

This is the result of the so-called 5% defect classification, i.e. the testers have been trained to reject a certain quantity. On occasions, variations in the actual reject quantity are not taken into account.

Efforts must therefore be made to objectivize the subjective tests by introducing an evaluation standard and also making it reproducible. This evaluation standard may be simply an accepted/rejected assessment, for example. Some of the evaluation criteria are defined numerically by entering defect limits. Training for the sound test is also a subjective test. However, this procedure takes place calmly for a small number of representative roofing tile samples.

Individual Measuring Techniques

The following are available:

- Sound test (cooling cracks)
- Visual surface and fold inspection
- Inspection of colour deviations
- Measurement of planeness

Sound test with Neural Networks

The main aim of the sound test is to detect fine cracks, i.e. in particular cooling or heating cracks. Since these hairline cracks are almost invisible at 'first' glance, a test using a camera system can be ruled out. On the other hand, these cracks can be heard relatively well by a trained ear. However, there are no 'exact' mathematical measuring techniques for this type of sound test. It is therefore necessary to use pattern recognition methods (sound patterns) in conjunction with artificial neural networks. Together with special signal analysis (spectrum) and the neural networks, the time and expense spent on adjusting and optimizing the classifier is reduced considerably. The aim of the sound test system is to adapt itself to the available expert knowledge in the factory by means of accepted and rejected sounds.

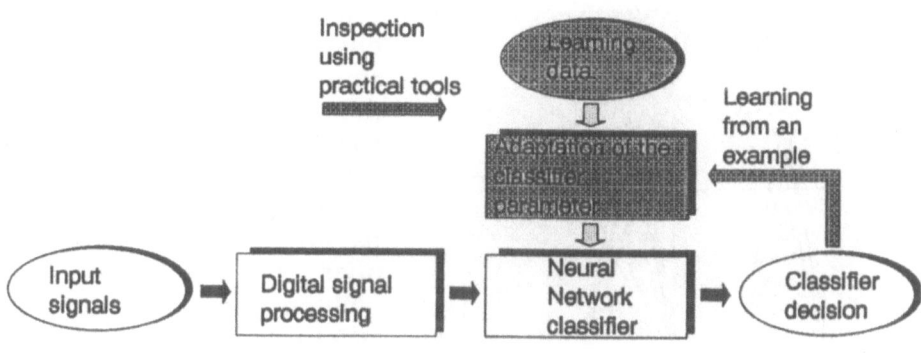

Figure 1: Signal processing in the sound testing computer

Understanding of Neural Networks

Unclearly defined classification tasks call for different solutions. Given the present requirements it is not possible to evaluate complex vibration signals via simple parameters, such as the power. Other characteristics are used, such as the frequency spectrum of the noise to be evaluated. Engineers are familiar with these and the results obtained permit good interpretation by a human operator. The frequency spectrum is a characteristic vector, which may, for example, comprise 400 individual elements. Additional characteristics from the time range lead to characteristic vectors with several hundred individual elements.

It can be seen that it is not generally possible for such areas to define model-based good/bad ranges at acceptable costs. The question arises as to how the various class areas can be determined here. The problem can only be solved if it is possible to learn from examples. Neural networks offer this capability.

306

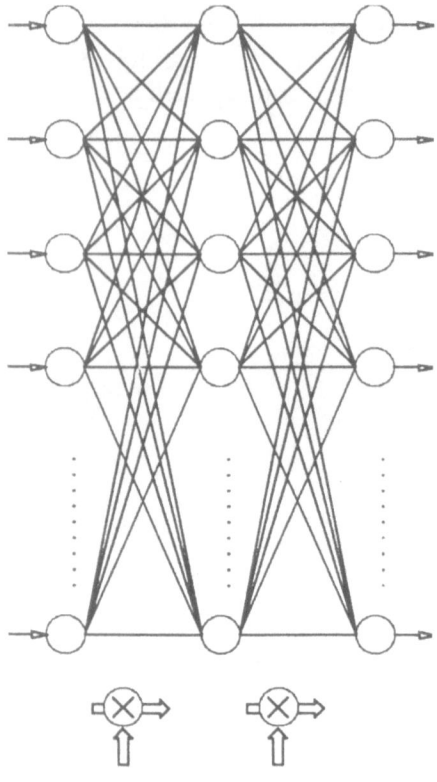

A three year old child can without any problem recognize the sound of an aeroplane, a car and a railway train. Visually handicapped people also learn to find their way about in their environment by interpreting sounds. These examples demonstrate the special cognitive capabilities of human beings in the assessment of complex sound patterns. Neural networks offer the possibility of imitating these capabilities within the limits imposed by present day technology. The name "neural network" stems from the function and structure of nerve cells and their interconnection in the brain. The figure shows the basic structure of a neural network.

Figure 2:   Structure of layered neural networks

A neural network mainly consists of three components:

- Arithmetic elements
- Topology of the network
- Learning rules.

At the input of the neural network we find the characteristics obtained from the random sample. From these are calculated the classification results in layered arithmetic networks with nodes and weighted links. In the learning process the learning rules control the change of the weighted links. These are thus in a position to dampen unimportant signal characteristics and to amplify important ones. The future function of the neural network is thus stored in the weighting.

Through training of the neural network the signal classes are thus automatically determined by the device. Such a classifier can therefore be used universally. A change or extension of the signal classes merely calls for renewed training of the neural network. Overall the following advantages are obtained:

- Reduced development costs
- High classification reliability
- Error tolerance in the learning phase
- Simple knowledge acquisition
- Quick application to production
- Cost reduction in operation
- Simple extendibility.

Neural networks are trained and not programmed. The learning process is based on random samples that have been previously classified by experts. The random samples consist of, for example, the actual sound pattern of fixed length and the associated assessment.

On the test stand the learning pattern is obtained in the manner laid down. It is expedient to clarify beforehand which sensors and which characteristics are suitable under the given environmental conditions.

Experts assign every random sample to a class. This classification assignment is an important input quantity in the learning process. The correctness of the assignment must be ensured and it must be comprehensible. The classification assignment can be oriented to previously practised methods. It is possible to obtain good/bad assessments or to introduce a system of awarding points. Where possible, the state variables that influence the learning random sample should be sensed on the test stand. This, may for example, be the speed at which the sound pattern was obtained. The knowledge base for the neural network is formed by random samples that have been classified by human experts. It must be ensured that this learning base is correct. It must be possible at any given time to check the random sample as a measuring data set with class assignment.

The learning random sample serves to train the network. The objective is to anticipate the decision of the expert automatically. The ideal situation is reached when the automatic classifier makes the same decisions as the human experts.

The test random sample constitutes the most important test. The trained classifier is checked by means of signals that have been classified by the human expert. These must not, however, have been used as training examples. Not until the classifier has proved itself may it be used in actual practice. By this means we are able to check the generalising properties of the classification system. That is, how will the classification system behave when a new noise occurs that did not form part of the learning random sample.

Neural networks thus permit the implementation of classification systems without it being necessary to resort to costly and risky analytical modelling. A trained network is rather a statistical description for the monitoring function. A prerequisite is that representative random samples should be available which then permit the modelling of class fields in the characteristic area through automatic training. Thus new methods are being created to pave the way to automatic monitoring systems that meet the requirements of modern monitoring and can be obtained at acceptable cost.

Procedure of automatic sound test in the tile factory

The following procedure is used during the sound test: An electromechanical hammer taps each tile which comes onto the conveyor belt. The sound is recorded contactlessly by a microphone acting as a sensor for data recording. The sound pattern and classification for each tile can be displayed in real time. Pattern recognition itself uses the frequency analysis of the sound (fast Fourier transformation) as a characteristic. Suitable, i.e. representative learning samples are required for successful classification. To this end, a selected learning collective (tile) is defined by the expert.

The selected sample roofing tiles are simply recorded as a learning pattern as they pass along the conveyor belt. The learning program is then started immediately afterwards and the sound computer can classify sounds live after a few minutes.

In order to attain maximum classification safety, the operator must "learn" the sample tiles for accepted and rejected for every individual type of tile (shape). The required number fluctuates here between 50 and 80.

A second learning method is also possible as an alternative. Training for this method is only normally carried out using one learning class, usually accepted samples.

If the routine verification of the classification produces an inadmissible error rate, the system can be simply optimized by relearning other tiles.

Other non trainable inspections for tiles

Plainness

Permanent measurement of plainness is important, not only if the average reject quota of the tiles is high with this test criterion!

One unplane tile (e.g. greater than 5 mm) will provide rain-leakage in a roof with some thousand tiles altogether.

Plainness can only be defined accurately by measuring. People can certainly detect very pronounced deviations from the target value when visually inspecting the tiles as they pass along the conveyor belt. Reliable and reproducible quality control for this dimensional accuracy therefore means that the entire production must be measured. We recommend the use of laser distance sensors to measure the tiles from above contactlessly as they pass along the conveyor belt. A computer evaluates the results.

Referring to accuracy the resolution of the sensors used lies within the range of < 0.1 mm. However, when specifying the measuring accuracy, it should be noted that this is a rough ceramic surface which is tested from the top on passing along the normal conveyor belt. A reproducible accuracy of the measurement of +/- 0.3 should therefore be assumed.

Surface Test

A digital image processing system is provided for the visual inspection of the roofing tiles.

The classification is made without using of Neural Networks.

For the purpose of a final inspection, the tiles are divided into the following subsections in which tests are performed according to different standards:

- The visible surface of the roof
- Fold areas (in the case of pressed tiles)

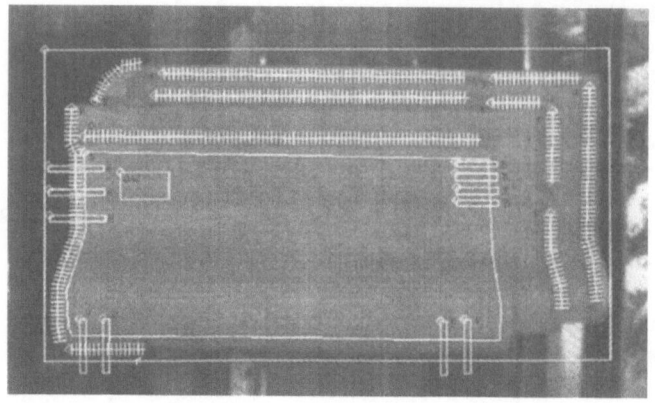

Figure 3: Inspection sections for image processing

Together with various customers, we defined the following smallest detectable defect parameters as practical.

Surface defects:

• Cracks	> 0.8 mm wide, simultaneously > 10 mm long
• Holes, Abrasions, Swellings, etc..	> 3 sq. mm, clearly visible.

Shape defects:

• External contour	> +/- 5 mm
• Fold defects	> 10 mm long

The relatively large assumed specification for cracks was chosen by special combination with the separate acoustic test. The aim of the acoustic test is to detect "hairline cracks". In particular, this reduces the risk of false classification during the visual inspection. If possible, no accepted tiles should be erroneously classified as rejects. The complete image-processing system for surface testing can also be used as a stand-alone measuring station before baking. Typical pressing and engobe defects are then sorted.

A low-cost variant of the visual inspection directly downstream of the press is also feasible to assist the press operator. This early sorting of rejects helps to save baking costs. The rejects can also be returned to the raw material cycle through recycling.

Inspection of Colour Deviations

If it is necessary to perform an actual colour inspection, there are two basic methods in technical terms:

- Extension of the surface test system to include a colour camera together with a computer and software

- An additional measuring system comprising a highly accurate colour sensor with separate evaluation.

The first method can be used if a surface inspection is installed anyway. A colour camera cannot actually resolve the colour as precisely as a colour sensor, but it is a cheaper and more compact solution. We offer both solutions which can be discussed beforehand depending on customs applications.

Mechanical Design

Integration in existing sorting lines normally means that the mechanics of the test system have to be adapted to the available space. The system is then planned on the basis of the available space. Due to the modular system of the individual test components, integration is possible even under difficult conditions.

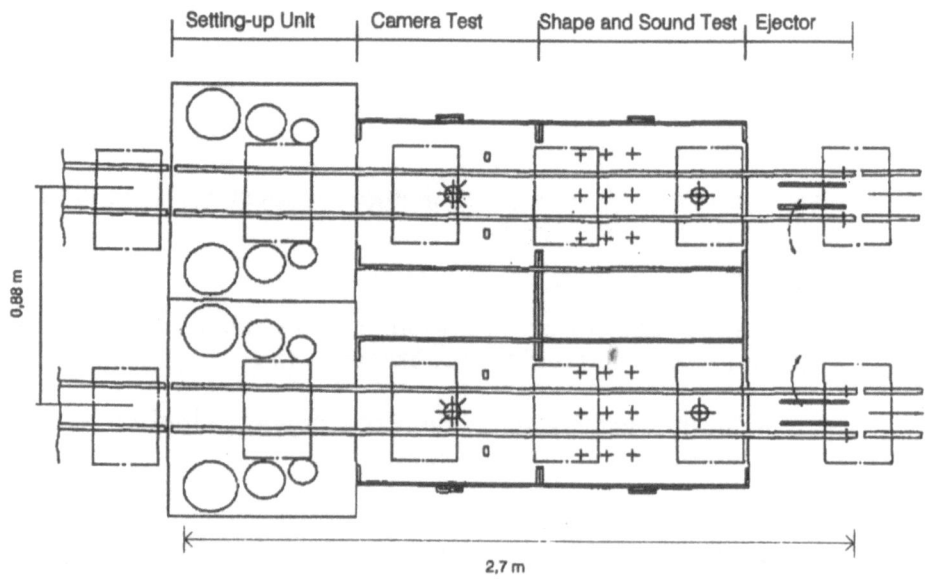

Figure 4: Layout proposal for 2 parallel sorting lines each with 3 test methods

Cost-Effectiveness

Various arguments should be taken into account when considering cost-effectiveness. An important factor, which can be expressed directly in monetary terms, is saving on labour costs. In the tile industry, multi-shift working often takes place on several parallel sorting lines. With two shifts on three lines, for example, labour costs amounting to 6 man-years can be initially saved each year. However, this example must be stated more precisely so that, if necessary, an employee from the factory must still be deployed to supervise the entire test system. Furthermore, a certain amount of time must be spent on regular maintenance and adjustment of the test systems.

The setting of limit values and learning patterns should, however, only be carried out by qualified machine operators (in cooperation with works management) and not by semi-skilled staff. Another potential area for savings is sorting before baking. In this case, considerable baking and raw material costs can be saved (through recycling) by separating the rejects with a reject quota in percentage terms. In addition to the economic aspect, this is definitely a contemporary, ecological task!

With regard to the costs of the described systems, efforts are normally made to obtain depreciation over a typical period of 2 to 3 years.

Summary

All measurements are performed as a fully automatic 100% inspection during throughput. The tests have become objective and reproducible. The modular system presented can perform all the necessary tests and inspections on roofing tiles. In order to save on sorting staff, at least two test systems must be used in combination: a sound test and a visual inspection. The following applies in this case: Hairline cracks can be easily classified in the sound test while large cracks and surface damage can be easily detected during the visual inspection.

Automatic Sorting of Pot Plants with a Neural Network Classifier

ir. Toine Timmermans
ATO-DLO, Agrotechnological Research Institute
Wageningen, the Netherlands
e-mail: a.j.m.timmermans@ato.agro.nl

Abstract

ATO-DLO has built a flexible and universal sorting system for pot plants. The system is composed of a colour camera, image processing hardware and specially developed software. It can be applied for sorting of several types of pot plants because of the implementation of learning techniques using statistical discriminant analysis and a neural network classifier.

1 Introduction

1.1 Product Quality

Actual topics in plant nursing are quality monitoring and improvement of the process efficiency in the greenhouse. A high quality product with a low cost price is the solution to handle increasing competition. In future greenhouses the role of human labour will be limited, because of further mechanisation and computerisation. Sorting of the plants on quality is one of the few tasks that still has to be performed by humans. Labour cost for sorting and product handling are a substantial part of the cost price.

Besides the high costs another big disadvantage of human grading is the subjectivity of the judgement. The human eye is very good in recognising different patterns but has limited capabilities to perform objective estimates of for example size and shape. When grading on size humans have a tendency to compare the plants in a group and select the extreme plants and put them in the class of small and big plants. The grading criteria can be changed completely when a group with another size distribution is examined. Considering the fact that quality of plants is determined by numerous plant features, like size, height, colour, shape, symmetry, etc., it is not surprising that humans have problems to maintain the same objective decision criteria. In most situations several people sort plants on quality and then it is practically impossible to hold on to the requested constant grading criteria.

1.2 Computer Vision

Computer vision techniques have proven to be successful for objective measurement of several agricultural products. Computer vision encloses the capturing, processing and analysis of two dimensional images. In general a video camera is used to acquire the images. The images are digitized into a matrix of picture elements (pixels) that each have a certain grey or colour level. Relevant features are extracted from the images using specific software algorithms. Finally the measurement data are used to classify the object. A few research projects focused on machine vision applications for pot plants have been described [1,2]. In applied horticultural journals the latest developments in technology and the suppliers of the sorting systems are reported [3,4]. At this moment about ten different grading systems for pot plants with a different degree of complexity are operational in dutch greenhouses. The field of computer vision for plant analysis has evolved in less than ten years from relatively simple laboratory projects to successful on-line applications.

2 Description of the Plant Sorting System

2.1 System Configuration

In the past three years the Systems Research and Development division of ATO-DLO has developed a universal system for sorting pot plants on several quality aspects. The intelligent part of the plant sorting system is a computer vision unit. The hardware of the machine vision unit consists of a high resolution 3CCD colour camera, an Intel 486 computer and a special image digitizing computer board (framegrabber). An illumination chamber with high frequency tube lightning is used to illuminate the plants. In cooperation with Visser ITE b.v., a mechanisation company that has its main activities in the horticultural sector, a first operational processing line is built with electronics and mechanical parts to transport and separate plants. The system has a capacity of about 5.000 plants per hour. For subsequent operational systems the selection, design and construction process of the camera arrangement, the colour of the conveyor belt and lightning configuration depend on the type of plant.

2.2 Product Knowledge

The knowledge about grading plants is partly described in quality rules that are defined by the dutch auction organisation. Most information however is stored in the minds of growers and product quality experts. In research projects dealing with building expert systems based on human knowledge the bottleneck was in many cases the knowledge acquisition process. In general it is difficult and laborious to extract explicit knowledge rules from experts. This is definitely true for expert knowledge on ornamentals. It is difficult for experts to explain what are the relevant criteria for inspecting plants and what is their relative ranking. Another factor is that the consistency in decisions of experts individually and correlation between different experts for judging pot plants is not high [1,2].

To translate product knowledge into information that can be processed by a computer it is necessary to derive the set of features that is relevant to classify a type of plant and to define the discriminating values for the different classes. Numerous different plant types exist and a major functional demand was that the system could handle as many types of products as possible. Because of the difficulty to extract knowledge from experts and the number of different plant types a flexible tool is necessary to incorporate product knowledge. A software platform based on techniques of learning by examples was developed to include product knowledge.

2.3   Software Platform

In figure 1 the universal software platform is displayed. The left column shows the normal measurement procedure. In the right column the selection and training moments are indicated. The process of determining the quality group of the plant is composed of two stages, a feature extraction stage and a classification stage.

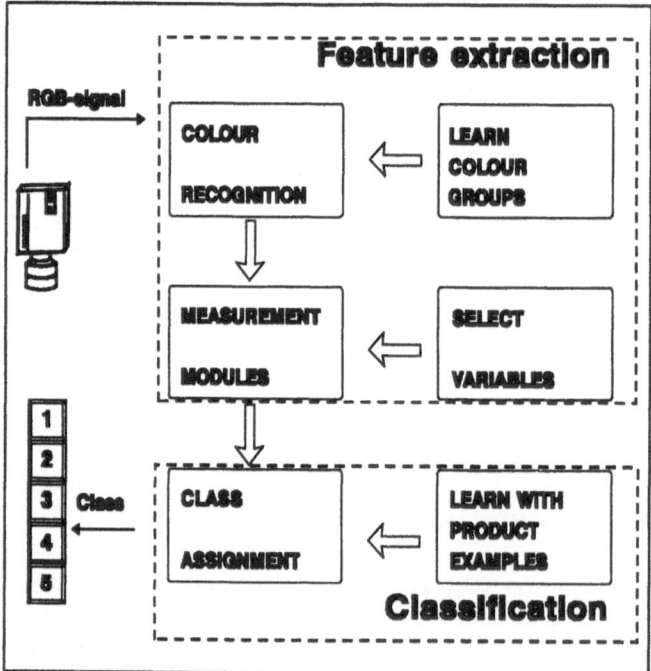

Fig. 1.   Measurement procedure and training modules.

The feature extraction process is split in two successive operations. The image taken with the colour camera has a resolution of 768*512 pixels. Each pixel can have over 16.7 million different colour levels. In the colour recognition module the image colours are reduced to meaningful groups. For a flowering plant the groups are for example: background, leaves and flower area. The colour recognition module can be trained by selecting pixels in example images. Using a pattern recognition technique the colour space is separated into non overlapping clusters.

After the data-reduction step the relevant features of the segmented objects can be measured. The software contains in total over 30 features that can be measured for each object. These features include area, size, colour, width, diameter, height, and a number of shape and symmetry descriptors. For each type of plant an optimum subset of variables has to be determined. This is done by a combination of using general product knowledge and a stepwise variable selection procedure [5,6]. In the classification stage the quality group of the plant is determined by applying pattern recognition techniques in the multi-dimensional feature space.

3   Supervised Learning

For both the segmentation in colour clusters and the classification of the objects into a defined number of groups supervised pattern recognition techniques are used. Training sets for the colour classification are built using images of example plants by selection of typical colours using the computer mouse. Training sets for product classification are built by showing examples from the different quality groups to the camera system. Product experts build the training sets.

Several techniques are available for performing supervised classification. The methods can be divided into parametric and non-parametric classification. A parametric classifier assumes a functional distribution of given samples. Examples of parametric classifiers are the Bayes classifier and the discriminant analysis techniques. Non-parametric classifiers do not assume any functional distribution of the samples. The K-nearest neighbour rule, multi-class partitioning algorithms building decision trees and a neural network classifier are examples of a non-parametric classifier [7,8]. For the plant classification application statistical discriminant analysis and neural network techniques have been implemented.

3.1   Statistical Discriminant Analysis

Statistical discriminant analysis construct separating surfaces in the multi-dimensional variable space, such that different groups are separated. With discriminant analysis, discriminant axes are calculated so that the projections of the data points of the classes on the axes are separated maximally. Two different versions of this technique are used: linear discriminant analysis (LDA) and quadratic discriminant analysis (QDA). With LDA linear separating surfaces are calculated, with QDA the surfaces are curved. With the LDA technique exactly the same results can be obtained by using multiple linear regression [9].

3.2   Neural Networks

A simulated neural network is created by scientists as a result of an objective to develop a computer model that matches functionality of the brain in a fundamental manner. Neural networks offer several potential advantages over statistical classification techniques. Neural networks are relatively robust and fault tolerant. Neural computing systems possess the capability to generalize information and get reasonable results with noisy or incomplete input [10,11]. Practical advantages of neural networks are easy implementation and prototyping. For most applications neural networks are better than or at least equal to statistical classifiers [1,12,13].

4   Experiments and Applications

Numerous experiments with several types of plants have been done to examine the performance of the developed prototype. Ten types of flowering plants, three green plants, one half grown plant and five types of seedlings have been analyzed with the camera system. In this chapter two standard applications are described. A more detailed description of the experiments is reported in a previous article [6].

4.1   Sorting on Colour, Size and Flowering Stage

This chapter describes an application of grading full grown Saint Paulia plants on colour, size and flowering stage. The final quality inspection of this type of plant is still done by humans, because manually bad leaves have to be removed and the flowers have to be positioned above the leaves. This means that the vision system is only used for objective measurement of the flower colour, the size and flowering stage.

For colour recognition the RGB images are segmented into 7 clusters. Three different techniques were tested to separate the colour clusters: LDA, QDA and the NN (neural network) classifier. For the NN several different structures were tested. A fully connected feed forward network with one hidden layer containing 8 neurons gave the lowest error. Figure 2 shows the network structure. The neural network was built with the NeuralWorks Professional II/Plus shell. The RGB (Red, Green and Blue) values were used as input, divided by 256 to get values in the range from 0 to 1. For each colour class an output neuron was defined. When classifying, the pixel was assigned to the class connected with the output neuron having the maximum output value. A back propagation learning schedule was used to train the network. For this application QDA is superior to LDA. The NN performed better always better than QDA, but with both QDA and the NN classifier less than 1% of the total image is misclassified.

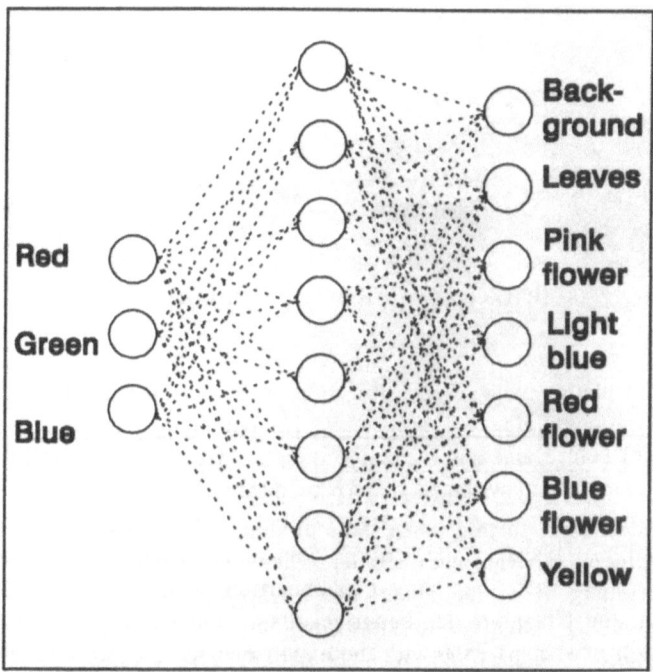

The plants are sorted into three quality groups based on the total area and the flower area of the plant. In the two-dimensional classification space the quality groups can be separated easily. For a classification problem with low complexity there is hardly no difference in performance between the tested pattern recognition methods.

4.2   Sorting on Colour, Size, Flowering Stage and Quality

In the previous application plants are manually inspected on quality. For some applications like pot-roses, azalea and begonia this is not an acceptable procedure. For these plants the bad quality plants have to be recognised by the camera system and have to be put on a separate belt. There are several reasons for a plant to be a B-quality plant. It can be a problem with the thickness, the leave structure, the position on the pot, the height, the shape, the symmetry, etc. Multiple camera views are necessary to be certain that the quality defects can be detected. In figure 3 a picture of 2 good and 2 B-quality plants is shown.

The software for this application is based on the system described in chapter 4.1. The same procedures and techniques for recognising the colours and classification on size and flowering stage are used. The recognition of the B-quality plants is a relative difficult problem. A procedure of supervised learning is practically the only way to realise this task. A feature set of about 8 parameters is selected that

Fig. 3   Two good (left) and two B-quality (right) pot roses.

have a high correlation with all possible quality defects. Examples of parameters
are the width ratio, height deviation, leave fill percentage, distance between pot
centre and plant centre and the vertical symmetry. Good and B-quality plants are
shown to the camera and with the pattern recognition techniques the segments in
the multi-dimensional feature space are calculated. This classification problem is
relatively complex, because the groups can not easily be separated in the
classification space. Their are many non-linearities in the discriminating planes.
For this reason the NN gives far better results as the LDA and QDA. With the
NN a correlation of at least 95% with the expert decisions is accomplished.

5.   Conclusions

ATO-DLO has built a flexible sorting system for pot plants. The brain of the
machine is a colour image processing system. In cooperation with Visser ITE b.v.
the system is integrated into a plant processing concept and is commercially
available. Product knowledge is incorporated in the feature extraction and
classification step. In the first step of the feature extraction the colour images are
compressed into meaningful object groups. This classification procedure is trained
by showing product examples. The quadratic discriminant analysis method is
chosen to separate colour groups. In the second step of the feature extraction
relevant features are measured of each object. Based on the measurement data the
object is classified using pattern recognition tools. Statistical discriminant analysis
(linear and quadratic) and neural network techniques are used for pattern recognition. The system is trained by showing product examples to the camera.

The system is successfully tested on several types of full grown plants, half grown
plants and seedlings. The selection of the best pattern recognition technique
depends on the complexity of the application. LDA and QDA have limited
capabilities to segment the feature space into clusters. Deformed or split groups
can not be separated with the selected discriminant analysis methods. A NN can
construct deformed clusters and therefor a NN is superior for more complex
problems. For these type of classification problems complexity can be associated
with the distribution and shape of groups in the multi-dimensional feature space.

The decision on what learning technique to prefer in practical situations is not only based on the relative performance of the classifier. Robustness of the learning technique, easiness to adjust and learn, verification and visualisation possibilities and education level of the end-user are important aspects to consider when taking the decision what classification technique to choose. A disadvantage of applying NN classifiers is the optimisation process of the structure. There is no universal network structure or general recipe that can solve all possible classification problems. The construction of the training set is for a network more important than for discriminant analysis. The structure of a neural network is more or less a black box. For objects that are not covered by the product examples, classification results are sometimes unpredictable and not acceptable with a NN. The operator of a sorting system is in many situations also the local product expert. It is important that whenever the operator does not agree with the decisions of the camera system he has options to adjust the classification performance. Because interpretation of the knowledge that is stored in the network is impossible, on line adjustment is difficult without further training of the network.

References

1.   Brons, A., Rabatel, G., Sévila F. Neural network techniques to simulate human judgement on quality of potplants, On machine vision systems for the agricultural and bio-industries, 1991, 153-161
2.   Dijkstra, J., Meuleman, J. Automatisch sorteren van potplanten, Agro Informatica 1994, 7:22-24
3.   Vegter, B. Beeldverwerking vergist zich niet, Vakblad voor de bloemisterij, 1992, 32:42-43
4.   Vegter, B. Wonderen duren altijd wat langer, Vakblad voor de bloemisterij, 1994, 34:68-71
5.   McCabe, G.P. Computations for variable selection in discriminant analysis, Technometrics, 1975, 17:103-109
6.   Timmermans, A.J.M., Hulzebosch, A. On-line sorting of pot plants with a neural network classifier, AI in Agriculture, 1995, 2:203-208
7.   Talmon, J.L. A multiclass nonparametric partioning algorithm, Pattern Recognition Letters, 1986, 4:31-38
8.   Belkasim, S.O, et al. Pattern classification using an efficient KNNR, Pattern Recognition, 1992, 25, 10:1269-1274
9.   Voet, H. van der, Hemel, J. Multivariate classification methods and their evaluation in applications, PhD thesis, University of Groningen, 1988
10.   Masson, E., Wang Y. Introduction to computation and learning in artificial neural networks, European Journal of Operational Research, 1990, 47:1-28
11.   Neuralware, Neural computing, NeuralWare, inc. Pittsburgh, PA, 1991, 356
12.   Lo, Z.P., Bavarian, B. Comparison of a neural network and a piecewise linear classifier, Pattern Recognition Letters, 1991, 12:649-655
13.   Rath, T., Artificial neural networks for plant classification with image processing, AI in Agriculture, 1995, 2:197-202

Neural Networks
Applied to Direct Marketing

R.P. ter Heide

Sentient Machine Research B.V.

Amsterdam, The Netherlands

Abstract

Direct Marketing is an area with large amounts of data and complex dynamic relationships. Neural networks help explore and model customer profiles and product relationships. Applications in business to consumer and business to business marketing are discussed.

1 Introduction

The application of neural networks in direct marketing is starting to breakthrough in the Netherlands. Sentient Machine Research (SMR) is one of the companies that has pioneered in this field. This paper will discuss why this breakthrough is happening and what kind of applications SMR has helped to implement. It will also enlighten the key elements to successful applications. Current developments providing better marketing tools will be discussed briefly.

2 Necessity for Adaptive Systems?

2.1 Database Marketing

Database marketing starts with information identifying your clients (name, address, telephone number, and account number). This enables you to communicate with your clients on a large scale, using e.g. personalised mail. To enable you to improve your personal approach, additional marketing orientated information (social, financial, consumer information) is needed. One of the problems is that good information is expensive, and available information has complex, and often unknown or circumstantial relations to marketing goals. On top of that, businesses operate in a dynamic market with fierce competition, and are confronted with scale enhancement. Quick decisions must be made, based on large amounts of information, which is not in a suitable form for management decisions. A database marketer needs tools for information analysis and decision support that can cope with the conditions mentioned here.

2.2 Adaptive Systems Support

Information analysis starts by the description of variables through simple statistics, like means, frequency tables and cross tabulation. This tells you if there are any main factors describing your client database. Interesting consumers however, are more often described in terms of typical profiles than e.g. single variables. Decision support systems need to go one step further because they should relate data to target variables which can be easily used for decision making. So, we can distinguish between data profiling, which finds and describes interesting patterns, and data modelling, which is concerned with mapping relationships between patterns and target variables. Tools we have used for data profiling are minimum variance cluster techniques, neural networks, and machine learning techniques. For data modelling we have investigated (linear) regression, neural networks, genetic algorithms, and machine learning techniques.

The advantage of using adaptive techniques, in particular neural networks, is that they are good in pattern discovery, can model non-linear/complex relationships, are non-parametric, and are robust for noise. There are several alternatives that can compete with neural nets if you single out certain aspects. However, in view of the growing complexity of marketing tasks, neural networks provide promising low-cost solutions [1]. SMR has learnt that in data modelling backpropagation networks show the best results. Where as for data profiling associative networks like SMR's DataDetective are more adequate. Several application domains in which SMR has implemented neural networks will be discussed in the next paragraph.

3 Applications

In the discussion of the following applications a distinction is made between consumer marketing and business marketing. Although the techniques applied are generally the same, the conditions and requirements differ.

3.1 Business to Consumer

The business to consumer market is characterised by large customer databases. The strategic marketing approach is group orientated, and based on dealing with large segments in the customer database. That is why performance is measured on a group level. This is reflected in the fact that marketers often regard their target group as an ideal case (they treat customers as a homogeneous identifiable group, which, of course, is a simplification). Only on an operational level do the individual's characteristics become important (for selection, personalised mail, billing).

In co-operation with the Dutch Association for Direct Marketing, Distance Selling and Sales Promotion (DMSA), SMR has investigated applications of adaptive techniques in direct marketing. Two main tasks for consumer database marketing were reviewed: scoring models for customer selection, and data profiling to describe interesting groups. These two tasks are difficult to optimise simultaneously, so they will be discussed separately.

3.1.1 Customer Selection

A large consumer database contains a wide variation of customer types each with their own interests. By relating these interests to the willingness to buy a certain product marketers can select profitable groups for contact by mail.

The task of a neural network is to model the relationship between consumer profile and product interest. The input can contain information on age, gender, lifestyle, consumer history. The output for training is based on known results. These can be the results of a previous mailing, or historical buying data from a sales database. In most cases the network output is an ordinal score used to rank all consumers in terms of high or low chance of sale. Before use the resulting model is evaluated on an independent validation set to provide an independent quality measure.

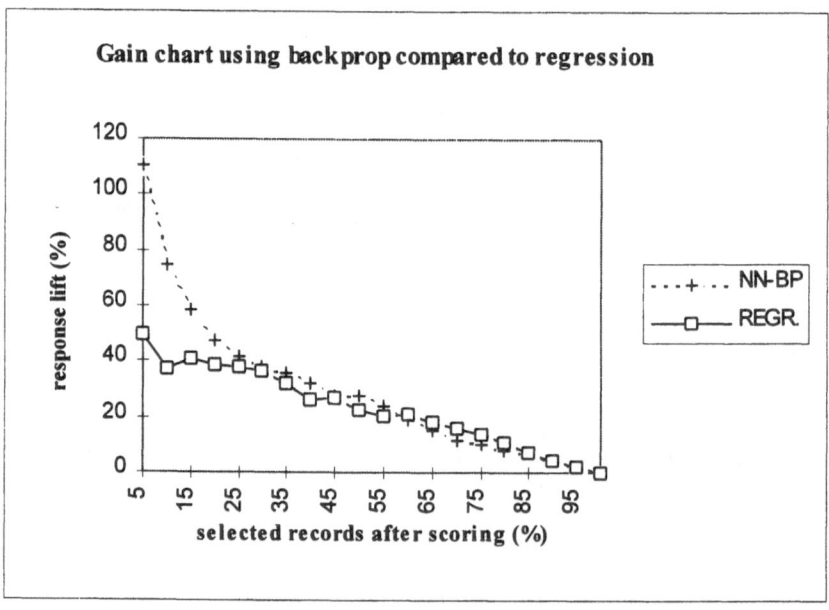

fig. 1 Typical cumulative gain chart. A cumulative amount of records is selected based on the highest score. The response for the selection is calculated and then compared to the base level response to obtain the response lift.

In the DMSA-project most cases concern customer selection. Default response rates before selection are typically between 0.2% and 10%. Typical volumes are between 50.000 and 1.000.000 consumers. Model performance is often evaluated by calculating the response lift in relation to the model score based ranking. In fig. 1 a typical cumulative gain chart is shown based on a backpropagation network model. A standard regression model is used as a benchmark reference. It can be seen that the main advantage of the neural network is a good discrimination of the high rated prospects. It is often difficult to discriminate among low chance prospects. Response level estimates for low response groups are based on a small number of observations. This means that estimates have relatively large fluctuations.

Comparable modelling results have been obtained for different consumer databases. The relevance of the input data in relation to the target data is the main cause for differences in performance level. Generalised profiles coupled to postal code are commonly used for prospect mailings. In these cases input information is only casually related to a specific consumer. When using individual consumer history and personal lifestyle information models reach higher performance levels. This is reflected in fig. 2 where profit gains for 3 typical cases in the DMSA-project are shown. The highest gain is related to a cross-selling activity where more specific information was available. However even using generalised data, profit gains can be interesting.

In direct marketing neural network models are sometimes disposable, only used for a one shot activity. Reuse is restricted to understanding the main input factors which are important for selection. In other cases mailing specific consumer groups is a continuous activity. Model reuse is then common practice. Performance is continuously monitored by the use of test mailings based on random selection. When a decrease in performance occurs it may be necessary to retrain, or, if additional factors come into play, even re-engineer the model.

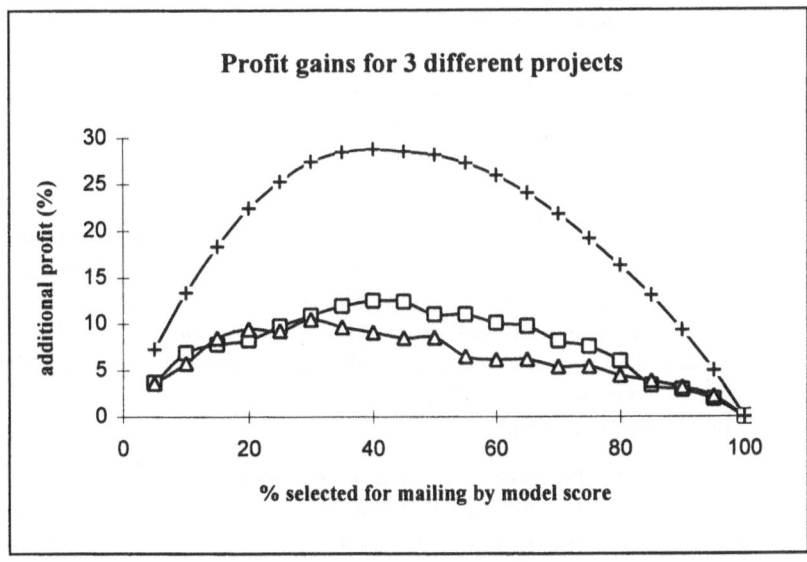

fig. 2 Profit gains for different mailings in relation to the percentage mailed. Additional profit gain is expressed as a percentage of the total possible profit.

3.1.2 *Customer Profiling*

Customer profiling is concerned with providing adequate descriptions of interesting groups in the database. Interest is reflected in terms of mailing response level, product interest or turn-over. Backpropagation neural networks are poor in providing insight in the important patterns (groups). Here, associative clustering techniques can be used to provide a solution. SMR has developed it's own propriety tool based on an associative neural network DataDetective and a visual cluster map called Looking Glass. Looking Glass maps relationships discovered by DataDetective from a high dimensional input space to a two-dimensional visual plane. The mapping is one-to-one in that all the input records are explicitly represented in the visual plane. The user can easily select groups of consumers and analyse group characteristics.

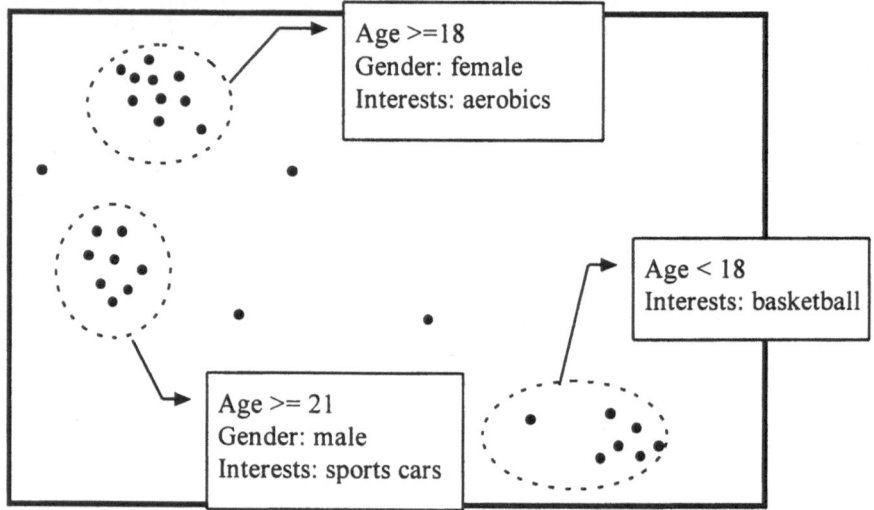

fig. 3   Example of a clustering of sports shoe clients with the Looking Glass Cluster visualisation tool.

An example taken from a media planning application is shown in fig. 3. Buyers of a typical sports shoe have been selected and clustered. Clustering shows three distinct groups with distinct profiles. The tool lets the user determine the groups interactively, based on the number of groups the user finds relevant. Decisions on the number of clusters and the relevance of describing variables need not be determined in advance. They can be made while a projection of the actual database structure is in view. The tool is now being further tailored for media planning by a major publisher in the Netherlands.

In the DMSA project we have used this method to inspect a fund-raiser's database. It was shown that better discrimination is possible than with traditional statistical methods like K-means and CHAID. Another area where the DataDetective / Looking Glass tool has been successfully applied is in the description of top prospects for a bank. These were previously scored by a backpropagation model.

The Looking Glass tool divided these top prospects into three interesting subgroups. Currently SMR is using the same tool to investigate supermarket buying patterns.

3.2 Business to Business

The business to business market is characterised by a large turn-over per customer. Marketing is more orientated towards the individual client. The number of clients in a database can vary from a few hundred to up to millions. In most cases the marketing budget can not support a full fledged approach of every client. One of the methods used to tackle this problem has been jointly developed by MSP Associates and SMR. The Customer Marketing® Rating method tries to optimise marketing effort [2]. The well known rule of the thumb that 20% of the customers provide for 80% of the turn-over is extended to a further subdivision of the top 20% in a customer pyramid (fig. 4). By concentrating marketing effort on the Top, Big and Medium clients a better return on investment is obtained. Neural networks can be used to rate a client's potential return. Clients are then evaluated as stable, or having booming or declining expectations. This can be used as a guideline to direct marketing effort. The SMR propriety neural network DataDetective has been successfully applied to over 10 different cases in Customer Marketing® (from banking to office equipment providers). DataDetective not only rates different customers but can also show users why (showing associated clients used for prediction). This is an important feature because it enables user acceptance as well as user intervention.

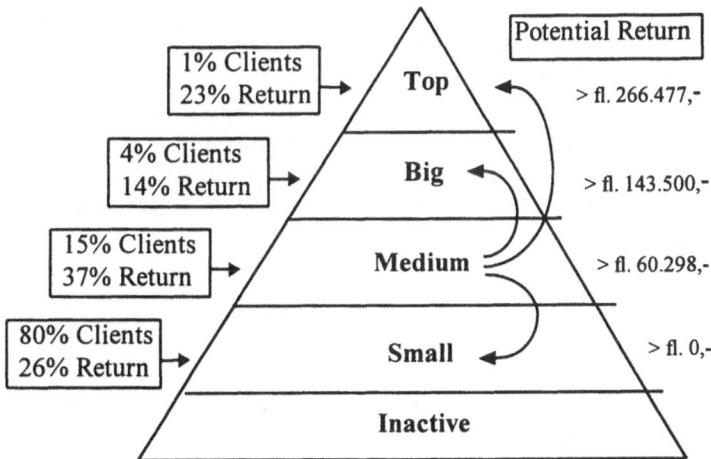

fig. 4 The Customer Marketing Pyramid with per group the percentage of clients and the percentage of return. Neural nets identify potential for migration for individual clients.

Predicting return is a common task. Other prediction tasks like cross-selling and prospect rating have also been implemented. Plans exist to extend the use of neural networks to other marketing management areas like human resource management and media planning (specifically business to business).

4 Key Elements to Success

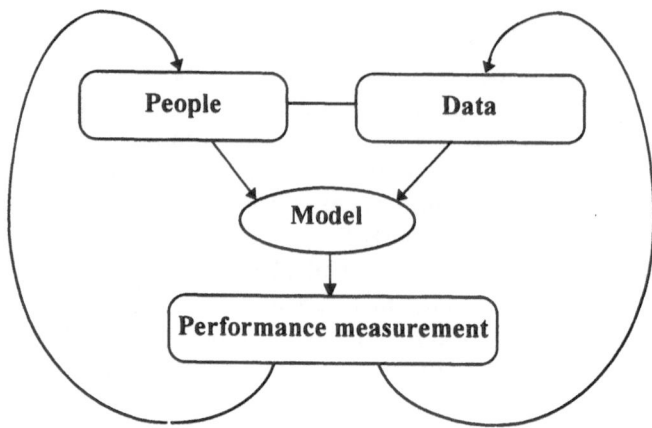

fig. 5 The three key elements to success in building models for direct marketing.

The past years SMR has seen that there are three key elements that determine the success of model building (fig. 5). Commitment of staff and management is necessary to test and integrate methods and techniques in an operational environment. Good data is always important. Better input data can give you more performance improvement than optimising a technique alone can. However it is a fallacy to think you should first concentrate on the database. In most cases the notion of good data is a result of using it in market models. Performance measurement provides feedback for marketers and database managers. They can use this to improve their tactics, their data and the quality of their models. The marketplace is a dynamic environment requiring people, information, and models to be continuously updated on what is happening. This all may seem trivial from a management point of view. That is maybe why these items do not always get the attention they need.

5 Current Developments

5.1 Hybrid Systems

The use of hybrid models is becoming common practice. One reason to use them is to decrease the number of badly modelled records (large errors). Backpropagation networks for example, often suffer from certain sub-domains that are badly learnt. Different starting conditions often give different subsets of bad records. SMR has learnt that combining the results of several backpropagation networks with different starting conditions can reduce the number of large errors. Other parallel concepts using alternative techniques (regression, DataDetective) have also been investigated and have shown improved results.

Hierarchical systems, using a conquer and divide approach, can be useful to tackle large databases. Specific trends may be hardly different from background noise if you are looking at the entire database. Once you've divided the database in smaller groups it's often easier to distinguish relevant factors from noise. Segmentation in smaller groups can result from a clustering process, but it can also be based on the scores provided by a scoring model. For example, we have seen that we can profit from an optimal scoring technique like backpropagation and still understand what is going on by investigating the scoring results using a descriptive segmentation tool.

5.2 Visualisation

Data visualisation is becoming the next important step in data modelling. Marketers need to do more than just optimise their daily work. Their daily work is constantly changing because of the dynamics of the market. Being the first to recognise changes and react adequately is very important. The problem is the large information overflow. How can you recognise deviations, if you don't know what to look for? Systems may be able to detect new groups or patterns but it doesn't mean these are relevant. Marketers must explore these new patterns interactively and decide for themselves whether they are important. Tools must support quick inspection and evaluation of interesting data. Visual tools are preferred because they can provide complex abstractions and still remain easy to use. We have seen that SMR's Looking Glass cluster map is a step in this direction. Developments in virtual reality will provide further tools.

6 Conclusions

Neural networks have been successfully implemented in marketing activities like customer selection, customer profiling and customer rating. Success mainly depends on managing people, data and performance feedback. Current developments show promising results for hybrid systems and visual tools.

Customer Marketing® is a registered trademark of MSP Associates.

References

[1] van der Veer RCP, Wierenga B, Kluytmans JCHW. Neurale Netwerken in Marketing. In: Recente ontwikkelingen in het marktonderzoek, Jaarboek van de Nederlandse vereniging van Marktonderzoekers 1994-'95. Uitgeverij de Vrieseborch, Haarlem, 1994, pp 9-24.

[2] Wurtz W, Curry J. Customer Marketing Ratings. In: Conference Workbook on the 10th National Conference/Exposition on Database Marketing, pp 204-215.

Modelling Market Dynamics in Food- and Durables Markets

Reint Jan Schüring
BrandmarC B.V.
Amsterdam, The Netherlands

Abstract

The organization structure and information supply are the largest obstacles for an effective marketing communication. Accountability of marketing communication is today possible through simulation of market dynamics with the use of natural computation models, if we are prepared to draw organizational consequences from them.

Introduction

The last few years effectiveness of marketing communication have received much attention. This is due to the social-economical developments, changing relations in competitions, less stable and more dynamical markets, and the increased power of retail chains. The brand seems to have lost in power. An obvious candidate to blame is marketing communication. Indeed the brand is assumed to be carried by image, reputation and most of all advertising. Loss of a brand must be related to less influence of these factors than previously had been assumed. Besides, statistical studies have shown that the influence of advertising is minimal. So, what is the contribution of marketing communication anyway ?

The question to the effectiveness of marketing communication is in fact the question to the added value of marketing communication, or the accountability of marketing communication.

Proposition 1
Marketing communication is a crucial part of successful marketing operations

The results of longitudinal statistical studies, performed several years ago, have shown that advertising hardly contributed to turnover and sales of a brand. Recent studies, which either used innovative nonlinear computation models and integrated analysis of marketing- and communication-data or which made use of so-called single source data gathered in panel studies, show that marketing communication can have a large influence on the success of a market operation (e.g. ARF 1991, Jones 1995). However, timing, dose and context appear to be crucial boundary conditions.

ARF indicates that news value is very relevant: a new campaign, new information about the product, a new media strategy. News value acts very positively on the success of marketing communication.

Jones indicates that a commercial has to show its effects on sales within a week. Long term effects of advertising can only emerge if the commercial has short term effects.

Jones denies the existence of long term effects of advertising independent of short term effects. If a commercial will not lead to more sales today, then it will also not lead to more sales over half a year. Jones also indicates that seeing the same commercial more times in a short period will lead to a strong decrease of marginal returns.

These results were obtained on the basis of so-called single source studies; Each week the advertisements which a consumer sees as well as the products he/she buys are recorded.

BrandmarC operates on the basis of computer simulations of market dynamics which are performed on the basis of data from various sources. The mean relative significance of variables as well as the point elasticity of variables is computed with the use of a factorial design, using a neural network.

Other techniques which are meant to read off nonlinear computation models used by BrandmarC - like "what-if" analyses and "scenario" analyses - provide insight in possible optimal strategies of the own brand as well as the ones of the competitors.

BrandmarC has shown that e.g. advertising, under certain circumstances and in a certain area, can give a powerful stimulance to the market share. Too much advertising can negatively influence the market share. With the use of BrandmarC we could investigate for the first time the possible effects of e.g. advertising. Curiously, Kotler mentions in Marketing Models (1992) two operating models of advertising which show only a positive or indifferent effect of advertising. It shows again that linear statistics is not able to analyse markets properly.

The crucial point of marketing communication is that it effect can be positive as well as negative. This depends on the position of the brand in the market, and the competition. Quantity is not only not always good, it can sometimes be very harmful. To conclude, marketing communication is a typical nonlinear phenomenon.

Proposition 2
The organization of marketing communication is insufficient to recognize the increasing dynamics of markets

The problem of effectiveness of marketing communication has mostly organizational back grounds. Nowadays, marketing communication is mostly "keep what you have got", "keep on going", "abdominal feeling" "extend" and very operational colored (instead of strategical). Marketing communication is too strongly oriented to reach the objective which stem from marketing communication itself. Marketing communication is seldomly defined as change and dynamics, much though the image of marketing communication may let us suspect otherwise.

Changes in the marketing communication often involves crises, changes in personnel and/or suppliers.

This can be explained by the following field of tension:

Markets behave nonlinear. Marketing communication is organized in a linear way.

If the price of a brand is risen twice without a significant loss in market share, then this implies by no means that a third rise will result in the same effect. If advertising show positive effects on the market share, then this implies by no means that a doubling of the budget will have a double effect. On the contrary, the effect could even be negative. Large advertising expenses can work out positively. With another campaign, or even with the same campaign such high expenses can cost market share. The market share can increase if the spontaneous product familiarity increases, but this relations is not always linear to 100 %.

The spontaneous product familiarity can even be too high. Brand loyalty is important, but an increasing brand loyalty can be the start of a decreasing market share. Marketing communication does not work out in the same way and in the same quantity over all the distribution variables. In other words, goals which have resulted in a positive effect yesterday, might show negative effects tomorrow.

All these knowledge, but more precise and concentrated on a specific market can now be obtained. The knowledge and technology is available, but is there an organization which can transform this knowledge into profit?

The organization of marketing communication is linear in the sense, that it aims in particular at the linear developments between the turning points of the market. The pursuit is stability in marketing communication as well as in the organization of it. Turning points mean crises, ad-hoc policy, changes in personnel and in advertising agencies. Somebody must be blamed. But the blame is on the insufficient focus on the moments of change. Which parameters can tell us **in advance** about the necessity to a change in strategy.

Proposition 3
The most important factor in the success of a brand (in marketing communication) is a high score on the price/value ratio.

Price is an overestimated variable, marketing communication is underestimated. More important than price is price perception, at least in those markets in which BrandmarC had both variables at its disposal. Also for food-products it is possible that the price is too low. Market leaders seem often to be on two third of the total price range of the market. This means that markets are carried by a right balance between image and price perception.

Proposition 4
The price/value ratio is continuously threatened by external and internal processes.

The news value of communication around a product increases or decreases, the competitor introduces new products, new campaigns, etc. , a competitive market is rising, introductions of Own Brands undermine A-brands, the age of a brand is high, a price rise leads to loss of spontaneous product familiarity, distribution problems, strikes, etc.etc. With the help of BrandmarC the price/value ratio is measured through the relations of price or price perception and image (quality) with market share or sales. Even if the price of a brand is constant, this implies by no means that the relation between price and market share is constant. After all, there is competition and an image that might have been changed. With BrandmarC we can follow all these elasticities in time. For instance we can indicate if the price elasticity is increased or decreased in time. The elasticities are mutually independent, like all explaining variables in the BrandmarC models.

To influence these elasticities is the key in marketing communication. This means that not the organization should prescribe what happens with the marketing communication, but the changes in elasticities should be determining.

Proposition 5
The organization of marketing communication should be aimed to recognize possible changes (the moments of change) in price/value in advance to the actual change. (To take action with this knowledge, either to undo the potential change, or to use it in an optimal way.)

To deal with nonlinear problems, i.e. market problems, means that the organization of marketing communication should be more aimed at the moments of change than at the stable periods before and in between. The added value of a marketeer, and hence the added value of marketing communication is just in the early recognition of the moments of change and hence the need of a strategic change. A crucial part of effective marketing communication is timing, dose and context: act correctly on the right moment and the right place. With the use of nonlinear computation models, like neural networks and genetic algorithms, we can obtain a good insight.

Proposition 6
The organization of marketing communication is completely insufficient in the acquisition of information needed for a correct timing and dosing.

Marketing communication is subdivided in many segments: corporate, sales, marketing, public relations, sponsoring, direct marketing, sales promotion etc. Each department, each supplier, knows its own goals, routines, truth/untruths, own studies/data etc.

Each segment is a bureaucracy on its own. Studies are aimed to prove that certain involved activities are correct, whether or not within a defined objective. Evaluation criteria and objectives are determined by the segment itself, and often are skipped. The probability on positive results of all these ad-hoc studies is big, the information on movements in the market is small. The BrandmarC analyses show that much redundant data is gathered by companies. On the other hand, crucial variables, like image and advertisement appreciation is completely insufficiently recorded. Information about below-the-line activities is very scarce, public relations is hardly recorded or analyzed, even in markets in which they clearly play an important role.

Proposition 7
All information about the market must be integrated, analyzed and evaluated in a central manner

A necessary condition to make marketing communication accountable and to give a good insight in the context, the right timing and the dose.

Besides this, other important goals are served: everyone has the same information at their disposal (decreasing the time spend at meetings), research will be tuned to each others (no unnecessary overlaps between studies, no unnecessary or to specific research questions), research can structurally be tuned to the information question of a marketeer, and the question whether or not the study is able to describe the dynamics of the market in a sufficient way can be investigated.

Besides there will be no confusion about the relevance of the variables. The discussion can aim itself to the choice of an optimal strategy in relation of the possibilities of the brand.

Proposition 8
Marketing communication research is mostly not aimed to make changes in the price/value ratio visible in an early stage.

One measurement is no measurement! The state of a brand is known by looking at the sales figures.

Market research must show most of all the trends and developments. Structural, frequent and efficient market studies are therefore indispensable. Without structure no possibilities for comparison, without frequency no developments and without efficiency much unnecessary data and at the same time too little information and too high costs. Ad-hoc research is equivalent to no research, or to a political document, but it will seldomly produce information. Many ad-hoc studies, if not all, are superfluous if the structural research is good. Much more information can be obtained from data of structural research. Studies aimed to find the truth in the head of the respondent (psycho-social research) tells us mostly about the status quo, not about the moments of change.

The analyses in BrandmarC show that analyses of structural research data produces more information, better information and more precise information than the so called qualitative studies. This does not mean that qualitative studies are nonsense, but rather that this type of studies should be embedded in a structural research program.

Recently I gave a lecture for the department of advertising and research. I claimed that 10 researchers asked about the measuring of price/value ratio would be stuck for an answer. The attendant market researchers, including representatives of the large market research institutions, just nodded. It would not be appropriate to say that everybody is just fooling around, but often it seems like that. Nevertheless I think that the researchers would be stuck for an answer for a good reason. Measuring the price/value ratio through a study seems very difficult if not impossible to me. The price/value ratio can only be simulated by model building in which all the relevant variables have a place.

Proposition 9
Marketing research institutions and clients do not ask them selves often enough whether they do research on relevant subjects.

Under the pressure of time and more important under the pressure of budget, most elementary demands on market research is violated. Seldomly the validity of a questionnaire is studied. The set-up of a study, the huge number of questions, the words which are used, etc. change with each study and with each customer without too much reflection. Image is sometimes put into operation as product property, sometimes as image (values of the brand) and most of the times somewhere in between.

Often it is not clear why just those image statements are put to the respondents. It could be just as well 20 other statements. How do we know which statements we have to put to really measure the image. Or the price/value ratio ?

The pretest research is thriving and although American studies have shown that there is no relation between the outcomes of pretest research and the real performance, and although we all know that a good pretest is no guarantee for good results, we do not look for other methods to measure the possible performance. We are attached that much to stability that we maintain our routines even against our better judgement !!

New mathematical techniques are able to simulate nonlinear processes. Provided that the right data is available. Sometimes these techniques are denoted as "artificial intelligence" or "self learning systems". Personaly I rather call them "natural or nonlinear computation techniques". After all, the computer remains stupid, even with these techniques. The suggestion raised by the words "artificial intelligence" put many people on the wrong track. Neural networks and genetic algorithms are sometimes called "natural" because of the way of computing (reasoning) is related to our model of evolution and of the nervous system. Provided the correct methodology is used, these techniques are able to give us reliable insight in the dynamics of nonlinear processes like markets. A condition is that a good monitoring study is available.

Frequent measuring of the most important variables like spontaneous product familiarity, price (perception), image, media, distribution, advertisement appreciation and of course sales or market share are therefore necessary. Not only of ones own brand, but also the ones of the competitors.

Proposition 10
Every large marketing organization is with the use of natural computing models and under the condition of another organization of marketing communication, able to make the effectiveness of marketing communication visible and even to influence it, in other words, to make marketing communication accountable.

The techniques are now available, as well as the knowledge about efficient market research. Now we have to wait for the organization which will make the profit. The information column in companies, now often horizontal, must become vertical.

The top management, now often the strategic motor of an enterprise, must make a clear statement about the information which is necessary for a right timing and dosing of marketing communication. This information is of crucial importance for the further existence of the enterprise. A marketing computing engineer can play an important role as intermediary between market and mangement. Product management can be integrated in sales. marketing communication must be equivalent to strategic thinking and deciding about communication on the basis of information, obtained by analysis of good monitor research with the use of natural computing models.

Anyway, marketing communication is a matter of the top-management. Recent studies, also data of BrandmarC, show that communication in many markets is a decisive factor. Context, timing and dosing determine the rate of efficiency.

Top down, bottom up. Strategic decisions, data transformed into information. In this way, a restricted investment in efficient market research and analysis will be pay out itself double.

References

Advertising Research Foundation (ARF). *Transcript Proceedings.* NY, 1991.

J. Ph. Jones. *When Ads Work.* Lexington Books, 1995.

Lilien, Kotler, Sridhar Moorty. *Marketing Models.* Prentice-Hall, 1993.

Handwritten character recognition
using neural networks

A.C.R. Hogervorst, M.K. van Dijk, P.C.M. Verbakel, C. Krijgsman
Document Access B.V.
Rotterdam, The Netherlands

Abstract

The use of character recognition in automated data-entry applications is described. The processing of the documents on which the characters to be interpreted reside, starts with making electronic images of them. At the end of the process, the information read from the forms is present in a database. Neural networks are used to recognize the individual characters in the form images. The confidence of each recognition, which is provided by the neural network as part of the classification result, is one of the things used to customize the application to the demands of the client.

1 Introduction

Document Acces B.V. is a growing Dutch company, specialized in software projects and consultancy on document management systems, automated data entry and workflow automation. The company was founded in 1991, and employs about 45 persons.

This paper describes how Document Access B.V. uses neural networks in their projects for automated data entry. The result of such a project is a computer system which makes electronic images of the documents, interprets the handwritten characters on them, and puts the results in a database. In most projects this computer system replaces (part of) a data-entry system, in which typists enter the data manually.

The remainder of this paper will start with a description of the components in a typical automated data-entry application. Then the core of such an application, being the single-character recognizer, will be discussed in more detail. Section 4 will show how an automated data-entry application is customized in order to meet the demands of individual projects. Before ending with some conclusions, one project, performed for a Swiss insurance company, will be described as an example.

2 Automated data entry

The basic scheme of an automated data-entry application is given in figure 1. The processing of documents (forms) starts with scanning them, in order to obtain electronic images. In the electronic images, the area's (fields) containing series of characters to be interpreted are then cut out for further processing. These images are enhanced by removing noise pixels, removing lines (e.g. originating from the box around the characters), etc.. A segmentation procedure then selects the areas

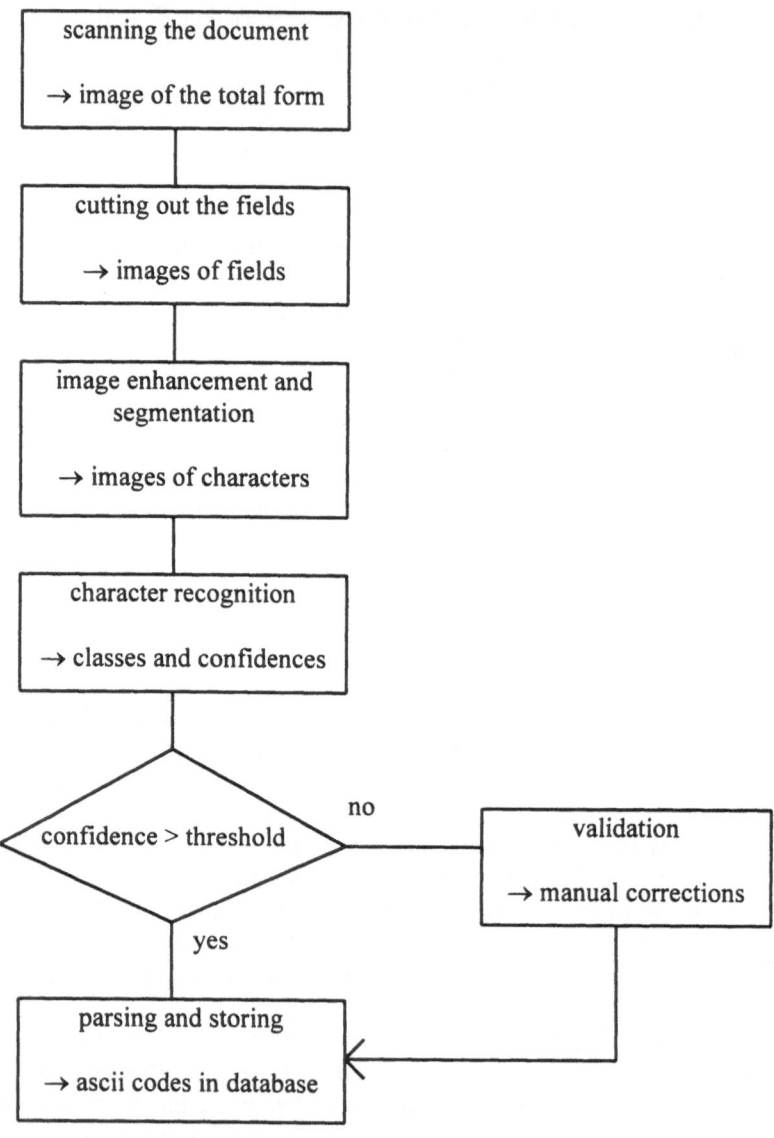

Figure 1. Basic scheme of a typical automated data-entry application.

containing the individual characters.

These images of individual characters are the ones which are processed by the character recognizer. The result of the recognition is for each character the class (e.g. whether the character is a '5', a '6', or an 'A', etc.) and a measure of the confidence. The confidence values of the characters in a field are used to decide whether the recognition results for that field are accepted or rejected. In the latter case, the recognition results for that field are sent to a validation terminal. There they are presented to a person together with the image of the field. The image provides the context of the character which is recognized with a low confidence. The person checks the recognition results, and makes corrections if needed. The end result of the automated data-entry application is that the handwritten information is stored in a database in the form of ascii codes, as would have been the case for manual data entry by typists.

Often, other verification methods can be used. For example, if a form contains several numbers and their sum, the recognized sum can be compared to the sum of the recognized numbers. If a number contains an internal redundancy, as is often the case for identification codes, this redundancy can be used as a check. If a field can only contain words from a limited set, it can be checked whether the recognized word is a member of this set.

Figure 2 gives an example of a hardware configuration. This client-server configuration contains two stations for scanning the documents, two stations in which the electronic images are processed and the characters are recognized, and four stations for the validation of the recognition results. Of course, the optimal number of each type of station depends on the actual project demands.

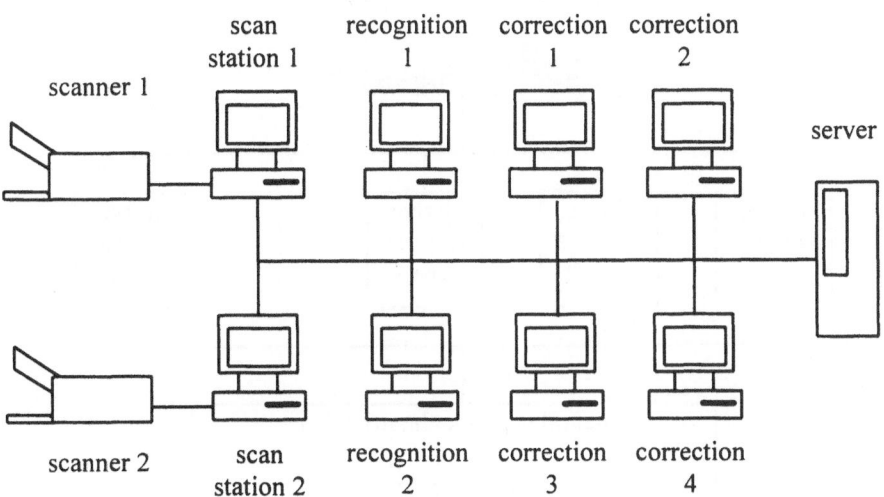

Figure 2. Example of a hardware configuration for an automated data-entry application.

3 The character recognizer

The character recognizer is the core of the automated data-entry application. Its performance is crucial for the performance of the complete automated data-entry system. In order to comply to the increasing performance demands, Document Access continues to put effort in the further development of this software.

As shown in figure 3, the character-recognition process consists of two steps. The first step is the derivation of features from the bitmap. Depending on whether recognition speed or recognition accuracy is more important in an actual project, the types of features to be used are chosen. A feature set may consist of geometric features or of coefficients resulting from a transform. The calculated features are used as input for the neural network, which performs the classification task.

In theory, the step of deriving features can be skipped. The pixels of the (normalized) character image can be used directly as input to a neural network. However, the task the neural network has to do is easier if some preprocessing in the form of the feature calculation is performed. Therefore, and because of the extra flexibility, we have chosen for the intermediate step of the feature derivation.

The second step is formed by the neural network calculations. Again depending on the project demands, a choice can be made for one network or for a combination of networks. In the first case, a choice can be made out of different network types.

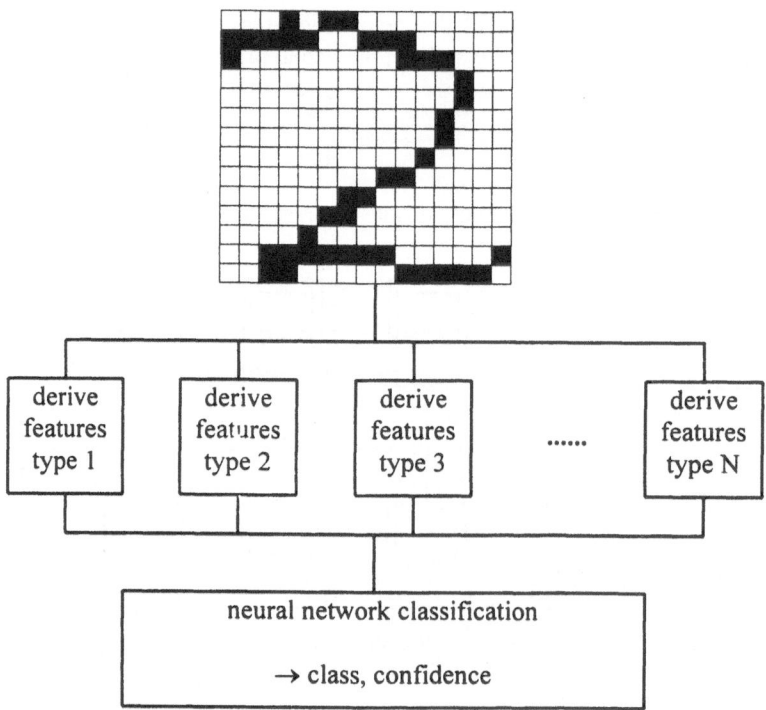

Figure 3. Basic scheme of the single-character recognizer.

In the second case, networks of different types are chosen in order to get the best synergy.

The neural network not only delivers the class of a character on the image (e.g. a '2', or an 'A'), it also delivers an indication of the reliability of that classification in the form of a confidence value. A low confidence may be caused by a poorly written character (e.g. a '6' with a short tail, resembling a '0'), a low quality image, or extra information in the character image (e.g. part of a signature).

4 Customization

Above, the choice of the feature types and the choice of the neural network, are already discussed. Another important facet concerning the customization of the automated data-entry application is the confidence threshold above which the character classification is accepted without further validation. A higher value of this threshold will cause more rejects, and hence more work for the persons behind the validation/correction stations. A lower value will result in more classification errors passing uncorrected. These errors are called substitutions. For applications in which the consequences of substitutions are very costly, as is e.g. the case for the interpretation of the amount on a cheque, one would choose a high threshold, in order to reduce the number of substitutions.

Figure 4 gives an example of a reject-substitution curve, which shows how the number of substitutions and the number of character rejects (both expressed as the percentage of the total number of tested characters) vary as the threshold changes. The curve has been made on the basis of recognition results of correctly segmented characters (digits in this case) which are not used in the training. The shape of the curve depends on the quality of the character images as well as on the quality of the

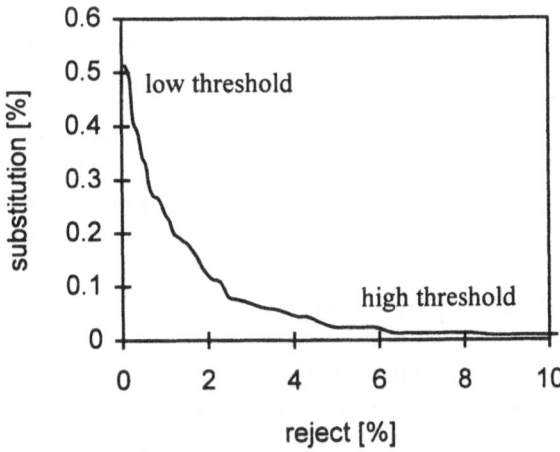

Figure 4. Example of a reject-substitution curve for the recognition of handwritten digits.

character recognizer. The closer the curve is to the axes, the more favourable is the situation.

For a point on the reject-substitution curve for individual characters, the number of rejected fields and the number of fields containing one or more substitutions can be estimated, using the average number of characters in the field, assuming that the classification errors and the low confidences occur randomly, and that no segmentation errors occur. A field is rejected if the confidence for one or more of its characters is below the threshold. For a small number of characters in a field (e.g. 5), and for a small character-reject percentage, the reject percentage for the field can be calculated by multiplying the character-reject percentage by the average number of characters in the field. The same approximation is valid for calculating the percentage of the fields containing a substitution.

A factor which influences the quality of the image of the handwritten text enormously (from the point of view of the application designer), is the design of the form. A form designed especially for the purpose of automated data entry gives much better results than other forms. To start with, it is advantageous to use forms printed in a color which is invisible for the scanner. In that case the boxes around the character and the accompanying text do not appear in the electronic image. It is further important to encourage persons to write the characters clearly separated from each other, e.g. by providing for each character a separate box on the form. The segmentation procedure can cope better with clearly separated characters than with characters connected to each other. Unfortunately, it is not always possible to change existing forms. For example, if the forms are not the property of the organization which has to perform the data entry, as is often the case for cheques, the software has to be customized to the peculiarities of those forms.

Another issue is the data on which the neural network is trained. In order to get the best results for an application, the data for training the neural network should be as good as possible representative for the data the network must process in practice. Therefore, the neural network is trained on the images of characters originating from the forms to be used by the client, filled in by as many different real form authors as feasible, and scanned with the scanners to be used by the client. If the forms are to be filled in by a limited number of persons in practice, the neural network should be trained (also) on characters written by those persons.

For the same reason (achieving optimal results), in most cases different recognizers are used for different fields on a form. For fields in which only digits are expected, a recognizer for digits only is used. For fields in which only capitals are expected, a recognizer trained only on those types of characters only is used.

5 Example of an application

Helvetia Krankenkasse is the largest Swiss health care insurer. The information on 25,000 claim forms has to be entered in the computer each day. Originally, about half of this work has been done by 18 data processing clerks employed by Helvetia (each processing 150 claims per hour), and the rest by an external data-entry agency at the cost of $ 0.75 per claim.

Using the automated data-entry system built by Document Access B.V. in cooperation with AT&T/GIS Switzerland, Helvetia was able to process 400 claims per hour per clerk. As a result, the company no longer has to turn to other data-entry agencies for claim processing. This saves the organization approximately $ 2,300,000 each year. The investment costs have been recouped in just eight months.

The system processes forms which have been designed by Helvetia in cooperation with Document Access. The double-sided forms are printed in a color which is not visible for the scanner. For each character to be filled in, a separate box exists on the form. This encourages the person, who fills in the form, to write the characters at the indicated locations, and as a result, the characters do not touch each other. The form contains 168 data fields. In 75% of these fields only digits are expected, in 15% of the fields only capitals are expected, and in 10% both type of characters can appear.

For this application, three neural networks have been trained: one for digits, one for capitals, and one for the two types of characters together. The software performs various correlation tests to check the correctness of the recognized characters. The only manual actions to be performed in this application are the operation of the scanners and the validation of the rejected fields. At the end, only three characters out of 10,000 end up incorrectly in the database.

6 Discussion

We have described the basic set-up for automated data-entry applications. The core of such an application contains one or more neural networks for the classification of the handwritten characters in the image of a form.

Automated data-entry applications, as described above, are profitable in several situations. As long as other techniques, such as EDI (Electronic Data Interchange), has not ousted the large scale use of forms from the market, there is a market for the automatic interpretation of forms. Many project opportunities exist with a return on investments within two years. As the performance of the recognition software is improving, due to our continuous research efforts, the number of such projects is even increasing.

Neural Networks - The Future of Forecasting in Finance?

Hans Georg Zimmermann

Siemens AG

Corporate Research and Development, ZFE T SN 4

81739 Munich, Germany

email: Georg.Zimmermann@zfe.siemens.de

Abstract

The most well-known forecasting technique in finance is chart analysis, which only evaluates data from a specific time series in the past. In contrast, a fundamental analysis attempts to describe the actual dynamics of market processes. The success of chart analyses is handicapped by the low volume of input information, while that of a fundamental analysis is limited by the complexity of the market and the fact that it disregards the psychological factors influencing decision-making. Neural Networks can be interpreted as an interacting decision process with the ability to extract a high dimensional nonlinear structure from observations by learning.

1 The significance of forecasts for financial applications

The ability to forecast developments is a central requirement for rational decision-making, since the merit of any decision is always measured by its consequence in the future.

This applies in particular to the financial sector, for example in exchange rate trading or capital investment. The most well-known forecasting technique in this context is chart analysis, which only evaluates data from a specific time series in the past. In contrast, a fundamental analysis attempts to describe the actual dynamics of market processes. The success of chart analyses is handicapped by the low volume of input information, while that of a fundamental analysis is limited by the complexity of the market and the fact that it disregards the psychological factors influencing decision-making.

The use of neural networks opens up new possibilities for forecasting complex economic dynamics. For the prediction of interest rates, exchange rates and stock prices, for the analysis of risks and predicting the demand of goods, this mathematical tool shows a new way to extract a dynamical structure from data of the past. Today it is becoming apparent that the discipline of neural computer science will be able to supply the banking industry specifically with better decisional base.

2 Steps toward developing a forecasting model

In principle, any forecast is based on observational data from the past from which a structure is extrapolated. It is assumed that this structure is invariant or at least approximately constant over time, thereby allowing the calculation of data of the future. The hypothesis of structural constancy in a still unknown future cannot be proven, but can be tested at known parts of the time series. The mathematical core of this task consists in assembling data from the past into a structure which may, for example, assume the form of difference or differential equations.

Conventional mathematical techniques concentrate on two specific types of structure: Linear regression techniques consider a large number of variables, but permit only linear interactions among the variables. Nonparametric regression methods permit complex behavior on a small number of variables.

In many cases, it is not possible to assign economic problems to one of these two special cases. Economic systems are by nature multivariable systems which can only be separated to a limited extent and whose nonlinear dynamics can only be observed under the superposition of a large number of influencing factors. In contrast to physics, it is not possible in the field of economics to construct a laboratory experiment in which almost all the variables are kept constant and a very few are observed in their mutual interactions. Nor is it possible to repeat an experiment under the same conditions. Consequently it is not possible to change individual variables and analyse various scenarios to observe their influence on the dynamics of the market and thereby develop a perfect model.

Neural networks are a chance for econometrics. Due to their suitability for con-structing models for a large number of influencing factors and at the same time nonlinear interactions, they represent a mathematical tool which is able to provide a framework for constructing complex economics models.

3 Neural networks as decision models

For the long-term fusion between the disciplines of neural networks and economics, it is important not only to investigate the formal use of the mathematical formalism, but also to be able to provide an interpretation of the neural network basics in the language of the application field. This especially applies to the mathematics of neural networks, the historical roots of which were not associates with economic questions.

The mathematical description of a single neuron can be understood as a model representing the concept "decision". To explain this, let us refer to the situation of a stock exchange trader. The trader is continually confronted with hundreds of items of information. either by the data supplier on a screen, by telephone or from conversations. As a result of this deluge of information, the trader has no other choice but to focus his or her attention selectively. Concequently, some of the information provided is given greater emphasis, while most of the other indicators are pushed to the background. In a second step, the indicators that the trader has registered must be combined to form an overall impression, for example the statement that a certain exchange rate will rise or fall. A verbalization of this step would be, for example, the statement: "The mood on the floor is that the dollar will go up". In the third step, this assessment of the market situation motivates an action, that is a switch-like behavior: A call option is either placed or not.

The basic mathematical model of a neuron reconstructs precisely these three steps. The input information is filtered by multiplying the input signals by weighting factors, and an overall signal is formed by summing the distorted input variables. This value is then compared with a threshold value. If the latter is exceeded, the output signal is a one. If not, it is a zero.

A neural network in its totality can be interpreted as a complex model of the input/output behavior of a single agent (individual model) using inner neurons to model subdecisions of the thinking process. Or it can be interpreted as an interaction model between a large number of decision-makers (model of market process). In a feedback system of this type, the superposition dynamics of a large number of decisions towards an equilibrium is reconstructed, similar to normal stock market behavior. Even the non equilibrium processes in the market have its analogies in special neural networks dynamics.

4 Learning from observational data

Neural forecasting systems are also particulary suited for constructing ecomomic models because they are based on experiences. From given observational data, they can reconstruct a slice of reality by using learning processes. A vector of quantitative and/or qualitative data is presented to the input neurons and the network computes the output by its weighting and threshold processing. If the neural network supplies an incorrect forecast, the inner state of the network (i.e. the size of the weighting factors and threshold values) is modified by the error deviation from the past, with the result that the quality of the prediction improves.

However, a large quantity of sample data is required in order to adequately adapt a complex model. Unfortunately, economic investigations often have to be carried out under certain constraints, for example that the training data are incomplete. Typically they have a high noise level and there is a lack of training examples relatively to the complexity of the structure which we are looking for. In recent years a large number of mathematical methods were proposed to detect a structure in observational data, even under this more difficult conditions.

5 Neural networks versus expert systems

Due to their ability to extract a structure out of observational data by a learning algorithm, neural networks differ from expert systems, which in principle would also be suitable as forecasting systems. Here an expert explains his understanding to a computer specialist who constructs an expert system in form of a body of rules. At best, the computer system can only reflect the expert's understanding of the structures involved. And experts frequently contradict each other in the field of economics, for the very reason that it is difficult to grasp economic interrelations with their multiplicity of overlapping variables which display opposing tendencies.

The methodology of neural networks is more strait forward in this situation. The choice of important time series as input variables for the training process should be done by experts of the application field. But if different experts give different sets of important timeseries, these can be put together and the learning algorithm can be used to evaluate the ranking of importance. The answer to the question 'What indicators are really important to understand market behavior? is indeed an essential part of the analysis of the market dynamics. And a neural networks study is able to give an answer in a nonlinear, high dimensional framework based on actual data.

6 Research trends

Fuzzy Logic is another attempt to model aspects of human thinking. The fuzzy logic description of a dynamical system can be given in three steps. Let us start with a numerical state description of the system. In a first step this description is transformed to a verbal description. This is done by membership functions, which evaluates the validity of the verbal description. By this transformation it is possible to model the fuzzyness of language. Based on the verbal description we can model in a second step the dynamic of the system by logical rules. An example may be: If the exchange rate is increasing and the future is higher than the rate then the exchange rate will continue to increase. In a third step we have to transform the verbal forecasts of the different rules to a numerical forecast. One should have in mind that fuzzy logic allows the superposition of contradicting rules.

In the SENN framework (SENN = Simulation Environment for Neural Networks) we have developed a compiler that transforms all three steps of a fuzzy logic description into a neural net environment. Now it is possible to optimize the net by training data. After a back compilation of the optimized network to a fuzzy structure the improved rule description of the dynamic is given in a transparent way.

The combination of expert knowledge and learning from data allows an improvement in model building. We have shown this in exchange rate and stock index applications. From a mathematical viewpoint the problem of the combination of logic and data is focused in the point that the knowledge of structure and learning does not harmonize all the times.

7 The future of neural net methods

An increasing effort in basic research and applications can be observed worldwide. In this competition the Simulation Environment for Neural Networks (SENN) from Siemens is the most advanced timeseries analysis tool. It supports the model building procedure from the data handling and preprocessing to in depth network tools and an integrated postprocessor.

A lot of neural models for representing complex processes in the financial market have already left the development laboratories. With appropriate engagement on the part of financial institutions, their number will probably increase rapidly - the scientific foundations, that is the development of flexible neural function systems, their understanding as an interacting decision process and supervised learning algorthms are availible.

COUNTERMATCH: A Neural Network Approach to Automatic Signature Verification

Graham Hesketh

Applied Neurocomputing Centre, AEA Technology
Oxfordshire, UK

Abstract

Countermatch is a neural network algorithm for Dynamic Signature Verification (DSV). Its inherent tolerance for natural variations in peoples' signatures enables it to outperform alternative DSV methods which are based on feature extraction. The successful transfer of this technology into the marketplace was demonstrated by the UK Employment Service in the world's largest public trial of a signature verification system.

1 Introduction

Signatures and signature verification play a large part in our daily lives. For hundreds of years the placing of a mark or signature on a document has been the accepted method for authorisation. As we approach the 21st century, the ease with which goods or money can be obtained on the strength of just a signature has led to an explosion in fraud.

Now, when we sign a credit slip we are doing two distinct things: (1) we are legally sanctioning the payment of a specified sum of money, and (2) we are providing evidence that we are the legitimate user of the credit card. Whilst signatures will continue to be used to sanction our transactions, other means of providing evidence of legitimate use are being sought.

The traditional method of visually comparing the signatures on the slip and the credit card is fraught with problems. The person required to do the check may (1) forget to do it, (2) deliberately avoid doing it, (3) not care about doing it properly, (4) not be sufficiently skilled at recognising forgeries, (5) not have sufficient time to do an accurate comparison, or (6) not be sufficiently confident to challenge the signer, even though they actually believe the signatures are different.

An automated means of positively verifying the identity of the credit card user would eliminate all of these problems. A variety of biometric systems such as fingerprints, hand geometry, voice patterns or retinal scans have been proposed for this purpose. At present they are either too expensive, intrusive, socially unacceptable or have not been proven to work well enough.

Signature verification by computer, on the other hand, is a socially acceptable biometric. It is not costly and it has the advantage that the client is already required to provide a signature, so the verification process involves no further change to the established procedure. The only real obstacle to its widespread uptake in the marketplace is the performance of the biometric in the public domain.

In this paper I will describe a neural network-based signature verification program called Countermatch. This software has successfully completed rigorous testing at two offices of the UK Employment Service, in what was the largest public trial of a signature verification system anywhere in the world.

2 History of Countermatch

This section chronicles the development of the Countermatch program from its inception to its validation in a public trial.

2.1 Concept

Early in 1992, Professor Colin Windsor FRS of the Applied Neurocomputing Centre at Harwell Laboratory was working on an elastic matching neural network for use in matching gas chromatograms[1]. These are essentially one-dimensional "signatures" of quantity versus time, indicating the particular mixture of chemical compounds in a sample. He had the inspiration to see how the method could be applied to matching hand-written signatures.

A simple BASIC program was constructed, and with a primitive pen-input device to capture the signatures, Professor Windsor demonstrated the feasibility of the approach. This led to funding for an internal project to develop the program into a robust demonstration vehicle.

2.2 Initial Development

The program was rewritten in C and modified to use a specialist signature capture device (an NCR 5990 Signature Capture Tablet) which had an LCD screen to display the ink as it was collected. The development team and co-workers supplied signatures to train the neural network model.

An internal trial was conducted at the Harwell Laboratory in late 1992 to evaluate the current model and to provide subsequent training material for an improved model. A summary of trial statistics are given in Table 1.

---

[1] The algorithm he was working on is now a successful commercial product called MatchFinder. It is currently used to check the legality of fuels used by Formula One racing teams.

	Authentic Test	Forgery Test
Number of Individuals	79	82
Number of Attempts	205	388
Number of Single-Attempt Failures	5 (2.4%)	14 (3.6%)
Extrapolated 3-Attempt Failures	Less than 2 in 100,000	10%

Table 1

The program was performing well enough to be shown on live television. The BBC's "Tomorrow's World", a well-known programme devoted to technological innovation, included Countermatch in a section on neural networks in September 1992. The program performed faultlessly throughout rehearsals and, most importantly, during the live broadcast.

2.3 Exploitation

During 1993 the program was shown to a large number of interested parties, including banks, building societies, credit card companies and government agencies. The general response was that although the idea was good and the software seemed to work, they couldn't commit to using it at that time. Here are some of the reasons they gave.

1. They didn't like the signature capture device (if you have ever tried to write on glass you will know why).
2. It was a DOS program with limited functionality which would require significant development to integrate into their applications.
3. Introducing signature verification would be a large-scale project, and they did not want to be the first to try to implement it.
4. They were uncertain as to how the public would react to having their signatures verified by a computer.
5. There were no statistics available to validate the performance of the system in the public domain.

It is the last point which has been the major weakness in marketing any biometric system. Validation in realistic end-user conditions is essential to convince prospective customers of the true performance they can expect from an installed system.

A breakthrough arrived at the end of 1993 when Countermatch was seen by the UK Employment Service. They liked the program and agreed to host a public trial in two of their Unemployment Benefit Offices. A new version of Countermatch had to be written for the proposed trial, to meet the following requirements:

1.   Signatures must be collected from an inking pen writing on paper (to fit in with existing procedures and to meet legal requirements for auditing).
2.   The program must run on a network of signing stations, using a central database capable of dealing with thousands of clients.
3.   The program must have a very simple interface.
4.   The program must not disrupt the normal operations of the offices.

Early in 1994 a Windows version of Countermatch was written in C++. In conjunction with a commercially available graphics tablet (the Summagraphics Summascribe), it satisfied all the requirements for an extensive trial by the Employment Service. The systems were installed in Liverpool and Tyneside during August 1994 and went live in September.

2.4 The Employment Service Trial

The two pilot schemes each comprised six Countermatch terminals using 486DX PC's; three for enrolment of new clients and three for verification of existing clients.

Clients were enrolled onto a simple database by signing three times on the graphics tablet, and this entry provided their reference template. When they returned each fortnight they signed the appropriate Employment Service form which was placed on the graphics tablet. Within one to two seconds an acceptable/unacceptable diagnosis was returned by the PC. If after three such attempts Countermatch still identified an irregularity, then existing Employment Service procedures were brought into action, such as requesting additional identification or referring the client for interview.

From the beginning of September 1994 to the end of November 1994 around 8,000 benefit claimants enrolled onto the Countermatch system and provided over 36,000 signatures for verification, making this by far the largest public trial of a signature verification system anywhere in the world.

The size of the trial was not the only important factor. The people who used the system (the benefit claimants) were not volunteers. Genuine claimants had a vested interest in the system working properly for them, whereas people impersonating other claimants were risking a great deal[2] by attempting to beat the system. In short, it was as realistic a trial as it is possible to achieve. This is an important consideration for the evaluation of performance statistics, especially for behavioural biometric systems such as signatures. Voluntary trials lack the psychological stresses which occur in real-life situations, so their results may not be reproducible where it matters.

---

[2] It is a criminal offence to sign for unemployment benefit on behalf of someone else.

2.5 Results and Benefits

Although the trial involved members of the general public, the statistics from the trial are confidential to the Employment Service. However, they have agreed to release some details which are reproduced here.

- 98% of all sign-on cases were accepted, and of these 94% required only one signature attempt.
- With over 300 sign-ons daily, rejections amounted to around 7 cases per day.
- Of the rejections, only 2-3 per fortnight were attributable as failures of the Countermatch algorithm, giving a false reject figure of around 0.1%.

In addition to these statistics, the Employment Service people made the following comments.

- Using Countermatch had a deterrent effect with certain individuals being noted to have stopped claiming benefit after the equipment was introduced.
- It reduced clerical error by helping ensure front-line people double checked claimant file details. For instance, Countermatch identified an irregularity where a client with a similar name had been asked to sign the wrong unemployment benefit form.
- It boosted morale in the office as news of the system spread around the local community.
- Because it used existing forms and procedures it was easy to install and use.
- It worked smoothly because the clients were not obliged to perform any different actions from those previously undertaken.
- To a large degree client reactions were favourable.

Both offices were so satisfied with the performance of Countermatch that they continued using it, even though the trial had officially ended, and it is still in routine use today.

3 The Elastic Matching Neural Network

Countermatch is a neural network algorithm for Dynamic Signature Verification (DSV). DSV involves collecting signatures in real-time and comparing them with previously stored reference templates in order to determine the likely authenticity of the new signature. In effect it operates as a biometric test, verifying the identity of the signer.

Unlike other DSV systems, Countermatch does not attempt to extract features from the signatures (e.g. average speed of writing, maximum acceleration, number of zero crossings etc.). Instead, it attempts to match the entire specimen signature to a test signature. What this means in practice is that a potential forger must not only make the signature look the same but must also reproduce the way it was generated (such as the stroke order and speed of execution).

The stumbling block for DSV systems has always been false rejections of authentic signatures. It is very easy to create templates which forgers cannot reproduce, but genuine clients would find them too stringent for routine use. The neural network approach used in Countermatch effectively addresses this problem. The elastic matching process is very tolerant of natural variations in the signatures (e.g. longer/shorter strokes, missing/extra strokes), but remains sensitive to deliberate or "unnatural" variations (e.g. significant timing changes and scale differences) which are typical of forgeries. This leads to a more robust verification system.

4 Applications

There are many areas where automatic signature verification would be beneficial. This section outlines some of the obvious ones.

4.1 Attendance Registers

This involves positive identification of individuals at specific places and times. It could be used for benefit claims (c.f. the Employment Service), examinations (e.g. driving test, academic qualifications etc.), signing-on at the work place, or even for monitoring the whereabouts of minor offenders in non-custodial situations.

4.2 Access Control

This involves limiting access to secure areas (building or rooms), machines (individual computers or networks) or data (specific programs or databases). It would effectively replace (or augment) conventional passwords or PIN numbers.

4.3 Point-of-Sale

This is currently the largest potential marketplace for signature verification. Signature checking of credit, debit and cheque guarantee cards could be automated, and would significantly reduce fraud in the retail sector. Implementation details (smart cards, databases, compatibility of equipment, apportioning costs) have held back progress in this market. It is likely that small-scale self-contained credit card installations (i.e. for use within a particular store or retail chain) will be the first to take up this technology.

4.4 Electronic Document Interchange

As the computerised processing and interchange of data becomes commonplace, there are significant cost savings for large organisations if electronic storage and transfer can supersede the need to print, store and mail paper documents. However, certain types of documents require authorisation for technical, financial, quality assurance or legal reasons, and the major obstacle has been the inability to electronically authorise these documents. Computerised signature verification now offers a solution to that problem.

It is also interesting to note that signature verification offers a better solution than the Public Key / Private Key encryption method used in the U.S. Department of Commerce's Digital Signature Standard, since identification of the originator and non-repudiation of the signature involve physical evidence (the electronic signature) rather than knowledge (of the private key).

5 Conclusions

The historical stumbling block for automated signature verification has been the level of false rejections incurred whilst trying to detect fraud. The successful application of a neural network approach has shown that the problem can be overcome, and this has been amply demonstrated by Countermatch in the world's largest public trial.

Biometrics as a means of providing positive identification are very much on the increase. Of all the available biometric technologies, signature verification is likely to be the one which is most acceptable to the general public, since it involves existing, familiar procedures.

Signatures are all-pervasive in our society, hence the potential for automated signature verification is enormous. As we move towards the paper-less office, the processing and verification of electronic signatures will become part of everyday life. The only question is who will be the first to reap the benefits.

ZN-Face: A system for access control using automated face recognition

Jörg Kopecz, Wolfgang Konen and Ekkehard Schulze-Krüger
Zentrum für Neuroinformatik GmbH, Bochum
Universitätsstr. 160
D-44801 Bochum
Germany

June 13, 1995

Abstract

We present a biometric access control device which is based on the identification of human faces. The system combines a console for semi-automated image acquisition with the necessary algorithms for face recognition. Facial features are stored in a relatively compact data format (1.6 kB). ZN-Face runs on a Pentium 90 without any special accelerator hardware where it performs image acquisition, face localization and identification in less than 3 seconds. ZN–Face not only allows robust identification of stored persons (despite changes in facial expression or size), but also reliable rejection of unknown persons. With an acceptance criterion which safely rejects all unknown persons we achieve an identification rate above 99% (FRR< 1%). The ZN Bochum GmbH has sold more than 100 licences to various institutions and companies, among them the Kremlin in Moscow. The ZN Bochum GmbH holds the relevant patents for ZN–Face.

1 Introduction

Face recognition is a remarkable example for the ability of humans to perform with great reliability a complex visual recognition task. We are able to recognize thousands of faces which we have learned during our lifetime. Our visual performance is very robust against a variety of factors like changes in facial expression, head posture or size, illumination, background, facial aging or partial occlusion of a face. It is one of the challenges of artificial information processing to achieve at least partially a similar performance on systems for automated visual recognition. The ease with which the problem is solved by the human brain, i. e. a biological neural system, may serve as a guideline in the sense that neural information processing paradigms can be important also for the construction of automated face recognition systems.

In this contribution we will report on the development of the face recognition system ZN-Face which offers a turn-key solution for an access control system. ZN-Face is based on the neural algorithms of [1] which are briefly reviewed in Section 2 (Elastic Graph Matching). In Section 3 we describe the extensions necessary to migrate the algorithm into an automated access control system.

Numerous approaches for solving the face recognition problem can be found in the literature from which we can mention here only a few. A good and more comprehensive survey of the state of the art can be found in [2]. Early algorithms [3] use feature-based techniques (e. g. features like localization or thickness of eyebrows) or template matching for recognition [4]. Recently, a systematic comparision of feature-based vs. template-based algorithms has been undertaken by Brunelli and Poggio [5]. Gilbert and Yang [6] presented a real-time face recognition system using custom VLSI hardware. The hardware allows fast template correlation and the system is able to perform an identification from a 34-person database in 2 to 3 seconds. Another approach uses the decomposition of facial images into an 'eigenface' expansion (similar to Karhune-Loeve expansion) [7, 8]. These expansions achieve a very compact representation of faces, they are, however, rather sensitive against variations in imaging conditions.

Another group of algorithms (including this work) uses neural processing techniques for face recognition [9, 10] or gender recognition [11]. One of the few systems which does not only present a face recognition algorithm but also a complete access control booth has been proposed by Bichsel and Seitz [12]. This direction is also the aim of the current work, where a complete hardware-setup has been developed as a biometric access control device for real-world security applications.

2 Face recognition by graph matching

The basic recognition algorithm of ZN-Face is an extension of the Elastic Graph Matching Algorithm [1]. In contrast to most other face recognition algorithms which require distinct processing steps like localization, separation, standardization and finally recognition of faces, the current algorithm is more coherent in the sense that *one* basic principle is used to perform the above steps simultaneously. Here, faces are stored as flexible graphs or grids (see Fig. 1) with characteristic visual features attached to the nodes of the graph (*labeled graphs*). A data representation with labeled graphs has computational advantages:

robustness: The features are invariant against intensity changes and contrast changes in the image. Furthermore, the features are less affected by changes in head posture, size and facial expression than raw grey level features.

data compression: compared to the raw image which has in our case a size of 128×128 pixels, the size of the graph (about 1.6 kB) is smaller by a factor of 10.

scaling: a sparse graph can be readily adjusted to changes in the geometry (size, perspective, see Fig. 1B,C). Change in size in a pixel-based data representation, for example, would require a more complicated transformation.

distribution: Compared to a high-level feature description (e.g. 'eye', 'nose') the graph contains sufficient information distribution on simple, but numerous features. Even if the information at a single graph node is missing due to occlusion, recognition is usually still possible due to the information at the remaining nodes.

In summary, the face recognition algorithm consists of three basic steps:

(1) Convolution of the image with feature extractors used in the labeled graph representation.

(2) A stored graph is matched to the image by an optimization procedure: Position, size and inner structure of the graph are varied in order to maximize the similarity between graph node features and corresponding image features. Penality terms inhibit too large deformations of the graph. As a result, a new graph with the actual image features as node labels can be extracted.

(3) In order to decide, whether stored graph and image show the face of the same person, the extracted graph is compared also to a number of reference graphs. Only if certain significance conditions are fulfilled the match is accepted.

3 Access Control System ZN-Face

ZN-Face offers an automated access control system which performs a biometric identification of persons from their facial images. To achieve this, several extensions are necessary to transform the Elastic Graph Matching algorithm of Sec. 2 into a system which is easy to use in real-world security applications:

- *Fast computation on standard hardware:* In former times the graph matching was performed on a transputer-based platform and required about 12 seconds for the identification. Now, the algorithm has been optimized and ported to a standard PC (Pentium PC/90 Mhz *without* special accelerator hardware). Computation time for a full identification is 3 seconds.

- *Semi-automated image acquisition:* No operator is available to adjust the camera (persons may differ in their size and their distance to the camera) and to trigger the image acquisition. An appropriate semi-automated procedure is described below.

- *Easy-to-use system administration:* A Windows-based[1] GUI has been developed which allows simple control by an authorized system administrator (e.g. adding or deleting access rights, monitoring the access protocol).

- *Possibility to handle large databases (1000 persons and more).*

---

[1] Windows is a registered trademark of Microsoft

In order to achieve the last point without unacceptable increase in computation time, we perform with ZN-Face *verification* instead of recognition. Verification has the additional advantage of higher security because in addition to a standard personal identification via secret number (PIN) or card, the face is used as independent control of the person's identity.

The basic verification procedure can be described as follows: The hardware setup of the system consists of a standard PC with framegrabber and a ZN-Face console (a prototype is shown in Fig. 2) containing camera, ID acquisition device (card reader or PIN pad) and LC-display. The camera is positioned behind a semi-permeable mirror which is tilted about $30 - 50°$ against the vertical line. Camera and mirror are tilted in order to allow users of different height to view and position themselves centrally in front of the camera. (at the expense of some size variation in the acquired facial images). The user triggers the image acquisition and identifies himself (PIN pad or card). Now ZN-Face starts the verification whether the graph stored under the given PIN in the database fits to the acquired image. The final decision 'yes' or 'no' is based on the significance conditions mentioned above under step (3). This does not require to search the whole database, but only the stored graphs to the given PIN and a fixed number of reference graphs have to be searched.

It has to be mentioned that even with the user's cooperation during image acquisition, there is considerable variation in the images with respect to head posture, size and position. This is due to the user-dependence of the automated image acquisition. Most of the variation can be handled successfully by the robustness of the elastic graph matching. Also, variations like different facial expressions or wearing glasses are handled very well by the algorithm. To scope with the remaining cases, the system offers the possibility to store more than one image for a given person ("teach-in", see below).

Adding new persons to the database is made very simple for an authorized system administrator by using the graphical user interface (GUI) under Windows. An image of the new user is acquired in the same way as described above, and the system automatically locates the face and extracts the new graph (Fig. 3). In addition, the GUI offers a facility to browse through the verification protocol (Fig. 4). This control panel also offers the possibility of "teach-in": If an image acquired during verification is rejected although it shows the correct person (perhaps because stored and acquired image differ too strongly in head posture, thus leading to a low significance), the authorized system administrator has the option to add the picture into the person's record in the database. The system can handle an arbitrary number of images in a persons's record, although normally 1-3 images are sufficient for reliable verification.

4 Results

Nearly all papers on face recognition algorithms usually report only on the recognition rate achieved by the algorithm, i. e. how successfully a person is accepted whose image is in the database. But this is only one side of the coin: Often even more important for biometric devices is the question, how successfully persons are rejected whose image is *not* in the database. Biometric research names errors of the former kind as *False Rejection Rate (FRR)* while errors of the latter kind contribute to the *False Acceptance Rate (FAR)*. Of course, there is a trade-off between both types of errors and consequently it is much more difficult to build a system with a high recognition rate *and* a low FAR than a system with a high recognition rate alone.

The system ZN-Face offers the possibility to balance optimally the trade-off between FAR and FRR. 'Optimal' is defined here as the simultaneous minimization of FAR and FRR. In the verification, the actual acceptance value is compared against an acceptance threshold. If needed, the user has the possibility to shift the acceptance threshold, e. g. in in the direction of lower FAR, usually at the expense of a somewhat larger FRR.

We conducted two experiments with a database of more than 130 persons: In the first one (Fig. 5A), the database contained one image per person. For each person, the verification was tested with a different image $\mathcal{I}$ of the same person. Unsuccessful verifications (acceptance value too low) contribute to the FRR in Fig. 5A. Then, the person was removed from the database and it was tested, whether the image $\mathcal{I}$ was now rejected, as it should. False acceptances contribute to the FAR. As can be seen from Fig. 5A, the combined minimum of FAR and FRR is achieved at a level of about 3.5%. In Fig. 5B the same experiment is repeated, but now with 2 images per person in the database. In this case, comparing with a person's third image $\mathcal{I}$ usually yields higher acceptance values. Here, it is possible to shift the acceptance threshold in such a way, that FAR and FRR simultaneously drop to zero. (This is of course only true for the specific database, more comprehensive tests with larger data material have to be undertaken to estimate with higher accuracy, how low FAR and FRR actually are).

5   Conclusion and outlook

Elastic Graph Matching has proven to be a powerful algorithm for the recognition of human faces, especially because it achieves at the same time a reliable rejection of unknown faces. This makes it an appropriate tool for a biometric access control device which has been presented in this work. In addition to the results presented in the preceding section, the system has also been tested recently 'live' on several exhibitions (VISION'94 and SECURITY'94) where it operated successfully under real-world conditions with 'untrained' visitors. The fast identification of faces on standard hardware opens new application possibilities on the biometry market. The ZN Bochum GmbH has sold more than 100 licences to customers of various fields (among them the Kremlin in Moscow). The ZN Bochum GmbH uses distributors proven in the field of access control and hold the leading market position in this field.

References

[1] M. Lades, J. Vorbrüggen, J. Buhmann, J. Lange, C.v.d. Malsburg, R.P. Würtz, and W. Konen. Distortion invariant object recognition in the dynamic link architecture. *IEEE Transaction on Computers*, 42:300–311, 1993.

[2] V. Bruce and M. Burton. *Processing Images of Faces.* ABLEX Publishing-Corporation, Norwood, NJ, 1992.

[3] T. Kanade. *Picture Processing System by Computer Complex and Recognition of Human Faces.* Unpublished Ph.D. thesis, Dept. of Information Science, Kyoto Univ., 1973.

[4] R. J. Baron. Mechanisms of human facial recognition. *International Journal of Man Machine Studies*, 15:137–178, 1981.

[5] R. Brunelli and Tomaso Poggio. Face recognition: Features versus templates. Technical Report TR 9110-04, Istituto per la Ricerca Scientifica e Tecnologica, October 1992.

[6] J. M. Gilbert and Woodward Yang. A real-time face recognition system using custom VLSI hardware. In *Proc. of Computer Architectures for Machine Perception Workshop*, December 1993.

[7] M. Kirby and L. Sirovich. Application of the Karhunen-Loeve procedure for characterization of human faces. *IEEE Transactions on Pattern Analysis and Machine Intelligence*, 12(1):103–108, January 1990.

[8] M. Turk and A. Pentland. Face recognition using eigenfaces. In *IEEE Proc. of CVPR*, pages 586–591, Maui, Hawaii, June 1991.

[9] T. Kohonen, E. Oja, A. Kortekangas, and K. Mäkisara. In *Proc. Intl. Conf. on Cybernetics and Scociety*, Washington D.C., 1977.

[10] G. Cottrell, P. Munro, and D. Zipser. Learning internal representations of grey scale images: An example of extensional programming. In *Proc. Ninth Annual Cognitive Science Society Conference*, Seattle, WA, 1987.

[11] B. A. Golomb, D. T. Lawrence, and T. J. Sejnowski. SEXNET: A neural network identifies sex from human faces. In D. S. Touretzky and R. Lippman, editors, *Advances in Neural Information Processing Systems 3*. Morgan Kaufmann, San Mateo, 1991.

[12] M. Bichsel and P. Seitz. Der elektronische Pförtner: Automatisches Erkennen und Identifizieren von menschlichen Gesichtern. In R.E. Grosskopf, editor, *Mustererkennung 1990, 12. DAGM-Symposium*. Springer, 1990.

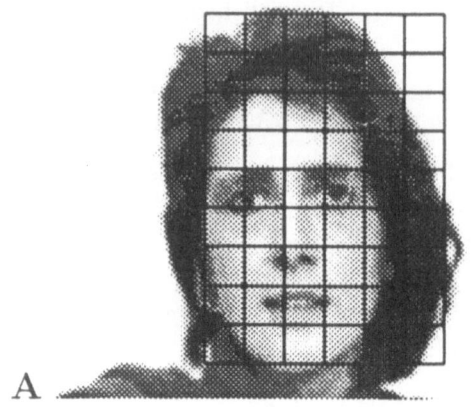

A

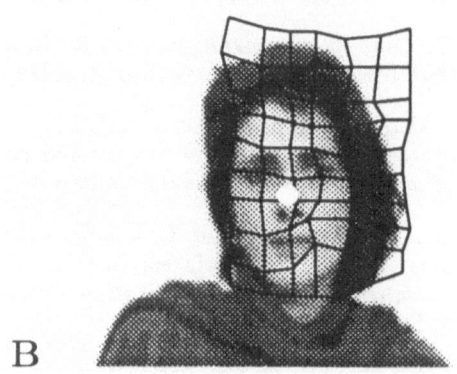

B

C

Figure 1: Data representation in *ZN-Face*. Such graphs can be shifted or scaled (**C**) efficiently in the image domain.

Figure 2: The access control system *ZN-Face* in a prototype realization, consisting of camera (behind semi-permeable mirror), LC-display and PIN pad.

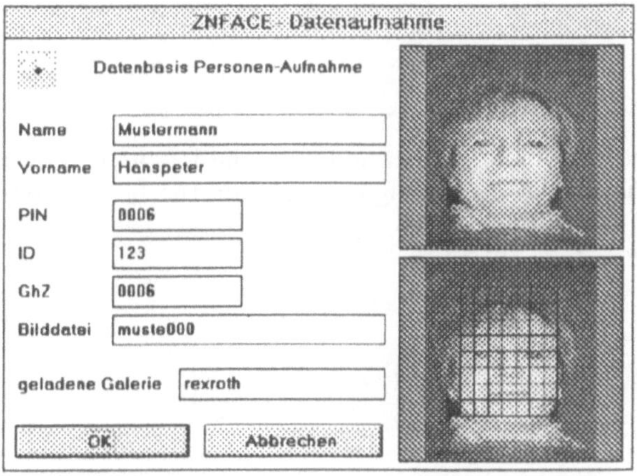

Figure 3: GUI for adding new persons to the database. The system automatically locates and extracts the new graph as shown in the lower image.

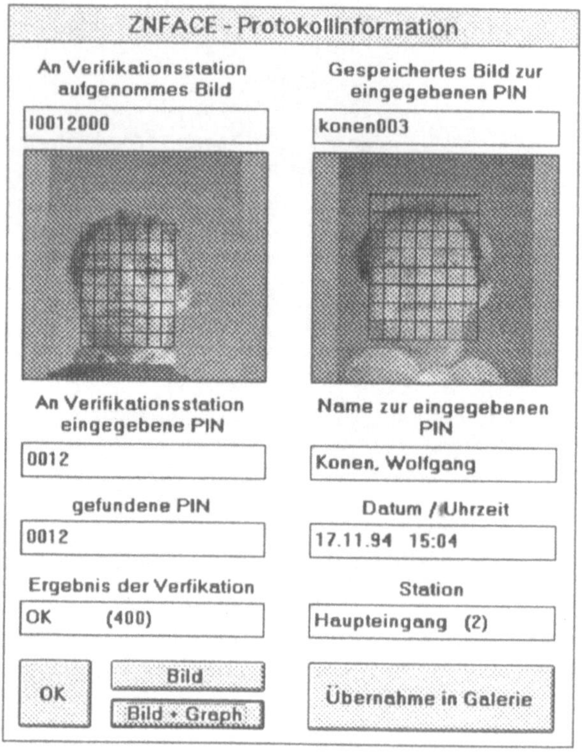

Figure 4: Protocol of the verification: On the left hand side the actually acquired image is shown, on the right hand side the best matching image from the database (to the given PIN).

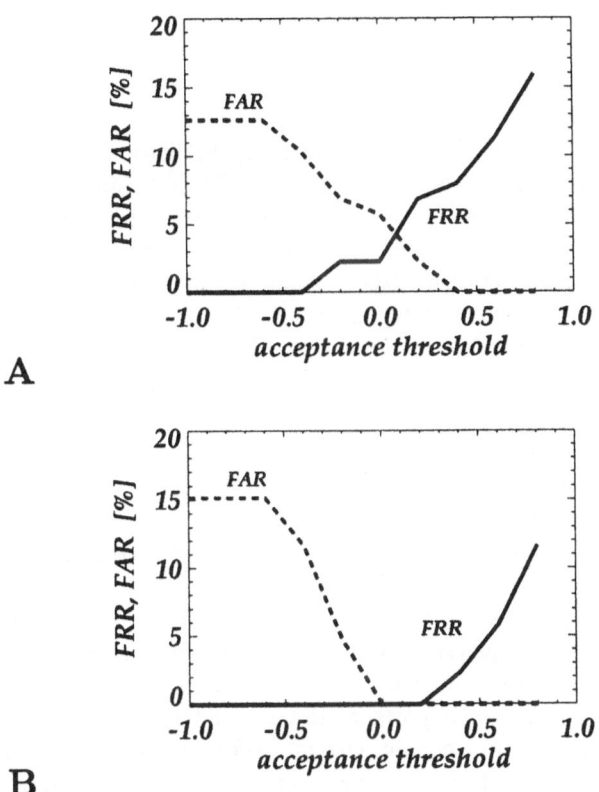

Figure 5: False rejection rate (FRR) and false acceptance rate (FAR) for recognition from a database of 130 persons as a function of the adjustable acceptance threshold. The threshold value 0 corresponds to the threshold learned by the neural net as optimal discrimination. The database contains one image per person in (A) and two images per person in (B).

Current Prediction for Shipping Guidance

J.C. Wüst
Rijkswaterstaat, North Sea Directorate
P.O. Box 5807, 2280 HV,
Rijswijk, the Netherlands

Abstract

Although the strong current in front of the harbour moles plays
a major part in the admittance of deep draught ships to the port
of Amsterdam, until 1994 no accurate current prediction could
be provided because numerical current models cannot yet be
used on an operational basis. Neural networks appear to be able
to fulfil the above need rather easily, using measured and
predicted wind and water level input data.

1 Introduction

One of the Dutch deep draught entrance channels is the IJ-channel to IJmuiden,
which makes the Amsterdam port accessible to ships of draughts of up to 16.5
metres. The eastern part of this 23 km long channel has a depth of 19.1 m. The
depth of the IJ-channel is not sufficient to allow safe channel passages under all
conditions. In order to guarantee safe channel passages an admittance policy is
carried out, which is based on an underkeel clearance criterion. Furthermore a
current based criterion is of importance in the admittance policy.

Until 1994 however no reliable current information was available; so the
current criterion could not be effectively applied. This paper briefly explains the
way in which a neural network has been able to fulfil the need for reliable
current information. But firstly a brief description is given of the admittance
policy and the production process of the necessary meteorological and
hydrological information.

2 Admittance Policy

The underkeel clearance as it is applied in the admittance policy is determined
by the water level, the ship draught, the draught increase of a sailing ship
(squat) and the draught increase due to vertical ship movements in waves as
mainly caused by swell.

Because high water levels roughly range between 0.35 and 1.45 m (neap and
spring tide respectively) the channel depth allows ships with draughts of 16.5 m

to sail on practically every high tide. This however is only valid for the so-called astronomical situation in which the water levels are mainly determined by the relative positions of the sun and the moon. The actual water levels may be lower as well as higher than the astronomical levels due to wind and air-pressure effects. This deviation can amount up to more than 1 m in extreme conditions.

For a given high water curve, the vertical ship movements in waves finally determine whether or not and if so, when a ship will be allowed to enter the harbour. If the vertical movements are too high, the ship will have to anchor at sea and wait for better conditions.

Apart from the underkeel clearance criterion, also a navigational current based criterion plays an important part in the admittance policy.

The relatively high current velocities in front of the harbour moles complicate the entrance of deep draught ships. A north-going flood current e.g. forces a ship to steer a few degrees more to the south in order to remain approximately on or parallel to the channel axis. When passing the harbour moles however, the cross-channel current suddenly drops and the ship has to perform a rather difficult manoeuvre in order to correct the abrupt course deviation. The upstream steering angle can simply be reduced by increasing the sailing speed. The sailing speed is restricted however because of the limited manoeuvring space within the harbour basin.

Given the harbour lay-out, the manoeuvrability of the ships, nautical experience and ship-simulator results, a criterion has been established that is considered to guarantee a safe passage of the harbour moles at all times. This criterion excludes a passage of the harbour moles for current velocities of more than 50 cm/s. Unfortunately the cross-channel current criterion is exceeded during almost every tidal cycle. During neap tide the (astronomical) flood current sometimes just does not reach the criterion, but during spring tide flood current velocities often exceed 100 cm/s. On top of this, winds exceeding force 5 to 6, depending on their direction, may cause significant wind-driven currents. Under stormy southwesterly conditions flood current velocities may increase with more than 50 cm/s.

Unfortunately the maximum flood current velocities occur about half an hour before high tide. Because of the relatively high flood currents, only under moderate neap tide conditions, an optimal use can be made of the high water levels. Under spring tide conditions however this is not possible (figure 1). If on top of this, wave and water level conditions coincide with each other in an unfavourable way, it is clear that the current velocity may be decisive with regard to the admittance of deep draught ships.

It is obvious that the decision whether or not to sail the channel can only be made using water level, wave and current velocity predictions. Accurate hydro-meteo predictions are essential for the realization of a safe and economically optimal admittance policy.

Until 1994 however accurate current predictions were not available. Until then the admittance policy was carried out, using a rough estimate of the astronomical

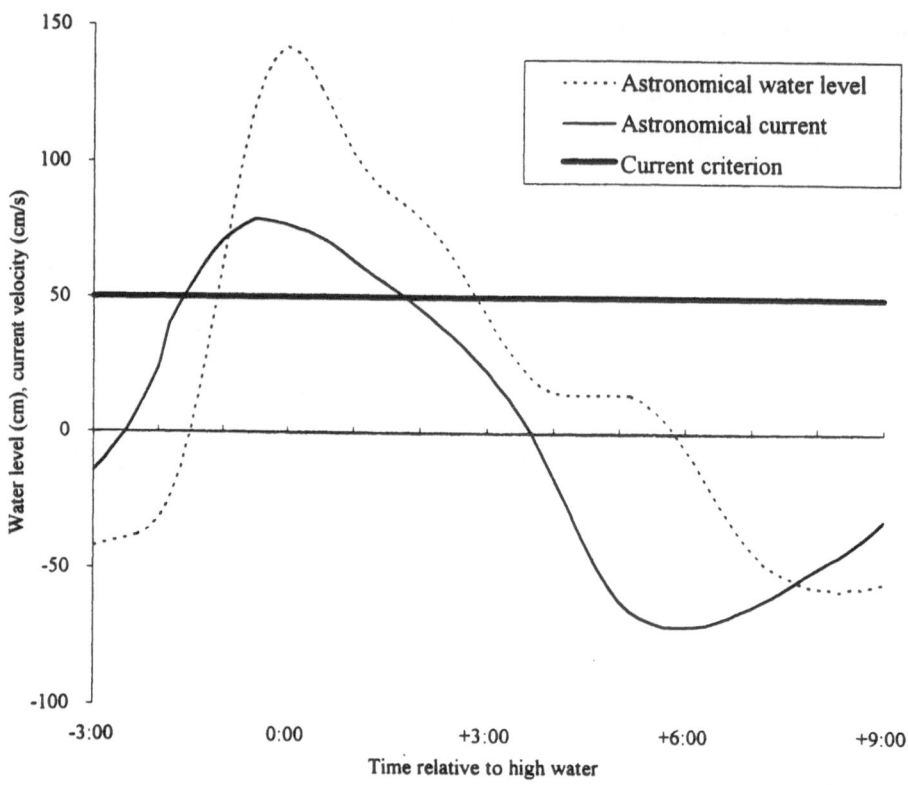

Figure 1. Astronomical water level and current curve.

current curve, which was based on an interpolation between three averaged astronomical current curves for neap, mean and spring tide. These estimates were not very reliable. Deviations of up to 50 cm/s could occur, due to errors in the current amplitude and particularly because of the impossibility of estimating the exact slack water moment, which appears to be rather variable in relation to the moment of high tide.

The absence of on-line current measurement information further hampers the estimation process. Although the wish for on-line current measurement information is an old one, it has not been fulfilled yet, because of the technical difficulty and the high operational costs of realizing and maintaining an operationally reliable measuring platform in the aggressive maritime environment. Moreover current measuring instruments are very vulnerable.

The limited accuracy of the estimated current curve forced the pilots to apply extra safety margins in planning their channel passage, sooner leading to (expensive) postponements of passages for at least one tidal cycle (12.5 hours). Under stormy conditions the situation was even worse, because the wind-effect could only be roughly determined on the basis of the pilot's experience.

3 Hydro-Meteo Predictions

As stated earlier, carrying out a safe and economically optimal admittance policy, requires hydro-meteo predictions of sufficient accuracy. This is even more important under adverse meteorological conditions when the admittance of a deep draught ship may easily be impeded. The importance of accurate predictions may simply be translated into economical terms, since an improvement of the prediction accuracy straightforward leads to a reduction of the waiting time of deep draught ships.

Hydro-meteo predictions for the guidance of deep draught ships are being provided by the North Sea Directorate (NSD) of Rijkswaterstaat from the Hydro Meteo Centre Rijnmond (HMR) at Hook of Holland. For this purpose the NSD processes the output of large scale numerical wind, flow and wave models, which are provided by the Royal Netherlands Meteorological Institute (KNMI), a division of which works together with the NSD in the HMR.
Accurate water level and wave predictions have been available for a long time, but no accurate current predictions could be provided until 1994. The main reason for this is the large spatial variability of the current, which requires a relatively fine grid numerical model, which simply demands too much computational time for operational purposes.

A conceivable solution to the computational restraints by means of nesting a small scale fine grid model into a large scale coarse grid model, is not yet feasible, because of the complexity of an operational implementation of such a system and because efficient and stable nesting algorithms are still being developed. Therefore the technical realization of operational numerical current prediction models will take a few years and a lot of effort.

Possible alternative solutions to the current prediction problem can be found in the field of statistical analysis, possibly combined with simple numerical methods. Both methods can be largely improved by the assimilation of measured data. The NSD uses such a method to predict waves for shipping guidance (ARMAX). Within the ARMAX model several model parameters are continuously tuned, according to a statistical analysis of the latest wave measurements.
An important advantage of this kind of models is that in critical situations when the wheather is rapidly changing and the 6-hourly distribution period of the large scale model outputs coincides unfavourably, new predictions can be made on short notice, using the latest measurement information.
Methods like these however, are not likely to be sufficiently successfull for generating a current prediction at IJmuiden, because of the lack of on-line current measurements and the complex non-linear character of the current phenomenon, which makes predictions difficult, especially in fickle weather conditions.

4 Neural Network Current Prediction

Given the above-mentioned limitations of the usual prediction methods and the promising capacity of neural networks to "understand" complex non-lineair processes, in 1993 the NSD took the initiative to investigate the merits of neural networks with regard to the prediction of current velocities. For this purpose nine months of simultaneously measured current, wind and water level data from IJmuiden were available. During the study [1], which was commissioned to All Fours NeuralTech, it soon became clear that a neural network was very well able to translate input water level and wind data into accurate current velocity output.

Special attention has been paid to the optimal way of presenting the water level and wind data to the neural network. Figure 2 gives a simplified illustration of the input presentation of the current prediction neural network. A typical current prediction neural network has about 100 input neurons, about 20 hidden neurons and one output neuron. Wind input consists of measurements up to 30 hours back in time and forecast wind data up to 24 hours in advance. Astronomical water level data are input for a period from 12 hours back in time until 12 hours in advance.

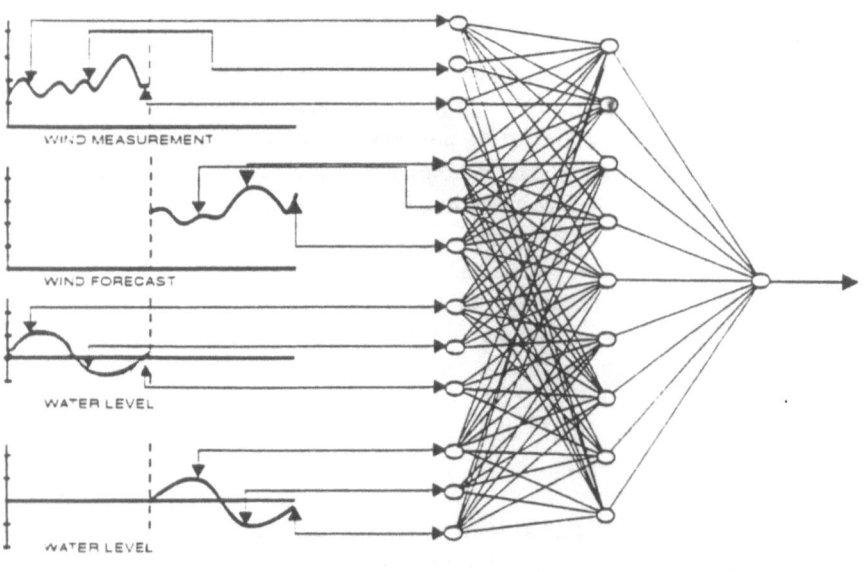

INPUT   NEURAL NETWORK   OUTPUT

Figure 2. Schematized current prediction neural network.

After a relatively short development time of a net three months in december 1993 an operational model has been implemented at the HMR. The neural network is implemented on a standard personal computer and generates a 24 hour forecast current curve in less than 2 minutes.

After a visual quality check and if necessary a correction, the required input data are automatically supplied to the neural network. The current prediction is routinely provided every 6 hours. A higher distribution rate is possible if desired.

5 Operational Neural Network Performance

After a three month introductory period the neural network current prediction system had been accepted by the IJmuiden pilots as a very useful and reliable aid. The neural network appeared to perform very satisfactorily, as compared to the accuracy of comparable numerical flow models.

The predictions were evaluated using both off-line current measurements and ship-felt current data [2]. Expressed by means of the root mean square error (RMSE) the accuracy was about 12.5 cm/s for predictions 6 hours in advance. If current velocities are hindcast using measured instead of predicted wind data the RMSE is about 11 cm/s. Under windless (near astronomical) conditions the RMSE falls under 8 cm/s.

In fickle wheather conditions with quickly veering winds and large wind speed fluctuations the neural network showed a good performance.

Under stormy conditions the network appeared to have some difficulty in predicting ebb current velocities; flood current velocities were predicted significantly better.

The neural network performed well in conditions up to and including force 9 winds, although it has only been able to learn a few examples of this kind. These results emphasize the well known large capacities of neural networks to generalize the presented examples.

Yet because of their relative scarcity in order to force the neural network not to neglect them, during the training process the most exceptional stormy periods had to be presented relatively more often to the neural network than more general conditions. Figure 3 gives an example of the operational network performance.

The performance of the neural network appeared to be less reliable in the following exceptional cases:
- When wind speeds exceed force 9, the results become less reliable because the network was not trained for such high winds. This result is not an unexpected one, because it is well known that neural networks have a limited capacity to extrapolate.
- A strong stratification may extend beyond IJmuiden, significantly affecting the current pattern, when an extremely high Rhine river discharge is followed by a period with little wind and little tidal mixing (neap tide).

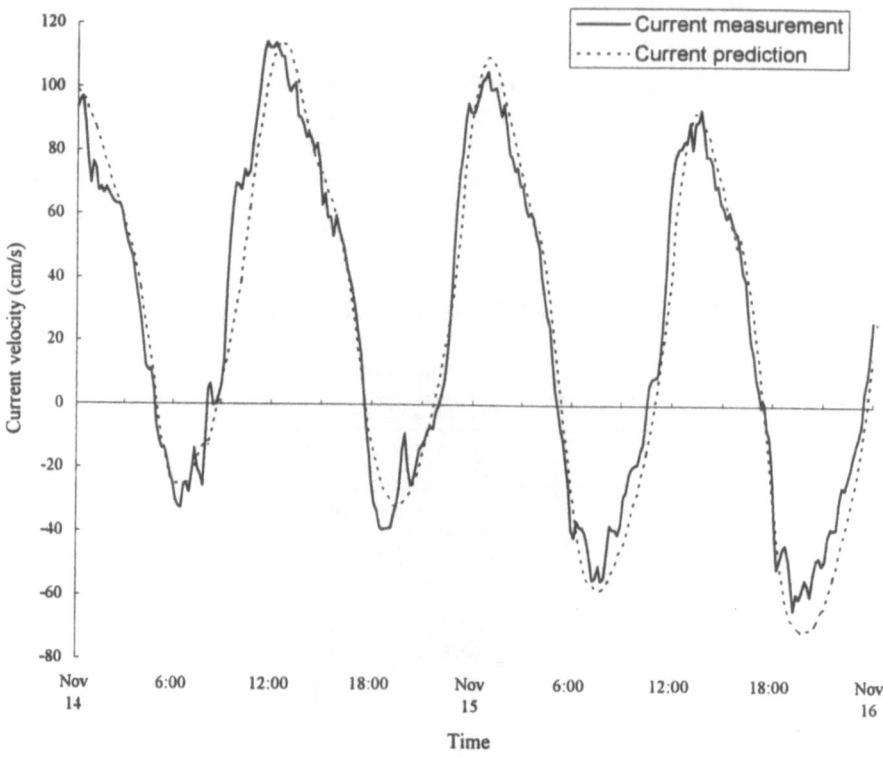

Figure 3. Network performance during a 1994 force 8 southwesterly wind.

An unexpected, close agreement between the network current and the ship-felt current throughout the entire IJ-channel, yields an important additional practical advantage. Because the pilot is able to use the ship as a current measurement instrument using the ship speed and the upstream steering angle, he is already able to assess the quality of the current prediction during the channel trip, well in advance of the actual passage of the harbour moles. This monitoring option is possible because of the homogeneous current pattern (as felt by the ship) and the constant tidal phase (simultaneous high tide) along the IJ-channel.

6 Conclusions

From the operational performance of the current prediction neural network it may be concluded that a neural network appears to be very suitable to "learn" the complex processes that rule the horizontal water movements along the Dutch coast.

Implemented on a personal computer the neural network makes it possible to generate quickly a current prediction at any moment, using the latest KNMI wind predictions and the latest on-line wind and water level measurements.

The neural network has shown a good performance in stormy conditions, emphasizing the large capacity of neural networks to generalize the information of relatively few examples.

The quality of the neural network output decreases in a number of well-defined exceptional cases, for which the network was not trained.

The neural network current predictions offer the pilots a better insight into the expected current situation in front of the harbour moles, so they may apply smaller extra safety margins, which leads to an economical reduction of the waiting time of deep draft ships.

At this moment in the HMR prototypes of neural networks for wave and water level prediction are operationally used for shipping guidance purposes. The fact that these fast networks also show a surprisingly good performance, emphasizes the value of neural networks for generating hydro-meteo predictions.

In the summer of 1995 another current prediction neural network will become operational for the guidance of deep draught ships to the port of Rotterdam.

A prerequisite for a successful application of neural networks will always be the availability of a reliable dataset which contains a good cross section of all relevant hydro-meteo situations.

Regarding the prospects and cost-effectiveness of the neural network tool and the operational drawbacks of other prediction methods, in the future neural networks may play a useful part in the field of operational hydro-meteo predictions.

7 Future Developments

Besides the above-mentioned "stand-alone" neural networks, alternative approaches in combination with numerical models, may yield very useful results. Some conceivable approaches will be stated below.

- Neural networks may be used to model sub-processes in numerical models which are not yet fully understood, or which cannot yet be effectively modelled.
- Neural networks have already proven to be very well suitable to interpret and correct the errors of numerical models [3].
- By training on its output, neural networks may be helpful for indirectly making operational use of a state of the art numerical model, which demands too much computational time for operational purposes.

References

1. van Noort GJHL. Een Operationeel Model voor Stroomsterkte-Verwachting ter hoogte van de IJgeul, All Fours, december 1993
2. Wüst JC, van Noort GJHL. Neural Network Current Prediction for Shipping Guidance. In: IEEE OCEANS 94 Conference Proceedings, Vol.1, pp. 58-63
3. van Noort GJHL. Waterstandsvoorspelling op locatie Antwerpen; een vooronderzoek, All Fours, june 1994

Applications of Neural Networks - Posters

Adaptive Nonlinear Control - Linearised models with Neural Networks

M.A.Hussain, J.C.Allwright and L.S. Kershenbaum

Process Systems Engineering Center,

Imperial College,

London

Abstract

A nonlinear control strategy involving a geometric feedback controller and adaptive approximation of the plant utilising a linearised model and a neural network model which approximates the higher order error terms is implemented. Online adaptation of the network is performed through a dead zone function. The proposed strategy is applied to a case study for output tracking of set point changes. The results are compared to that utilising the linearised model alone.

1 Introduction

Nonlinear control strategies involving geometric nonlinear control such as feedback and input-output linearisation have been one of the many strategies under considerable study in recent years. One of the main reasons for their popularity is that they provide the basis for an analogous development of nonlinear control theory with linear methods and are able to account for process nonlinearities as it directly utilises the model in the controller design. However they have a drawback in that the controller relies heavily on the accuracy of the model , which are prone to errors in many cases [1].

However at the same time, the modelling of nonlinear systems using neural networks have been widespread and found to demonstrate good results recently. Hence the incorporation of neural networks to model the uncertain nonlinear functions within this geometric control technique seems to be a promising strategy. In recent years neural networks have been shown to be able to approximate many nonlinear functions arbitrarily accurately [2]. For that matter, in this work we use linearised models in conjunction with neural networks to model the uncertain higher order terms. This is applied with adaptation of the neural network model online. This methodology is more reasonable as it takes advantage of a linearisation model when it is available and hence appears more robust than that of utilising the models or networks alone.

2 Nonlinear adaptive control formulation with neural networks

Consider a single input single output nonlinear system in its companion form , under closed loop control, given by :

$$x^{(n)} = -f(x) + g(x)u \tag{1}$$

where u is the scalar control input ($\hat{u} \, \varepsilon \, \Re$), x $\varepsilon \, \Re^n$ is the state vector and is assumed to be available for feedback. The functions f(x) , g(x) are nonlinear C^{∞} vector fields.

However in many practical cases, the nonlinear functions f(x) and g(x) are unknown or known only with some degree of uncertainty e.g. in linearised models.

In this study , we proposed using the neural network to model the uncertain part of the system's dynamic behaviour represented by f(x) and g(x) i.e. neural network models the difference between the true functions and that obtained from the linearised first principle model. In other words the functions , f(x) and g(x), are composed of the linearised term as well as the higher order terms(i.e. second order and above terms) which are modeled by the neural network. In mathematical terms, these representations can be given by :

$$f(x) = f(x_e) + f^{'}(x_e)(\delta x_e) + \hat{f}(x, w) \tag{2}$$
$$and$$
$$g(x) = g(x_e) + g^{'}(x_e)(\delta x_e) + \hat{g}(x, r) \tag{3}$$

where $\hat{f}(x, w)$ and $\hat{g}(x, r)$ are the neural network models of the higher order term (unmodelled term), x_e is the point of linearisation and $\delta x_e = x - x_e$ (w and r represents the parameters or weights of the network).

These equations are then substituted in the linearising control equation below to obtain a global linearising closed loop system,

$$u = \frac{y^n - L_f^n h(x)}{L_g L_f^{n-1} h(x)} \tag{4}$$

Adaptation of the weights for the neural network models are done using the steepest descent and the dead zone function , as proposed by Chen et. al [3], when near the minimum.

3 Simulation studies

Simulation studies were performed using the proposed strategy above for a second order SISO nonlinear plant involving level control for a two tank-in-series system. Simulations were performed for output tracking of the set point with a step change from the equilibrium position. The system is represented by the dynamic equations below :

$$A\frac{dh_1}{dt} = q_0 - k_1 h_1^{1/2} \tag{5}$$

$$A\frac{dh_2}{dt} = k_1 h_1^{1/2} - k_2 h_2^{1/2} \tag{6}$$

where A,area of tanks= $1m^2$, k_1 and k_2 , the valve coefficients inclusive of the conversion factors, are $1/2$ and $2/3$ m^6/min respectively. For this case the adaptive output tracking using the proposed procedure was performed for step changes in the controlled variable x_1(i.e. height of second tank, h_2) from its equilibrium value of 2.25 m, to a value of 4.0 m, by manipulating the control input, q_0. In this case the network estimating f_h has 22 hidden nodes while that estimating g_h has 10 hidden nodes. The coefficients of the feedback error equation used were $c_1 = 6$ and $c_0 = 10$.This result was also compared with the case of using the linearised model alone , in the controller equation. The result obtained for the proposed scheme showed good step change tracking with initial oscillations and a slight offset at the step change. However the scheme using linearised model alone showed instable control at the step change with the actual output diverging away from the required set point, which did not happen at a lower change in set point close to its linearised point (see Fig 1 and Fig 2).

4   Conclusions

The work outlined here demonstrates the viability of utilising linearised models in conjunction with neural networks. The results shows that this strategy gives results comparable to those obtained when utilising the actual nonlinear plant equations, which are rarely known exactly in practice. However use of the linearisation and neural networks gives better and stable performance than using linearisation on its own as seen in this particular case study.

References

[1] M. Kummel and H.W. Anderson, Controller adjustment for improved nominal performance and robustness - II , Robust geometric control of a distillation column , Chemical eng. Sc., 42(8), pp 2011-2023,1987.

[2] G. Cybenko, Approximation by superposition of a sigmoidal function, Math. Control Signal syst.,2,pp. 303,1989.

[3] Fu-Chuang Chen and H.Khalil, Adaptive control of nonlinear systems using neural networks, Int. J. Control, 55(6),pp.1299-1317, 1992.

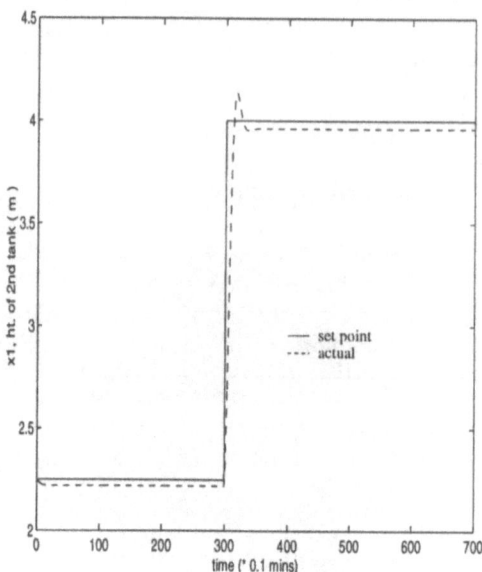

Figure 1: adaptive output control tracking of level in second tank - linearised-with neural networks

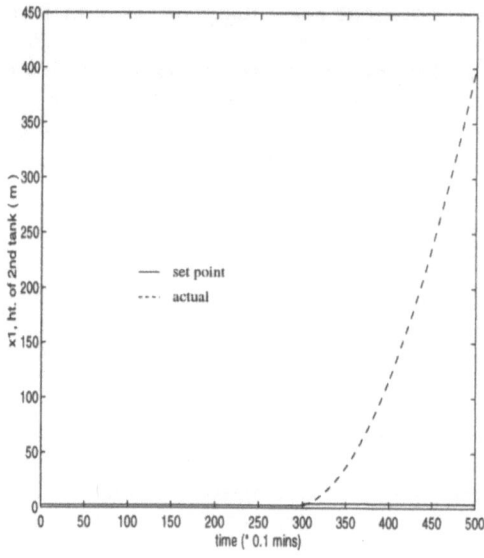

Figure 2: adaptive output control tracking of level in second tank - linearised model only

A Software Engineering Approach to Neural Network Specification

Erich Schikuta

Institute of Applied Computer Science, University of Vienna

Vienna, Austria

Abstract

An approach for the specification and implementation of neural networks within a software engineering framework is presented. We show the suitability of the formal methods of the OMT (Object Modeling Technique) introduced by Rumbaugh [2] using an object-oriented CASE (computer aided software engineering) tool.

1   Introduction

In the last few years many different methods for the specification of neural networks were presented. Most of these methods are based on the description of the processing elements, the topology, the adaption rules, and the net dynamics. It turned out that no general system exists to comprise all possible network paradigms. This led to the development of systems, which describe neural networks by defining the algorithmical aspects in detail and come down to a pure implementational description. These approaches show the deficiency of hiding essential neural network characteristics, as model paradigm, network categorization, etc. and hamper the view at the underlying concept.

We follow a different approach in proposing a model to specify what a neural network should do [4] and not, how its is physically implemented. Our model is based on an object-orientated (oo) view [3]. We introduce the oo software engineering framework as a suitable environment for neural network specification.

2   Design Concepts

2.1   Object-Oriented Design

One describing property of the oo design is the hierarchy of classes. A class comprises a set of objects sharing common functions. Generalization and specialization define a hierarchical class structure, which organizes the unique classes. Functions defined on a superclass are also inherited by all of its subclasses along the class hierarchy. The data model of the underlying oo system has to be based on three elements: classes, objects, and methods.

Objects represent unique entities and are classified by classes and described by methods. Classes are organized in inheritance hierarchies. Methods access and manipulate objects and can also be applied to other methods. This indirection allows a high degree of flexibility and expressional power.

2.2   Neural Networks Specification

The oo approach makes it easy to define, handle and administrate neural networks and seams therefore the most comfortable and natural design method for neural network models [1]. In the following we sketch an oo modeling approach for neural networks.

The basis of our design model is a generic neural network class, which comprises all possible network paradigms and models. The purpose of this rather generic class is to represent a unique network, which defines the net paradigm, network structure, dynamic behavior, etc. It can express complex neural systems consisting of different networks as well. Specializations of this class can be predefined classes represented by the network paradigm. Figure 1 gives a hierarchical structure of the neural network classes. Classes in lower levels represent subclasses.

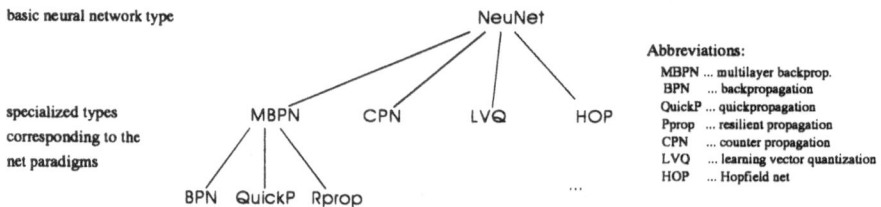

Figure 1: The neural network paradigm hierarchy

The class hierarchy expresses its power. A predefined system class provides all attributes and methods. If the predefined BPN class (see Figure 1) is chosen, the accompanied functions and tool set of the class can be used. A further expressional advantage is the employment of a predefined class and the overloading of its operators. This comes very handy by the hierarchy concept of the oo approach. Attributes or methods can be defined for an existing subclass and replace the original ones of the superclass. This allows the definition of highly complex neural systems and networks. For example, a system could be specified, which has different dynamics in different layers. As mentioned above a network can also be built from scratch. The hierarchical class system provides the model for the formal definition of a network family. For the realization of a specific network further classes are necessary, as classes for processing elements, connections, layers.

2.3   The Object Modeling Technique

The Object modeling Technique was introduced by Rumbaugh as a method for an oo software development approach. Its notation models the entities of the environment and describes them by classes and relationships throughout their life-cycle. Three basic model types characterize all aspects of the system in focus, the Object Model, a Dynamic Model and a Functional Model. The Object Model describes the static structure of the entities (objects) of the system. These are the identity, attributes, operations and relationships to other objects. The Dynamic Model shows the behavior of the system respective to time and the sequencing of the operations. The Functional Model depicts the transformations of data and catches the system's work.

The OMT method provides a tool to develop a comprehensive model of what the system in question is supposed to do, without specification of the implementation.

A complete description of the Rumbaugh's Object-Oriented concepts and methodology can be found in [2].

2.4 Comparisons of the Concepts

We set the three concepts against each other and map the describing components of each model.

Object-oriented view	Neural Network Model	OMT
classes	paradigms	Object Model, Functional Model
objects	neural networks res. network topology	Object Model
methods	network dynamics (training, evaluation, etc.)	Dynamic Model, Functional Model

The models of the OMT overlap and have no unique mapping. However they can comprise the whole specification process.

3 Specification Approach

In addition to the outlayed modeling via the network paradigms we can also use the oo concept to define other describing components of a neural network. In the following we model a simple neural network structure with the means of the OMT methodology. For simplicity we restrict the example to 3 describing network units, the generic neural network class 'NeuNet', nodes and links. 'NeuNet' is an aggregation of (consists of) nodes and links, which is expressed by the '◇' symbol. The nodes and the links are associated by a 2 to many relationship, i.e. a link needs 2 nodes and a node can have many links. Generally a class consists of a name, attributes and operations (Figure 2). In our example this is for the net links the identifier 'Links', the attributes 'Weight', 'FromNode', and 'ToNode' and the 'UpdateF' function for the learning phase.

This would be just the beginning for the specification of the static structure of a neural network. We can continue with the other phases of the OMT model to specify the learning and update phase of the neural network and the dynamic behavior at all.

4 From Specification to Realization

Many state of the art software tools available today allow to generate programming language code directly out of the specification. They provide the

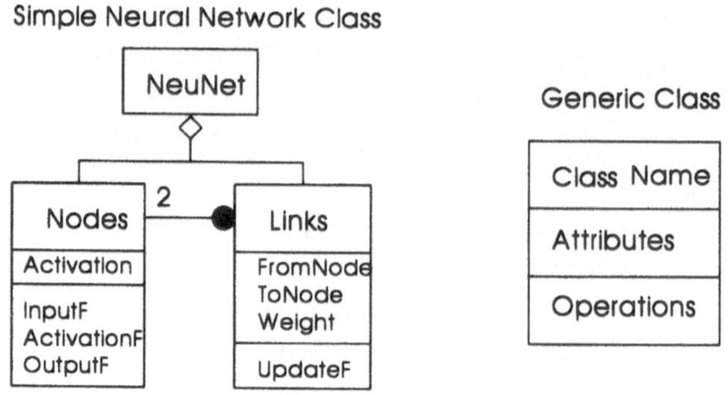

Simple Neural Network Class

Figure 2: Specification example

functionality to transform the OMT notation into a target programming language, as C++ or C. The final implementation can therefore be directly derived from the model without any concerns about implementational aspects.

5 Conclusion

We could prove that an oo software engineering tool, in our case the OMT methodology, provides an unchallenged platform in flexibility, expressiveness, but simplicity for the specification and implementation of arbitrary neural networks. It represents design information at a level above that of the physical implementation. This meets our goal stated at the beginning of this paper.

References

[1] G. Heileman, et al., A General Framework for Concurrent Simulation of Neural Network Models, IEEE Trans. on Software Engineering, 18, 7, pp. 551–562, 1992

[2] J. Rumbaugh, et al., Object-Oriented Modeling and Design, Englewood Cliffs, New Jersey, Prentice Hall, 1991

[3] E. Schikuta, Embeddment of Neural Networks into an Object-Oriented Database System, World Congress on Neural Networks, INNS, 1993

[4] L.S. Smith, A Framework for Neural Network Specification, IEEE Trans. on Software Engineering 18, 7, pp. 601–612, 1992

A Neural Network-based Software Tool
for Number-Plate Recognition

A. Frosini, M. Gori[*], *L. Pistolesi*

Dipartimento di Sistemi e Informatica - Università di Firenze
Via di Santa Marta 3, 50139 Firenze, Italy
* E-mail : marco@mcculloch.ing.unifi.it

Abstract

In this paper we describe a software tool running on standard platforms, including PC and several Unix Workstations for the number-plate recognition. The software is based on a hybrid model where a society of autoassociator-based neural networks are properly coordinated by modules charged of segmenting the number-plate and the single characters, respectively.

1. Introduction

In the last few years, the availability of impressive computational resources at affordable cost has stimulated the interest for several *unconventional* problems of information processing. A remarkable number of these problems are in the field of image processing, where one is mostly concerned with the extraction of information from pixel-based arrays. The problem of the recognition of the number-plate addressed in this paper has been receiving a growing attention in a number of different applications. When dealing with natural images, the problem is made particularly difficult because of the different environmental conditions that give rise to significantly different images taken from the same number-plate. A remarkable noise is introduced by the the light conditions and by the variable angle between the camera and the moving car. Basically the rectangle can have different dimensions, the colors, the fonts, and the characters' thickness can be different. All these problems make number-plate recognition difficult and, sometimes very hard, also for humans. In addition, the automatic approach of the problem cannot neglect computational contraints that are necessary in order to develop any commercial product.

In this paper, we give an overview of the solutions that we have adopted for approaching the problem and discuss the experimental results found in motorway toll environment. The recognition is based on a hybrid model that acts using a sort of hypothesize and verify technique. The extraction of the number-plate and the character segmentation are based on classical edge detection techniques while the character recognition is carried out by Multilayer Perceptrons (MLP) acting as autoassociators.

2. Recognition model

A model for problems like number-plate recognition can be based on forward steps only, or can act under the control of a feedback process which informs on the likelihood of the found hypotheses.

The first model is simpler to implement and maintain; all the efforts are on the effective development of the modules, that give rise to a sort of pipeline process. The single modules have a simple interface and the task to be solved is clearly identified. Unfortunately, the structure of a similar computational model is intrinsically weak since the failure of one module of the pipe results in an error in the number-plate recognition.

A model acting under the control of feedback processes informing on the likelihood of the found hypothesis is referred to as hypothesize and verify. It is likely to be more expensive than models based on forward steps only, but the structure seems to be more robust because of the intrinsic capability of recovering errors. The verification module is quite critical in this model since it controls the evolution of the whole recognition process. The system described in this paper is based on the second kind of model and on neural networks performing character recognition.

2.1 Segmentation step

In order to meet the requirement of using *reasonable* computational resources it seems reasonable to rely on segmentation steps aimed at detecting the contour of the number-plate and the characters. There are a number of different serious problems that must be addressed in order to detect the countor of the number-plate. The first one is that there may be several objects in the back of the vehicles that produce a contour similar to that produced by the number-plate. A possible way of facing this ambiguity is that of considering to take the ratio of the dimensions of the detected rectangle and compare with the ratios produced by actual number-plates. Unfortunately, in practice, most of the times, this seems to be plagued by the presence of tilted number-plates. Once the contour of the number-plate is detected, the process for the segmentation of the characters begins its operation. Unfortunately, at least in Italian number-plates, the detection of the first character does not allow us to predict the segmentations for all the remaining character since no all of them are located at the same distance. We have used segmentation algorithms based on thresholding techniques properly developed under a number of heuristic rules derived on the basis of experimental suggestions.

2.2 Character recognition

As already pointed out, the character recognition plays a crucial role, since it makes also possible the verification step with the subsequent feedback to the hypotheses provided by the segmentation modules. Hence, the recognition module cannot simply perform a character recognition, but must be capable of providing a reliable score of the likelihood of its presence in the hypothesized area. The very different environmental conditions make this recognition problem quite difficult if we consider that one is required to perform *very high* scores for the recognition of the single characters, since the recognition of the number-plate needs the contemporaneous successful recognition of many characters.

We decided to use connectionist models for their recognized ability of facing effectively similar problems [1]. The additional requirement arising in our problem, however, makes the use of MLP, acting as pure classifiers, not very well-suited. These networks perform in fact very well for pattern discrimination but, unfortunately, fail exhibiting a reliable score of the likelihood that a character of an expected class is present in an hypothesized area. In order to overcome this problem, in this paper we suggest to use MLNs as autoassociators. When using this learning mode, the input is forced to reproduce the outputs, and rejection criteria can be based on the way the input is approximated by the output. It has been proven that these criteria have clear theoretical foundations that legitimate their usage [2].

The main advantage of a classification model based on a society of autoassociators is that it is intrinsically modular but, unfortunately, it is not very well-suited for performing the discrimination of very similar classes. For this reason we have also introduced a set of discrimination networks with MLP architecture and classifier structure, aimed at discriminating very similar classes (2-7-Z, 0-D, 8-B, 5-S).

3. Experimental results

The results of our experiments refer to a significant number of samples of 1000 photos taken in real conditions and containing also many photos where the number-plate is not recognazible by humans either. In these conditions the percentage of number-plate recognized varies from 50% to 70% depending on the quality of the sample. If the sample was patiently cleaned up from the bad photos, these figures could be significantly higher. Another important indicator of the efficiency of the model is the error rate, that is the percentage of number-plates correctly recognized. From the beginning, as a matter of fact the philosophy of the project was *"rejections yes, errors not"*. The error rate is now about 2%. We are working to increase the percentage of recognized number-plates but particularly to reduce the error rate and we think to have still a good margin of improvement.

Fig. 1: Example of a correctly recognized number-plate

The figure 1 represents an example of a number plate correctly recognized. On the top-left there are overprinted the segmented characters before they were processed by the recognition module.

The figure 2 shows an example of a number-plate not correctly recognized with an error on the second character. The result of recognition is FTL00423 instead of FIL00423.

Fig. 2: Example of number-plate recognized with one error

At the moment, the program runs on MS-DOS PCs and several Unix platforms icluding Intel x86, IBM RS/6000 and DEC Alpha. The performance of recognition is quite satisfactory. On a 486DX4-100 PC with FreeBSD v2, the total execution time is about 3.2 seconds on the average, on a IBM RS/6000-250 is about one second less.

4. Conclusions

In this paper we have described a neural network-based system for the recognition of number-plates. The model relies on a hypothesize and verify paradigm. The system has exibited successful results in motorway toll environment. We are currently working to improve the robuteness of the system and make it available for number-plates of different formats.

References

[1] Y. Le Cun. *Generalization and Network Design Strategies*. Proceedings of Connectionism in Perspective, pages 143-155, 1989. Elsevier Publishers.

[2] M. Bianchini, P. Frasconi, and M. Gori. *Learning without local minima in radial basis function networks*. IEEE Trans. on Neural Networks, pages 749-756, Vol. 6, No. 3 1995.

Parallel Cross-Validation of Artificial Neural Networks

E.P.P.A. Derks, W. Melssen
L.M.C. Buydens
Laboratory for Analytical Chemistry, University of Nijmegen
Nijmegen, the Netherlands

Abstract

In this paper, a simple parallel cross-validation method for estimating the optimal number of hidden units for multi-layered feedforward neural network models trained by the generalized delta learning rule is presented. A neural network model is considered optimal when an ideal compromise between modelling accuracy (bias) and generalizing ability (variance) is found. The problem of training by means of gradient descent based methods, on data containing local minima is adressed. The cross-validation program has been parallelized to operate in a local area computer network. Development and execution of the parallel application was aided by the HYDRA

1 Introduction

In chemometrics, cross-validation has become a commonly used technique for estimating the number of latent variables (pseudo-rank) of a calibration model, especially in situations where only a limited number of measurements is available. Since no separate testset is used, cross-validation can be considered as an internal validation method for creating a model with optimal predictive abilities. A model is considered optimal when the best compromise is found between the accuracy of fit (bias) and generalization ability (variance).

The estimation of the pseudo-rank for a calibration model can be compared to the estimation of the optimal number of hidden units [1] for a neural network model.

Since neural networks have proven to be very powerful non-linear function approximators, the risk of overfitting, yielding a poor generalization ability, becomes even more critical. Non-random components in observational noise and fluctuations due to experimental conditions are easily modeled by neural networks when a superfluous number of hidden units is used, leading to poor predictive abilities.

Applying conventional cross-validation on a neural network based model, obtained from sparse data, will inevitably yield unreliable results since neural

---

[1] The modeling ability of a neural network depends as well on the choice of the activation functions as on the number of hidden units. However, in this work only sigmoidal activation functions are assumed.

networks suffer from the fact that the models obtained are, in general, unreproducable. This is a result of the fact that training involves a gradient descent search in an error hyperplane containing, apart from the global minimum, many local minima. Hence, considering the huge number of possible network configurations and ways to initialize the weight values of a network, it is highly unlikely that a set of weights (which determine together with the network achitecture the actual model) obtained from a single training session will correspond to the best fitting model (*i.e.* the model associated with the global minimum in the error hyperplane).

In this paper, an approach is presented to allow cross-validation on neural networks in order to estimate the optimal number of hidden units. Our approach is based on the assumption that the probability of finding the global minimum, corresponding to a unique set of weights, increases when multiple starting positions on the error hyperplane are used for the gradient descent search. Although no guarantee can be given that the global minimum will be found, at least the probability of getting stuck in local minima will be reduced considerably.

Since the method described above, requires a lot of computing time and administration (for example, scheduling and monitoring the progress of the training sessions), a parallel extension of the conventional cross-validation procedure has been developed.

2 Cross-validation

The principles of cross-validation have extensively been described [1, 2, 3, 4], so only the general concept will be described.

The cross-validation data is obtained by dividing the data (size N) into Q training and test sets (training set (size $N - A$), test set (size A), whereas $(A \approx N/Q)$).

Then, the network is being trained and the outputs of the A test samples are used to calculate the predicted residual sum of squares (PRESS) represented by Equation 1.

$$PRESS = \sum_{i=1}^{A}(y_i - \hat{y}_i)^2 \tag{1}$$

This procedure is repeated Q times. The prediction errors are accumulated for every test set yielding the total PRESS. Given the number of hidden units , the predicted residual sum of squares yields a statistic to test the predictive ability of the neural network. Cross-validation on an increasing number of hidden units generally shows that the PRESS of the trainingset continuously decreases whereas the PRESS of the test set increases with the number of hidden units.

Since the generalized delta rule does not produce reproducable weights, it can be imagined that faulty neural network models are obtained which are not fully trained, due to the presence of local minima in the error hyperspace, so special attention has to be paid during cross-validation that local minima are avoided.

Ending up in local minima can be avoided by re-initializing the gradient descent pathway from different locations in the error hyperplane, during the cross-validation. As can be expected, the computational effort becomes quite substantional. For example, when cross-validation is applied with Q subdivisions, on 1 up to L hidden units and R re-initializations, the neural network has to be trained $Q \cdot L \cdot R$ times. In general, this leads to long execution times accompanied by long lasting computer overloads. For example, the cross-validation of a neural network for 1 up to 10 hidden units, on data subdivided into 10 cross-validation sets, applying 5 re-initializations ($L = 10, Q = 10, R = 5$), requires 500 independent training sessions.

3 HYDRA driven parallel cross-validation

The computation time can be reduced, approximately by a factor Q by parallelizing the cross-validation procedure. This means that Q networks are trained simultaneously on the Q subdivisions of the data. One single parallel run yields identical results as a standard cross-validation procedure. As already mentioned in the previous section, it is highly recommendable to retrain the neural network R times, starting from different positions on the error hyperplane.

From the prediction errors obtained from the R neural network models, the minimum, the mean and the standard deviation are calculated. The minimum prediction error is used for estimating the number of hidden units whereas the mean and standard deviation are used as statistics to give insight in the ruggedness of the error-hyperplane and the reproducability of the various training processes.

Since the retraining cycles are completely independent for every subdivision of the data, parallel cross-validation can be performed by using the HYDRA parallel programming environment [5].

HYDRA driven cross-validation consists of the next steps

1. Assign the Q data subdivisions (training and test set) to Q computers and set the number of hidden units to 1.

2. Train and test the networks and for l hidden units and record the prediction errors. Then, repeat the training and test procedures on each computer R times (for the calculation of the statistics (minimum, mean, standard deviation) of the R prediction errors).

3. Increase the number of hidden units by one and repeat step 2. When all networks for L hidden units are trained and tested, the procedure will finish, yielding a cross-validation table containing all prediction errors.

By means of averaging the statistics of the prediction errors for all the cross-validation subsets, a general measure for the predictive ability, based on the data used, is obtained.

4 Conclusion

Problems arise when conventional cross-validation is applied on neural network models trained by the generalized delta learning rule, due to the fact that weight initializations yield unreproducable neural network models. Every new initialization can be regarded as a new start position for the gradient descent search for the global minimum. Although special learning parameters (*e.g.* noise injection, momentum factor) can help to avoid local minima, no guarantee of finding the global minimum can be given. The probability of finding the global minimum might be enhanced by selecting various random start positions for the gradient descent search. Consequently, there exists a bigger chance of "walking around" the local minima. Obviously, the chance of finding the global minium directly depends on the smoothness of the error-hyperplane and the number of local minima. Cross-validation by means of re-initializations and re-training the networks yields information about the smoothness of the error-hyperplane and probably a better 'neural' model can be established.

Finally, we conclude that the implementation of the cross-validation program in HYDRA, yields a considerable gain in execution time. Moreover, a robust and reliable execution in a computer network can be guaranteed.

Acknowledgements

The authors wish to acknowledge Geert Rolf and Ronald Pijpers for the development of HYDRA and the implementation of the parallel cross-validation program. E.P.P.A. Derks is supported by the Dutch Foundation for Chemical Research (SON) on behalf of the Dutch Organization for Scientific Research (NWO)

References

[1] Allen, D.M. The relationship between variable selection and data augmentation and a method for predictions. *Technometrics*, 16:125–127, 1974.

[2] Geisser, S. The predictive sample reuse method with applications. *Journal of the American Statistical Association*, 70:320–328, 1975.

[3] Stone, M. Cross-validatory choice and assesment of statistical predictions. *Journal of the Royal Statistical Society*, 36:111–133, 1974.

[4] Wold, S. Cross-validatory estimation of the number of components in factor and principal components models. *Technometrics*, 20(4):397–405, november 1978.

[5] Melssen, W.J., Derks, E.P.P.A., Beckers, M.L.M., and Buydens, L.M.C. Parallel processing in a local area network of computers. part i. hydra: concept, configuration and implementation of parallel applications. *Computers & Chemistry*, 1995. submitted.

Neural networks in large scale bank effect recognition

A. Sierra, C. Santa Cruz*, V. López*, G. Fractman, J. Dorronsoro*, C. Aguirre

Instituto de Ingeniería del Conocimiento
(* also in Departamento de Ingeniería Informática),
Universidad Autónoma de Madrid, 28049 Madrid, Spain

J. M. Soto, A. Medina, R. López
Keon, S. A.,
Orense 6, P.2 28020 Madrid, Spain

1 Large scale bank effect processing

Bank drafts are widely used as agreement documents for the deferred payment of goods and services. As such, they can be effectively negotiated, usually at a discount, by banks and other finacial intermediaries, as they represent money to be cashed at the arrival of a certain date. Since a huge amount of commercial transactions are financed this way, large institutions can turn a profit dealing with these effects before they are due for payment. However, no effective electronic recording of these transactions usually takes place when the drafts are issued. Thus, draft's information has to be put in electronic form before any dealing is to be done.

In this paper we will discuss neural recognition whithin the bank draft processing application of a major spanish bank (see [1, 2, 3, 6] for other references in neural character recognition). This may involve the processing of up to 100.000 effects on peak days, which in practical terms means that the system has to recognize 1.500.000 characters per hour, or roughly over 400 characters per second. If done manually, a team of about 60 persons is needed to perform this task. It is thus clear that in this situation an automated system for draft processing has a good chance of being economically effective, provided that it posses two usually competing requirements: 1) a not too high rejection rate, which would send back a large number of effects to manual processing, and 2) a very low error rate, which does not lead to costly and lenghty checking and improper charging remedying operations.

In the next section we will describe the major aspects of the neural recognition system. However, and although we won't pursue it now, we want to point out the crucial influence on the recognition process of several image processing issues. For instance, although Spanish bank drafts (see figure 1) have well defined zones where the transaction's information is to be placed, all too frequently that information either overflows its holding zone, or overlaps with the labels identifying it.

Moreover, factors such as the scanner speed, its lens–to–effect distance, or the height variations of the effect position caused by the carrying mechanism, distort the effect image. Furthermore, all sorts of fonts and font sizes are

394

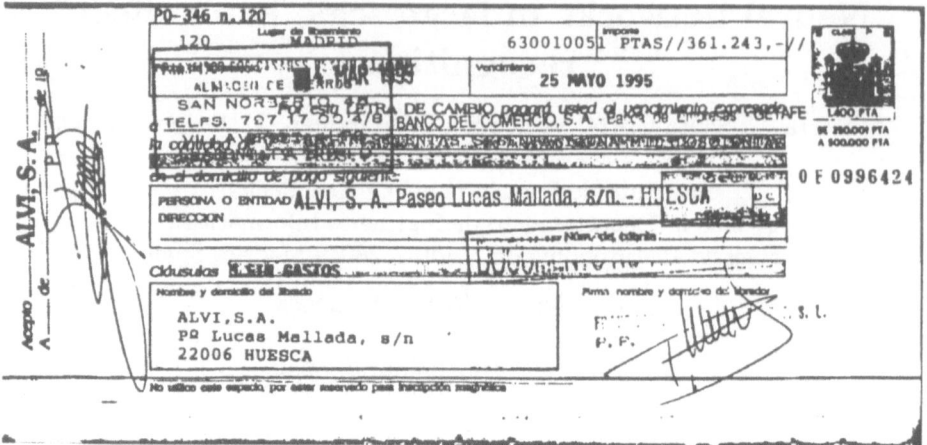

Figure 1: Typical problems encountered in effect recognition are the use of date stamps, the overlapping of printed information with zone boundaries, or noise due to uncontrollable factors.

possible and the above mentioned image scale variations cause that font size may vary, in an almost continous fashion, from under 10 points to above 20 points. Finally, printer quality has an extreme degree of variability, as could be expected if we take into account that the system is to process some 30 million documents per year coming from over 3 million potential issuers.

All the above circumstances will have a very strong influence during recognition, to the point that, in our experience, it is useless to run the image processing and character extraction tasks in a recognition independent fashion.

2 Neural recognition of bank effects

Once various extraction, segmentation and image enhancing phases have been completed, individual characters are compressed to a 40 parameter set giving mostly spectral information. That compressed information directly enters the recognition modules. Two different neural procedures are used for recognition purposes. The first one is a set of perceptrons [5] with several layers and complexity of about 20.000 free parameters (weights and bias). The second one is a set of radial basis function networks [4], each one attuned to an individual letter or number. While the perceptrons usually have better generalization abilities than the RBF networks, these offer faster training (more on that below). Furthermore, given their natural probabilistic interpretation, high values of the transfer function associated to a particular character greatly insure its proper recognition (in some sense, each of such a network's gaussian can be associated with a particular font).

These considerations are used in the recognition procedure. A character is first considered to be properly identified if its identification by any of the RBF nets gives a high value. If this is not the case, the character will be considered properly identified if the RBF and the perceptrons substantially agree. If even this does not happen, the system accept the perceptron's identification but only if it has a markedly high value.

Notice that the system is rather strict when deciding if a character has been properly recognized. Moreover, this procedure will unavoidably mark a sizable number of characters as undecided, and force therefore a number of rejections. This will lead to an equilibrium between the rejection and error rates, which can be modulated in terms of the settings used to decide when a character has been suitably recognized. Taking for instance the field where the draft's amount is written, the system presently processes correctly about 55 % of these fields, and rejects about 45 % of them. The per field error rate is here about 5 per thousand.

Notice that system performance has to be measured in terms of recognition rates per number of fields, rather than per number of characters. From a practical point of view (and certainly from the customer's one), rates per character are not too useful: what counts are fully recognized fields, or even better, fully recognized effects. In the amount field this means that not a single digit can be missed. On the other hand, some flexibility is possible: nonessential characters can be missed provided they are not placed inmediately before or after the actual amount digits.

Finally, let us briefly comment on training complexity. The huge amount of characters to be processed requires that rather large training sets have to be used for the building of the neural recognition modules. The present database contains over 700.000 characters, and keeps growing, altough at a fairly manageable rate. This large database size is not a crucial problem for RBF network building, since this is done in a per character basis. The situation is however quite different for the perceptrons. Perceptron training usually takes over 100 epochs to converge. If a full perceptron, capable of dealing with the upper and lower case alphabets, 10 digits and assorted punctuation signs and other delimiters, is to be built, the full training set has to be processed on each one of these epochs. Moreover each character implies adjusting over 20.000 weights. These figures imply that a full size perceptron training requires over 1.4 trillion floating point operations, counting just one operation per character per weight (something overly optimistic even for highly optimized training algorithms).

In fact, typical training time for a full size perceptron is about 9 days in an IBM RS 6000/370 machine (with a 20 MFlops performance). In contrast, training time for a perceptron in charge of recognizing the characters of the amount field, is much shorter (about two days), since it only involves numerical characters and a few letters and punctuation characters. In particular, this means that font adding, a pretty straightforward operation for the RBF nets, is highly time consuming when dealing with full alphanumeric perceptrons, since it involves retraining a whole network. Fortunately, these perceptrons able to deal with the full alphanumeric character set are only used in the bank and debtor fields, where lexical redundancy allows the interpreter module to make up for characters not recognized.

3 Conclusions

Neural network based character recognition is a viable procedure for large scale document processing applications. However, to be succesful, it has to be tightly coupled with the image processing components of that application: faulty bitmap extraction or segmentation can render a highly efficient network recognizer useless. This makes quite difficult that such a system can be set up using off–the–shelf independent components: even if their individual quality is excellent (as it happens with some well known line removal or OCR products), application tailoring and module communication won't reach the performance an the degree of flexibility that a more integrated system can achieve.

In any case, the very same large scale of the system, which on the one hand leads to recognition rates far from 100 %, makes economically viable a system with very low error rates even if its recognition rates are not truly exceptional. As our system shows, even a 50 % per field recognition rate, coupled with good processing speed and large document flow handling capabilities, makes an application perfectly viable from a cost effectiveness point of view.

References

[1] Y. Le Cun, B. Boser, J. Denker, D. Henderson, R. Howard, W. Hubbard, L. Jackel, "Handwritten digit recognition with a backpropagation network", in Advances in Neural Processing Information Systems 2, Morgan Kauffman, 1990.

[2] I. Guyon, V. N. Vapnik, B. E. Boser, L. Bottou, S. A. Solla, "Structural risk minimization for character recognition", in Advances in Neural Processing Information Systems 4, Morgan Kauffman, 1992.

[3] R.Y. Li, M. Xu, "Character recognition using a fast neural–net classifier", Patter Recognition Letters 13, 369–374, 1992.

[4] J. Moody, C.J. Darken, "Fast Learning in Networks of Locally–Tunned Processing Units", Neural Computation 1, 281–294, 1989.

[5] D.E. Rumelhart, G.E. Hilton, R.J. Williams "Learning Internal Representation by error propagation", in Parallel Distributed Processing, MIT Press, 1985.

[6] "Special Issue in Optical Character Recognition", Proceedings of IEEE, 80, 1027–1209, 1992.

Author Index